Paleoethnobotany on the Northern Plains: The Tuscany Archaeological Site (EgPn-377), Calgary

Evelyn Siegfried

BAR International Series 2691
2014

Published in 2016 by
BAR Publishing, Oxford

BAR International Series 2691

Paleoethnobotany on the Northern Plains: The Tuscany Archaeological Site (EgPn-377) Calgary

ISBN 978 1 4073 1338 2

BAR Publishing is the trading name of British Archaeological Reports (Oxford) Ltd. British Archaeological Reports was first incorporated in 1974 to publish the BAR Series, International and British. In 1992 Hadrian Books Ltd became part of the BAR group. This volume was originally published by Archaeopress in conjunction with British Archaeological Reports (Oxford) Ltd / Hadrian Books Ltd, the Series principal publisher, in 2014. This present volume is published by BAR Publishing, 2016.

Printed in England

BAR
PUBLISHING

BAR titles are available from:

BAR Publishing
122 Banbury Rd, Oxford, OX2 7BP, UK
EMAIL info@barpublishing.com
PHONE +44 (0)1865 310431
FAX +44 (0)1865 316916
www.barpublishing.com

Abstract

The Tuscany habitation site (EgPn-377) located in northwest Calgary was excavated between 1995 and 1997. The site stratigraphy of the large depression contained a series of buried paleosols situated between Mazama tephra above, dating to 6730 ± 40 ^{14}C years BP, and Glacial Lake Calgary sands below, dating to approximately 13,900 calendar years ago. These paleosols comprised the focus of this dissertation. One of the research objectives was to examine the site for spatial information via the processing of bulk sediment samples. Such samples had the potential to yield information on the distribution of small-scale archaeological remains throughout the site. Sediment samples representing 1% volumes were collected from each excavated level of each unit in the site grid. Through flotation processing an inventory of bone, lithics, insects, fungal spores, mollusks and charred macrobotanical remains were recovered. The charred macrobotanical remains were the focus of this research. Though the inventory is small, it provides a representative sample of the remains of plants that grew locally in the depression through the early Holocene. The charred botanical remains were compared with pollen and soil studies along with modern vegetation and climate records to develop a model for open parkland in the area for the early Holocene. The reconstructed landscape appears to have provided a habitat for a broad spectrum of fauna along with a diverse inventory of potentially useful plants for early Holocene peoples to exploit.

Acknowledgements

At last, I am at the point where I must end this great adventure. It has been an experience that has endured the entire range of possible emotions—from the euphoric highs, to the seemingly bottomless and endless lows, and deep, dark moments of true despair. However, it is done and for that I am truly thankful.

There are many people who have helped me in numerous ways throughout my time in program to the final completion of this research project. I must begin by thanking Dr. Gerald Oetelaar for giving me the chance to work on the Tuscany Archaeological Project during three summers (1995, 1996 and 1997) and develop a research project based upon the bulk sediment samples that were collected from the habitation site (EgPn-377). Gerry also provided logistical support through the Tuscany project, which gave a number of students the opportunity to work with flotation of the bulk sediment samples as well as the laborious job of sorting samples under the microscope. While the majority of the work was completed by me, an important contribution of labor was provided by students including Carmen Olsen, Debbie Meert, Murray Lobb, Beau Cripps, Ann Clegg, Amanda Dow, Candace Ayres and Leslie-Lynn Sinkey. The Tuscany project also covered the costs of radiocarbon dating carried out on selected samples from the site.

I wish to extend a deeply sincere thank you to my supervisor Dr. Brian Kooyman, who provided guidance and support through the duration of my time in program and played an instrumental role in the scholarships received. Extenuating circumstances led to an extension of time in program through which Brian's support continued to its conclusion. This was most appreciated. I must also extend a sincere thank you to Dr. J. Scott Raymond and Dr. Len Hills from my committee for their support, especially during the latter part of my time in program.

Time in program was funded in part through the Education Department of the Bigstone Cree Nation, Desmarais, Alberta, and many thanks are extended to the various people who provided administrative support with all the necessary paperwork, which took care of business over the years. Additional funding was provided through the Department of Archaeology, University of Calgary, through Graduate Research Scholarships and Teaching Assistantships, which were greatly appreciated. A Thesis Research Grant from the University of Calgary provided additional funding. Three scholarships including the Coutts Family Memorial Scholarship for Archaeology 1995/96, the Plains Anthropological Society – Native American Student Award 1998, and the National Aboriginal Achievement Foundation Scholarship through the Provincial Museum of Alberta, also provided additional support to my dissertation work.

I must extend a very heartfelt thank you to Dr. Alwynne Beaudoin of the Archaeolgical Survey of Alberta, Edmonton. Much of the final work with the identification of macrobotanical remains would not have been possible without the use of the comparative seed collection that is curated at the Provincial Museum of Alberta, Edmonton. Alwynne provided an incredible amount of patient guidance and advice with the identification of my charred botanical specimens as well as several extremely important working hours with the SEM facilities at the University of Alberta. Alwynne also provided instrumental support for the National Aboriginal Achievement Foundation scholarship that I received, which gave me an opportunity to work with the seed collection and develop both a working knowledge and appreciation for its value in future research. All my time spent at Alwynne's facilities in the Provincial Museum represents some of the best memories I have from this project as a learning experience.

I have to say thank you to the 'Ladies-of-the-Main-Office', Lesley Nicholls and Linda Berreth. They provide help and good advice when needed -- sometimes saving students like myself from desperate little predicaments. I will miss looking after the kitties, Tod and Tirrups and the poodles Jojo, Phoenix and Khepri.

I must also extend a great inclusive thank you to all my former and present office partners that have come and gone over the years like Roman Harrison, Jocyelyn Williams, and Peter Dawson. Tamara Varney is the last of the best and we have certainly solved a lot of the world's problems together -- if only the world knew! There are many, many others, presently in program and those long gone from these hallowed halls who have made my time in program an enjoyable adventure overall. Trevor Peck, who 'resided' a few doors down deserves a special thank you -- we too solved many 'world' problems, cried on each other's shoulders sometimes, and had a ton of fun conversations that kept us both on track for the prize. Those who have tolerated my angst and frustration in more recent years include Latonia Hartery, Liz Robertson, Judy Klassen, Joe Moravetz, Sandra Garvie-Lok, Frank Timmermans, Ceasar Appentik, Nancy Saxberg, Sean Goldsmith, Jason Gillespie, Jason Harris, Mike Turney, Christy de Mille, Luc Bouchet, Susan Tupakka, Janet Blakey, Deepika Fernandez, Andrea Waters, Michelle Shatz. Alison Landals gave me a couple of working contracts over the years that really helped me out and I am very thankful for those opportunities. Apologies to those I've missed -- there are too many great people to list!

I must also extend a hearty thank you to the Geology-Crater-Guys down the hall from 'us Arky types'. Mike Mazur saved me from several major computer-meltdown moments when I had nobody else to turn to and thought my little virtual world was coming to an end. Another kind of horrible meltdown may have happened without his help! We also enjoyed many soapstone-carving hours together creating some of the best works of art ever seen on this planet -- well, we think so, anyway! Wayne Edwards and more recently Matteo Niccoli have also been pretty good pals who have helped make the last of my time in program a lot of fun. Dr. Alan Hildebrand has given helpful advice on occasion and I leave the garden I created in the front hallway to the care of his capable 'green thumb'.

Finally, I must say thank you to my family. My mother has been very patient with me over the years. All the things I have forgotten to do when she has asked me have always been forgiven -- trust me, I've forgotten a lot! The special little containers of food so I could work into the evenings were so thoughtful and helped me out many times over the years. I must also give hearty thanks to my brothers Guy and John. We all help each other out in different ways and remain good friends. They have helped me fix my trucks on numerous occasions when I could not afford to pay mechanic's wages on my student salary. I am indebted to them for their skills, time and labor. My brother Peter has recently moved to China to teach english. I wish him well and hope he has an adventure and many tales to tell when he returns.

DEDICATION

TO MY PARENTS

JOHN PAUL SIEGFRIED
(1927-1980)

VICTORIA SIEGFRIED (*nee* AUGER)

"Miyawsin kohgkpiyhkiyhowasoohk"

(It is nice when you raise children)

Table of Contents

ABSTRACT....i

ACKNOWLEDGEMENTS....ii

DEDICATION....iv

TABLE OF CONTENTS....v

LIST OF TABLES....ix

LIST OF FIGURES....x

Chapter 1: INTRODUCTION

Paleoethnobotany: Perspectives and Problems for Understanding Sites on the Northwest Plains....1

Approaches and Issues in Paleoethnobotany....3

- Quantification....3
- Hypothetico-Deductive Approach vs Narrative Interpretation....3
- Meaning of Materials Recovered and Context....6
- Taphonomic Approach....6

Problems this Dissertation will Address....7

Site and Data Used....7

Bulk Sediment Sampling Strategy: Levels of Sampling....11

Summary of Chapters....11

Chapter 2: METHODOLOGY

Sediment Sampling and Standardization....13

Flotation Processing of Bulk Sediment Samples....15

- Introduction....15
- The First Flotation System of the Tuscany Archaeological Project....16
    - Problems with the Tub Flotation System....16

The Alternate Method: Bucket Flotation 17
Experimental Water Sieving: Alternative Methods to Derive Samples 19
The First Water Sieving Test 20
The Second Water Sieving Test 20
The Third Method of Water Sieving 20
Summary Discussion on Experimental Water Sieving 20
Conclusions on Water Flotation and Water Sieving 20
Sorting of the Light and Heavy Flotation Fractions 21
Identification of the Seeds 22
Pollen Versus Macrobotanical Remains 22
Charred Plant Fragments 24
Charcoal Research 25

Chapter 3: CALGARY SETTING, IMMEDIATE SITE SETTING AND SITE CULTURAL FEATURES

Climate 27
Climate Classification of the Calgary Area 27
Ecology 28
Fescue Grass Ecoregion 29
Aspen Parkland Ecoregion 30
Discussion 31
Tree Cover in the Calgary Region 31
Nose Hill Park 31
Coulee Environments 32
Soils and Vegetation of the Tuscany Habitation Site 32
Fauna of the Calgary Region Today 35
Geology/Topograpy 35
Drainage 35
Physiography 36
Glacial Lake Calgary 36
The Post Glacial Deposits and Cultural Chronology 37
Using Faunal Remains to Derive the Colonizing Vegetation 38
Detailed Description of the Site Specific Topography 41
Cultural Remains and Features 41
Definition of the Amalgamated Levels Based on Sediments and Soils 44

Geomorphology at the Tuscany Site 50

Summary 53

Chapter 4: THE MACROBOTANICALS

The Charred Wood 55

Charred Seeds and Seed Fragments 56

Juniper Seeds 56

Kinnikinnick Seeds 57

Chenopodium sp. Seeds 58

Rose Seeds 60

Picea sp. Seeds and Cone Scales 61

Type D Charred Seeds 63

Other Charred Seeds 63

Interpreting the Charred Seed Inventory 68

Chapter 5: THE SPATIAL ANALYSIS

The Excavation Grid 70

The Hearth and Structure 79

Spatial Distribution of the Charred Seeds 81

Planview Distributions 81

Distributions of Seeds Through Profiles 91

Fire Ecology 107

Chapter 6: LOCAL ECOLOGICAL AND CULTURAL LANDSCAPES IDENTIFIED

Ecological Landscapes 109

Using Fossil Pollen Studies to Derive Vegetation Influences 114

Cultural Landscapes 119

Seasonality 123

The Site Location 124

Chapter 7: POLLEN AND MACROFOSSIL STUDIES IN SOUTHERN ALBERTA

Introduction 125

Review of Holocene Vegetation Zones from Pollen Profiles in the Rocky Mountains: Postglacial to 6730 ± 40 years BP (Mazama Tephra Event 125

Review of Holocene Vegetation Zones from Pollen Profiles and Macrobotanical Studies East of the Rocky Mountains: Postglacial to 6730 ± 40 years BP (Mazama Tephra Event 128

Summary Discussion 132

Chapter 8: REGIONAL PALEOSOL STUDIES AND THE TUSCANY SITE

Dated Paleosols from Sites in Alberta ... 135

The Fish Creek Paleosols ... 135

The Crestmont Estates Site (EgPn-428) ... 137

Valley Ridge (EgPn-230) ... 137

Bowfort Road/85th Street Pit ... 137

Mona Lisa Site (EgPm-3) ... 137

The Hawkwood Site (EgPm-179) ... 137

The Gap (DlPo-20) ... 138

Rough Creek, Willow Creek and Bellevue ... 138

Summary and Integration with Observations at the Tuscany Site ... 138

Chapter 9: SUMMARY AND CONCLUSIONS ... 142

REFERENCES CITED ... 147

List of Tables

Table 1: Biophysical and climatic characteristics of Ecoregions surrounding Calgary ... 28

Table 2: Physiography and biophysical characteristics of Ecodistricts in the Calgary region ... 29

Table 3: Relative abundance of tree species per ecoregion ... 31

Table 4: Plants observed in the Tuscany habitation site and surrounding area ... 33

Table 5a: ^{14}C dated faunal records of the late Pleistocene and early Holocene in the Calgary area and southern Alberta ... 39

Table 5b: ^{14}C dated/relative-dated faunal records from sites in the Calgary region and the predominant vegetation associated for browse and habitat ... 40

Table 6: Amalgamation of the site sediment interpretations by sedimentary sections, pedogenic zones, soil horizons, and profile layers ... 45

Table 7: Comparison of the vegetation in kinnikinnick-juniper ecophases in the different Alberta Subregions ... 110

Table 8: Comparison of climate data from Northeastern and Southwestern Alberta ... 112

Table 9: Pollen and macrobotanical studies of the Rocky Mountains, Foothils and Plains in eastern British Columbia, Alberta and Montana ... 126

Table 10: Paleosol sites with radiocarbon dates from southern Alberta and Saskatchewan ... 136

List of Figures

Figure 1: Map showing the location of the Tuscany site within Calgary 8

Figure 2: This topographic map identifies the variability in the surface terrain around the Tuscany site 9

Figure 3: The Tuscany habitation site grid at the end of 1997 10

Figure 4: Koeppen's climatic zones (using data for 1931-1960) 27

Figure 5: The Ecodistricts surrounding the greater Calgary region 29

Figure 6: The Tuscany habitation site grid at the end of 1997 42

Figure 7: The Tuscany habitation site grid with the general parameters of the three activity areas outlined.. 43

Figure 8: Profile description of Unit 14 at the Tuscany habitation site EgPn-377 46

Figure 9: Legend of the layer descriptions for East, West and North wall profiles 48

Figure 10: Profile of 3 units along the east wall of the excavation 48

Figure 11: Profile of 3 units along the west wall of the excavation 49

Figure 12: Profile of 3 units along the north wall of the excavation 49

Figure 13: Wireframe DEM grid maps showing the trends of surface change through time 51

Figure 14: Three DEM landscape surfaces are represented here 52

Figure 15: An example of a charred plant fragment from Unit 14 Level T from the Tuscany habitation site.. 56

Figure 16: A magnified view of the cross section or transverse view of a piece of wood from Unit 14 Level U, from the Tuscany habitation site................................. 56

Figure 17: Another view of the piece of wood from Unit 14 Level U, at a higher magnification to view the pore and cell wall structure 56

Figure 18: *Juniperus horizontalis* seeds from the PMA comparative collection................. 57

Figure 19: Two views of two different Juniperus sp. seeds from the Tuscany habitation site (EgPn-377) 57

Figure 20: Modern kinnikinnick seeds from the PMA comparative collection.................... 58

Figure 21: Two different ESEM views are presented of charred kinnikinnick seeds from the Tuscany site 58

Figure 22: An example of one of the *Chenopodium* sp. seeds found in the flotation samples at the Tuscany habitation site.. 58

Figure 23: One other example of a Chenopodium sp. seed from the Tuscany site 59

Figure 24: Two views of *Rosa acicularis* (prickly rose) seeds from the comparative seed collection at the PMA 61

Figure 25: Two charred rose sp. seeds from the Tuscany site 61

Figure 26: The seed on the left is from *Juniperus horizontalis* 61

Figure 27: Two examples of *Picea* sp. seeds from the Tuscany habitation site 62

Figure 28: Two views (front and back) of a *Picea glauca* cone scale from an identified specimen from the PMA comparative seed collection 62

Figure 29: Two charred cone scales from the Tuscany habitation site proposed to be from *Picea glauca* conifer trees 63

Figure 30: A view of 11 charred cone scale structures from Unit 10 Level AA from the Tuscany habitation site 63

Figure 31: An ESEM view of a charred cone scale from the Tuscany habitation site (Unit 10 Level AA) 63

Figure 32: Two examples of the Type-D charred seed 64

Figure 33: An example of a *Potentilla fruiticosa* seed from a plant within 50 meters of the Tuscany habitation site grid 64

Figure 34: A charred *Symphoricarpus* sp. seed from the Tuscany habitation site paleosols 64

Figure 35: Two views of modern *Symphoricarpus occidentalis* seeds from the PMA comparative collection 65

Figure 36: A seed from a *Shepherdia* sp. from Unit 34 Level P from the Tuscany site grid 65

Figure 37: An example of a *Shepherdia canadensis* seed from the comparative seed collection at the PMA 65

Figure 38: An example of one of seeds of *Monarda fistulosa* (western wild bergamot) recovered from flotation analysis of the bulk soil sample from Unit 28 Level S at the Tuscany habitation site 66

Figure 39: Examples of seeds of *Monarda fistulosa* var. *menthaefolia* from the comparative seed collection of the PMA 66

Figure 40: Two examples of charred seeds tentatively identified as *Potentilla f ruiticosa* (shrubby cinquefoil) from the Tuscany flotation samples 67

Figure 41: Two examples of modern *Potentilla fruiticosa* seeds from a plant growing close to the Tuscany excavation grid 67

Figure 42: Two examples of charred seeds tentatively identified as *Galium boreale* (northern bedstraw) 67

Figure 43: Examples of modern seeds of *Galium boreale* (northern bedstraw) from the comparative seed collection at the PMA 67

Figure 44: Two examples of charred seeds tentatively identified as *Castilleja* sp. from the Tuscany flotation samples 68

Figure 45: An example of a modern seed of *Castilleja* sp. from the comparative seed collection at the PMA 68

Figure 46: Two halves of a charred baneberry seed from the Tuscany habitation site from Unit 50 Level S 68

Figure 47: Example of a modern baneberry seed from the comparative seed collection at the PMA 68

Figure 48: Density peak and post maps of the charred plant fragment sums for the Tuscany site (EgPn-377) 71

Figure 49: Greyscale density plot of the charred plant fragment count for Units 42 to 44 W-E in Northing 519 72

Figure 50: Greyscale density plot of the charred plant fragment count for Units 62 to 51 W-E in Northing 520 73

Figure 51: Greyscale density plot of the charred plant fragment count for Units 63 to 41 W-E in Northing 521 73

Figure 52: Greyscale density plot of the charred plant fragment count for Units 68 to 5 W-E in Northing 522 74

Figure 53: Greyscale density plot of the charred plant fragment count for Units 72 to 10 W-E in Northing 523 74

Figure 54: Greyscale density plot of the charred plant fragment count for Units 76 to 30 W-E in Northing 524 75

Figure 55: Greyscale density plot of the charred plant fragment count for Units 16 to 22 W-E in Northing 525 75

Figure 56: Greyscale density plot of the charred plant fragment count for Units 23 to 28 W-E in Northing 526 76

Figure 57: Greyscale density plot of the charred plant fragment count for Units 32 to 36 W-E in Northing 527 76

Figure 58: This diagram highlights the lower burn layers (K, L, and M) within the Tuscany paleosols 78

Figure 59: This diagram highlights the upper burn layers (H, I, and J) within the Tuscany paleosols 78

Figure 60: This figure summarizes information for Unit 14 78

Figure 61: Density peak and post maps of the charred seed sums for the Tuscany site (EgPn-377) 80

Figure 62: Peak and density maps of the distribution of whole/complete juniper seeds found within the paleosols at the Tuscany site 82

Figure 63: Peak and density maps of the distribution of juniper seed fragments found within the paleosols at the Tuscany site 83

Figure 64: Peak and density maps of the distribution of whole/complete kinnikinnick sees found within the paleosols at the Tuscany site 84

Figure 65: Peak and density maps of the distribution of kinnikinnick seed fragments found within the paleosols at the Tuscany site 85

Figure 66: Peak and density maps of the distribution of Chenopodium seeds found within the paleosols at the Tuscany site 86

Figure 67: Peak and density maps of the distribution of shrubby cinquefoil seeds found within the paleosols at the Tuscany site 87

Figure 68: Peak and density maps of the distribution of rose seeds found within the paleosols at the Tuscany site 88

Figure 69: Peak and density maps of the distribution of spruce seeds found within the paleosols at the Tuscany site 89

Figure 70: Peak and density maps of the distribution of spruce cone scales found within the paleosols at the Tuscany site 90

Figure 71: NORTHING 519 – The distribution of kinnikinnick seeds and seed fragments in Units 42 to 44 91

Figure 72: NORTHING 520 – The distribution of kinnikinnick seeds and seed fragments in Units 62 to 51 92

Figure 73: NORTHING 521 – The distribution of kinnikinnick seeds and seed fragments in Units 63 to 41 92

Figure 74: NORTHING 522 – The distribution of kinnikinnick seeds and seed fragments in Units 68 to 5 93

Figure 75: NORTHING 523 – The distribution of kinnikinnick seeds and seed fragments in Units 72 to 10 93

Figure 76: NORTHING 524 – The distribution of kinnikinnick seeds and seed fragments in Units 76 to 30 ... 94
Figure 77: NORTHING 525 – The distribution of kinnikinnick seeds and seed fragments in Units 16 to 22 ... 95
Figure 78: NORTHING 526 – The distribution of kinnikinnick seeds and seed fragments in Units 23 to 28 ... 95
Figure 79: NORTHING 527 – The distribution of kinnikinnick seeds and seed fragments in Units 32 to 36 ... 96
Figure 80: NORTHING 519 – The distribution of juniper seeds and seed fragments in Units 42 to 44 ... 97
Figure 81: NORTHING 520 – The distribution of juniper seeds and seed fragments in Units 62 to 51 ... 97
Figure 82: NORTHING 521 – The distribution of juniper seeds and seed fragments in Units 63 to 41 ... 98
Figure 83: NORTHING 522 – The distribution of juniper seeds and seed fragments in Units 68 to 5 ... 98
Figure 84: NORTHING 523 – The distribution of juniper seeds and seed fragments in Units 72 to 10 ... 99
Figure 85: NORTHING 524 – The distribution of juniper seeds and seed fragments in Units 76 to 30 ... 99
Figure 86: NORTHING 525 – The distribution of juniper seeds and seed fragments in Units 16 to 22 ... 100
Figure 87: NORTHING 526 – The distribution of juniper seeds and seed fragments in Units 23 to 28 ... 100
Figure 88: NORTHING 527 – The distribution of juniper seeds and seed fragments in Units 32 to 36 ... 101
Figure 89: NORTHING 519 – The distribution of lithic remains in Units 42 to 44 ... 102
Figure 90: NORTHING 520 – The distribution of lithic remains in Units 62 to 51 ... 102
Figure 91: NORTHING 521 – The distribution of lithic remains in Units 63 to 41 ... 103
Figure 92: NORTHING 522 – The distribution of lithic remains in Units 68 to 5 ... 103
Figure 93: NORTHING 523 – The distribution of lithic remains in Units 72 to 10 ... 104
Figure 94: NORTHING 524 – The distribution of lithic remains in Units 76 to 30 ... 104
Figure 95: NORTHING 525 – The distribution of lithic remains in Units 16 to 22 ... 105
Figure 96: NORTHING 526 – The distribution of lithic remains in Units 23 to 28 ... 105
Figure 97: NORTHING 527 – The distribution of lithic remains in Units 32 to 36 ... 106
Figure 98: Map of the Southwestern Alberta Subregions located in the greater region west of the City of Calgary ... 111
Figure 99: Comparison of the moisture and nutrient regimes and range of elevation of six species of shrubs and forbs associated with ecosites of Southwestern Alberta ... 113
Figure 100: Comparison of moisture and nutrient regimes and range of elevation of six species of trees associated with ecosites of Southwestern Alberta today ... 114
Figure 101: Comparison of moisture and nutrient regimes and range of elevation of shrubs, forbs and grasses associated with ecosites of Southwestern Alberta today ... 115
Figure 102: Diagram depicting the sediment profile at the Tuscany habitation site during the period immediately following the draining of Glacial Lake Calgary ... 117

Figure 103: The initial vegetation colonizing the landscape were grasses and forbs......... 118

Figure 104: Kinnikinnick and juniper begin to fill in the landscape along with willows, deciduous and conifer trees.. 118

Figure 105: Kinnikinnick and juniper fill the land surface again, along with willows, deciduous and conifer trees and herbaceous plants as undergrowth 118

Figure 106: Kinnikinnick and juniper remain in the landscape but in low numbers along with grasses, shrubs and forbs.. 119

Figure 107: The timeframe is now 6730 ± 40 years BP.. 119

Figure 108: This diagram represents the present day landscape, which is covered with buckbrush, grasses and forbs.. 119

Figure 109: Two views of the Lusk point found in the Tuscany habitation site excavation in Unit 37 Level S (N521, E526).. 120

Figure 110: A line drawing of the Lusk projectile point.. 120

Figure 111: Locations of the pollen and macrobotanical study sites discussed in Chapter 4 127

Chapter 1
INTRODUCTION

PALEOETHNOBOTANY: PERSPECTIVES AND PROBLEMS FOR UNDERSTANDING SITES ON THE NORTHWEST PLAINS

This dissertation explores an assemblage of charred macrobotanical remains derived from the archaeological excavation of a prehistoric site located on the northern Plains in Alberta, Canada. The examination of the charred plant remains provides information related to the paleoenvironment of the Tuscany habitation site for the early Holocene timeframe.

The nature of paleoenvironmental reconstruction studies determines the approach that is generally adopted by researchers today. Many variables are examined at the specific level and this provides a historical and descriptive aspect to this research. Synthesis of the information is thus inductive, moving from the examination of the variables themselves, to achieve a general understanding of the processes that may have formed them, how they are linked, and how they may have interacted during the past. At the same time, narrow and focussed questions can be explored with a Hypothetico-deductive approach, a process of questioning observations of the data to try to derive conclusions that may help resolve specific problems. The two approaches perform a "dualistic dialectic" (Mentis 1988) that leads to synthesis and understanding. These ideas will be explored further, below.

The most recent discussions about paleoethnobotanical research describe it as a "subdiscipline" within archaeology (e.g. Archer *et al.* 2000 and Hastorf 1999). Popper and Hastorf (1988:2) define paleoethnobotany as "...the analysis and interpretation of archaeobotanical remains to provide information on the interactions of human populations and plants." Ford's (1979:286) definition focuses more clearly on the relationship between people and the environment:

> Paleoethnobotany, then is the analysis and interpretation of the direct interrelationships between humans and plants for whatever purpose as manifested in the archaeological record. Its objective is the elucidation of cultural adaptation to the plant world and the impact of plants upon a prehistoric human population, not simply the recognition of useful plants, and its subject is all archaeologically known cultures, including so-called civilizations. Perhaps more than any other class of archaeological data, including artifacts, plant evidence expresses most aspects of past societies and their involvement with both external social and natural environments.

Hastorf (1999) notes that there is some confusion between the terms "paleoethnobotany" and "archaeobotany" which both apply to the study of plant remains from archaeological contexts. The term "archaeobotany" is defined by Miksicek (1987:211) as "... the art and science of recovering, identifying, and interpreting plant remains from archaeological sites." Hastorf (1999:55-56) points out that "archaeobotany" is a European term whereas "paleoethnobotany" is a North American, or New World term. For the purposes of this dissertation, a paleoethnobotanical approach is preferred for its definition and research goals for the charred macrobotanical data collected through flotation analysis of bulk sediment samples from the Tuscany habitation site in northwest Calgary. The research goals for this study are to develop an understanding of the environment that existed in the immediate area of the archaeological site under investigation and to relate that to human populations that occupied the site and surrounding region during the early Holocene timeframe.

According to Ford (1979:286) the theoretical foundations of paleoethnobotany are not explicit and methodology is borrowed from the botanical sciences. Ford (1979:286) differentiates six approaches to recovered data and how it has been presented: 1) scientific identification lists of plants; 2) crop plant evolution relative to human behavior; 3) domesticated plants and dispersion by diffusion; 4) ecological interaction between people and plant communities; 5) the use of ethnographic analogy; and 6) settlement and subsistence patterns and seasonality. As Ford (1979) notes, the use of more than one of these approaches is appropriate and determined by the research and the questions under investigation. In the case of this present study, approaches 1, 4, 5, and 6 will be combined in various chapters to investigate the charred macrobotanical data recovered from the Tuscany habitation site.

One of the problems for this paleoethnobotanical approach to the Tuscany site is the timeframe that is encompassed. The focus of this is the time period of 10,000-6800 radiocarbon years BP. Extrapolation of ethnographic analogy can certainly be accused of becoming problematic with such time depth. The ideas and suggestions presented must be viewed as exploratory and conjectural, 'best fit' scenarios intended to stimulate further questions and discussion of the models presented. The actual past human-plant relationships may not be explicitly derived from the evidence of this research, but a probable relationship can be presented based on the archaeological data in conjunction with knowledge derived from modern ethnographic studies.

The present research is focussed on a specific section of the Tuscany site, stratigraphically, and which falls into the time period 10,000-6700 years BP. The focus of this research is paleoethnobotanical. As such, the remainder of the site stratigraphy is summarized only as needed. This research is not presented as a site report and does not detail the excavation of the Tuscany site. The site report is pending. Information provided here on stratigraphy and excavation methodology is intended as descriptive background only. The author was not responsible for the methodology employed in the stratigraphic excavation of this site nor was the author responsible for the final set of profile descriptions and soil descriptions summarized in this dissertation. These were based on the collaborative efforts of employees and consultants of the Tuscany Project. Some interpretation of stratigraphy is related to the macrobotanical remains recovered and the stratigraphy and soil descriptions presented here are strictly intended to contextualize related ideas.

The Tuscany habitation site is located in the region identified culturally and environmentally as the Northwest Plains. This area in southern Alberta includes the plains and foothills bordering the eastern flanks of the Rocky Mountains. As noted by Markgraf and Lennon (1986) the paleoenvironmental history for this region is fragmentary. In Alberta, palynological research has provided most of the data used for paleoenvironmental reconstruction. More recent macrobotanical studies have also recovered materials from pond environments. The present research represents the first study of macrobotanical remains recovered from an archaeological site in Alberta that does not have a ponding history. The macrobotanical remains thus represent examples of the vegetation cover present in the general landscape, which will provide new insights to the paleoenvironmental reconstruction of this region of the northern Plains.

As outlined by Minnis (1981), one of the problems faced in the study of seeds from archaeological contexts is discerning the difference between modern deposits of an ongoing seed rain and the representation of a fossil seed rain that is the result of cultural occupation or a natural vegetation regime. Buried paleosols were the focus of the Tuscany site research and a completely charred macrobotanical assemblage was recovered. In paleoethnobotany a charred assemblage is normally assumed to be prehistoric (Minnis 1981), especially if the investigation is focussed on questions of domestication. The difference between this present study and most paleoethnobotanical studies reviewed in the literature is the time period (early Holocene) covered. This fact, along with the location in the northwest Plains, means there were no expectations that any domesticate crops would be represented at the site. In the case of the Tuscany site, the stratigraphic position of the buried paleosols that were the focus of study and the location of a capping volcanic tephra above them argued convincingly for an early Holocene prehistoric context. Subsequent dating of charred materials from the paleosols confirmed this irrefutably.

An issue that can affect the interpretation of a site is the possibility of contamination of the sediments that are the focus of research. Seed rains continue through time for archaeological sites located in natural settings. Any disturbance in a site (i.e. rodent tunneling) can introduce seeds of the present into the older site sediments. Minnis (1981) acknowledges that it is possible for a modern seed rain to become charred—definitely something to consider in a forested setting. However, charred seeds are generally considered the product of human activity. It is imperative that the modern ecological setting of the site is understood and an inventory made of plants in the site locale so that a comparison can be made with the items recovered in flotation analysis. This was one of the problems addressed in this study.

The focus on charred macrobotanicals remains at the Tuscany site is a product of the nature of the features available for study. Many archaeological sites contain features like hearths and pits. Examples of macrobotanical studies in the Plains demonstrate that most of this work has concentrated on the analysis of fill from hearth and pit features (e.g. Armstrong 1993; Drass 1993; Minnis 1981; Smith 1988). Seeds are burned when they fall into a hearth by natural means (blown in by the wind) or when they are introduced by human activity around a fire. Processing methods like parching can also lead to charred seed remains being introduced into pit features. Seeds recovered from hearth features may not have resulted from the use of that feature. They could be the by-product of discard practices when the stalk or leaves of plants were utilized and the remainder was tossed into the fire. Charred seeds can be introduced to features unintentionally after use when weedy plants grow in the area and then burn off when a fire sweeps through. Differentiating between cultural introduction of macrobotanical remains and introduction by other means is vitally important in the analysis of pit and hearth features from archaeological sites.

The fundamental difference between the macrobotanical studies reviewed and the Tuscany habitation site is in sampling methodology. There was only one feature found within the buried paleosols at the Tuscany site. It was sampled and processed by flotation. The rest of the bulk sediment samples from the Tuscany site were from the southwest corner of each level of each unit in the excavation grid (except where there was disturbance). This sampling methodology is very different from that of the other sites reviewed where only the fill of pit features and hearth features is taken for flotation analysis. However, the macrobotanical remains recovered from the Tuscany site were all charred, the main reason why they preserved in the sediments through time. This macrobotanical assemblage contains the remains of a number of plant species. The relationship between the charred macrobotanical remains and the cultural occupation of the site is an important subject of discussion in this dissertation.

APPROACHES AND ISSUES IN PALEOETHNOBOTANY

Quantification

The general approach to quantification of macrobotanical remains recovered from flotation analyses, has been the use of simple tallies and summations—how many of each item can be counted (e.g. Smith 1988, Armstrong 1993, Drass 1993). Individual plant species are identified by remains such as charred seeds or by identifiable parts that have unique structures. Lists are drawn up and seeds or structures are counted and summed. Macrobotanical remains from the Brewster site, Iowa were summed by weight to the nearest 0.1 gram (Conrad and Koeppen 1972). Since all the remains were identified as charcoal from a number of different tree species, this method of quantification of the remains yielded useful results.

Summation tallies can be used to generate lists of plant species represented and a total of identified pieces. These lists can be broken down by a time period approach, stratigraphically, or by cultural component. Quantification lends comparability to the data. Cultural components can be compared for potential differences in use or importance based on the relative amounts of charred seeds. For example, changes from low quantities of macrobotanical remains in lower levels (early Archaic), to higher quantities in upper levels (late Prehistoric) are interpreted to indicate greater usage or reliance on plant foods during the late Prehistoric in Smith's (1988) study of a number of sites in Wyoming. The appearance or disappearance of a seed type within features may indicate changes in environment or subsistence. The summed weights of charcoal at the Brewster site in Iowa demonstrated declining amounts of charcoal between the lower and higher levels in the site and differences in species of wood used as fuel (Conrad and Koeppen 1972). Simple quantification methods can provide information about activity areas and subsistence within site boundaries through time.

As noted, the sampling methodology at the Tuscany site led to the collection of a very different suite of sediment samples compared to those collected from other archaeological sites on the Plains. The sampling at Tuscany provides material from each level of each unit in the site grid, rather than being restricted to the cultural components or features of the site. This methodology allows the use of alternative approaches to understanding spatial and possibly environmental trends that may have been taking place within the site during the past. The tallies and summations of data have been used for a simple spatial analysis based on changing quantities of recovered macrobotanicals from each bulk sediment sample. The spatial analysis reveals trends in changing quantities of the macrobotanical remains that appear to reveal the original depositional surfaces. This kind of approach to flotation data is not possible when features are the only areas sampled within an archaeological site.

Hypothetico-deductive Approach versus Narrative Interpretation

Archaeology fits into the realm of the social sciences. It is not a pure science like physics or a biological science like botany, but is rather an interdisciplinary science, which borrows methods and ideas (Thomas 1998) in the pursuit of understanding the place of humanity on the earth through time. During the 1960s and 1970s, a movement that has come to be known as "the New Archaeology" or "Processual Archaeology" challenged the traditional approaches of archaeology (Gibbon 1989; Kelley and Hanen 1988). Fundamental to the movement was the desire to make archaeology more scientific (Kelley and Hanen 1988). Researchers realized that literature on the philosophy of science could be drawn upon to develop new approaches to attain that goal. The hypothetico-deductive approach to data was originally born out of the desire for science and scientific methods to be part of archaeology's means of inquiry to understand and explain. The hypothetico-deductive approach was adopted as a method for confirmation, a proceedure to test general statements that connect two or more variables (Kelley and Hanen 1988). As stated by Gibbon:

> Basically, however, the HD approach is a view of the process of theory formulation and evaluation – a blueprint of basic scientific procedure. A scientist begins a project by formulating a hypothesis (a statement of constant conjunction between two or more variables), or a provisional theory, and proceeds to test the hypothesis or theory through the process of the logic of confirmation… …i.e. the process that 'secures' the relation between a hypothesis and an observation report (Gibbon 1989:23-24).

While there are problems with the approach and often a fundamental misunderstanding of deductive versus inductive methods (Kelley and Hanen 1988), many archaeologists and other researchers frame their research using the hypothetico-deductive approach. Problems have been encountered in other disciplines such as ecology where Mentis (1988:6) notes that "The application of the hypothetico-deductive scheme in ecology is faced with certain conceptual, logistic and ethical problems." One of the deficiencies in hypothesis testing that Mentis (1988:13) views as problematic is that "Powerful testing demands focus on one or only a few hypotheses... ... that risks blinkering our science, particularly in a field that espouses greater or lesser interconnectedness and holism." It would be fair to say that archaeology espouses similar ideals and should be deeply concerned about problems with this methodology as perceived in other disciplines.

An example of a hypothesis driven inquiry is provided by MacDonald (1993), in an examination of a paleoecological problem concerned with the decline of birch during the early Holocene in western Canada. In this article, each key idea is prefaced with the phrase, "It can be hypothesised that.....", followed by a proposed idea which is discussed and which then provides a verbal 'test' of the hypothesis (MacDonald 1993). While providing a relevant example, this article also demonstrates some problems in the testing of the hypotheses that leave it open to criticism. For example, in examining the possibility of a pathogen being involved in the decline, birch pathogens are not investigated. The argument is made by analogy, drawing upon statistics available for *Castanea dentata* (chestnut) trees (MacDonald 1993). On the basis of this analogous argument, MacDonald rejects the hypothesis that a pathogen is involved. This is not a compelling argument. A more compelling argument would examine modern birch trees for pathogen related problems. A second hypothesis is proposed offering interspecific competition as a possible cause of birch decline. In two of four comparative sites examined there is a concurrent rise in pine pollen during the decline of birch, which is dismissed as coincidence and the hypothesis is rejected (MacDonald 1993). Again, the argument is significantly weakened because it is not proven that competition was not a factor. MacDonald (1993:94) accepts the final hypothesis that the decline of birch is due to an individualistic response to climate change, based on the rejection of three hypotheses including the two mentioned. It is not clear that MacDonald's final hypothesis should be accepted based on these problematic arguments. The hypotheses have not been falsified. As outlined by Thomas (1998:46):

> Once the hypotheses are defined, the scientific method requires their translation into testable form. Hypotheses can never be tested directly because they are general statements, and one can test only specifics. The key to verifying a hypothesis is simple: You don't. What you verify are the logical consequences of hypotheses.

In recent years it has become apparent that some of these methods/approaches may be inappropriate at certain levels of presentation and reasoning because of flaws that were not recognized when these ideas were adopted during the 1960's and 1970's. Archaeological theory has developed alternatives to a strict Hypothetico-deductive approach, which has roots in Logical Positivism. For example, Whitely outlines an alternative approach that moves theory forward without losing science:

> The moderate position also maintains that there is a true and objective past, although we may not be able to recognize it (in the sense of developing singular scientific tests that will reveal it). The point of science is then not necessarily to discover truth (an objective past), but to attempt to move increasingly closer to it. The way this is done is not through the critical tests of positivism (processual archaeology emphasized falsification as the preferred means of testing theories). Instead, it is with a procedure that Kelley and Hanen (1988) have labeled "inference to the best hypothesis": using empirical evidence to select the best among a series of competing hypotheses. This is an effort to employ a method of science that is more sophisticated than positivism, not a rejection of science in the general sense. It is then possible to accept this important critique of positivism – that fact and theory are not independent but are related – without rejecting science. A rejection of processual archaeology therefore does not require a rejection of science (Whitely 1998:11).

In ecology an inductive approach is described as ampliative, "implying that the content of the inference drawn is not present either explicitly or implicitly in the premises" (Mentis 1988:11). The patterns that may be present in the data and for which hypotheses or models can be proposed are not bound or constrained by preconception (Mentis 1988). In sociology the methodology of "Grounded Theory" has been developed and utilized for research, where data are gathered with a research question generally outlined and the data are coded afterward (Strauss and Corbin 1994). The theory is thus grounded in the data itself and arises through the coding, examination and exploration of those data (Strauss and Corbin 1990, 1994).

The next question those inclined to a stricter Hypothetico-deductive approach might ask is where is the testing of the data? Repeating the experiment of the flotation analysis undertaken in the present analysis is impossible. All the samples have been processed, sorted, and examined. The test of the data and the model presented can be undertaken with another analysis on a new and independent data set collected from another site with similar stratigraphy. Each new study completed on material collected from similar sites serves to test preceding studies (Mentis 1988). The original model(s) can be corroborated, rejected, or modified as required.

There are aspects of narrower or specific topics that may be addressed employing a Hypothetico-deductive approach where ideas are explored for 'best fit' answers. This would take place in a manner similar to that presented by MacDonald (1993) on declining birch populations during the past, criticisms notwithstanding. Such an approach is similar to the focused hypothesis testing method discussed by Mentis (1988). More appropriate for the paleoenvironmental data explored in this dissertation is a process which Mentis (1988:13) calls a "combination of or dualistic dialectic" between inductive and deductive approaches. It is a process of questioning-answering, a dialectic that is part of the exploration to find answers from large data and information sets that appear to be linked.

An initial and very general research hypothesis for the bulk sediment sampling program was developed by the Tuscany Project in 1995: "It is hypothesized that flotation analysis will help derive a better understanding of the spatial distribution of cultural material with the acquisition of vertical and horizontal data from the bulk sediment samples" (G. Oetelaar, pers. comm. 2001). At that point there was no indication that any macrobotanical remains might be recovered for the site. However, previous flotation work at the Carrier Mills Project, Illinois (Oetelaar 1982) and the Strathcona Park site, Edmonton (Oetelaar 1990) had demonstrated the probability that some botanical remains could be found (G. Oetelaar, pers. comm. 2001).

During 1995/96, a number of the bulk sediment samples were processed by flotation and examined with a low-powered microscope. Charred macrobotanical remains were discovered along with other remains including bone, microdebitage, insects and mollusks. The original hypothesis developed for the present research was proposed at that time, based on the realization that there were indeed botanical remains of interest to be recovered from the site. The original hypothesis proposed the examination of pollen, phytolith and macrobotanical samples to derive a vegetation history for the site. This vegetation history would then be compared to the paleoenvironmental reconstruction for southern Alberta based on other studies. The use of three lines of evidence (pollen, phytolith and macrobotanical) was intended to provide a self-testing set of data. One line of evidence could be compared to the others to see if the same or similar information was being derived from the same spatial contexts within the site. Unfortunately, the pollen and phytolith aspects of the research had to be abandoned due to time constraints. The remainder of the project was devoted to the processing and analysis of the charred macrobotanical remains recovered from 1043 samples from the buried paleosols at the Tuscany site. Testing and comparison of the results of the research was no longer possible within the site because of the lack of a phytolith and pollen analyses. Comparison with other southern Alberta data sets remained an objective of the analysis.

Nobody has attempted such an analysis in this region before. The charred botanical remains were not anticipated and the charred seeds were not actually recognized until all the samples had been processed by flotation. These remains were discovered through the process of sorting the samples. Initially it was not clear whether it was possible to identify the structures that appeared to represent seed remains. Once it was clear that some were seeds and were identifiable, it was possible to begin looking at them in greater depth. Problems with the Hypothetico-deductive approach increase with the complexity of the problems addressed. One of the main problems in the present research is that of pattern recognition. The charred plant remains present in the site that may be due to a number of processes, both cultural and natural. As noted by Reid (1985:172) discussing pattern distributions:

> The methodological issues are in how we isolate and interpret patterns from these distributions. Although workable programmatic strategies are lacking, good advice is available to indicate that pattern search and interpretation are inadequately handled by the hypothetico-deductive mode of hypothesis confirmation… … it is too narrow an approach, one that is inherently myopic in its scan of noisy archaeological data for the reliable recognition of multiple and often overlapping patterns.

This is one of the problems faced with attempts to derive an understanding of the paleoenvronment. While there may be patterns, they are difficult for us to recognize because of the 'noise' of the many variables that contribute to the picture.

> Paleoenvironmental reconstruction is generally an inductive process: the nature of the whole—in this case, the paleoenvironment—is inferred from that of one or more of its parts, the proxies. Examples of direct, unimpeachable links between proxies and the environment of the past are rare, and those that do exist are analogous to the "smoking gun" in criminal investigations. In the absence of such compelling information, a detective must painstakingly reconstruct the crime from clues at the scene, amassing a body of more circumstantial evidence. By defining a credible sequence of events, the investigator may eventually reach a conclusion (or verdict in this analogy) that is supported by a preponderance of the information at hand (Caran 1998:113-115).

Much of the material that is dealt with in this dissertation really has the feel of 'circumstantial evidence'. This is one of the problems that goes with the research field. A number of variables are summarized and examined—the train of thought and evidence amassed progresses from the specific to the general. In the present study this has led to discussions about sediments, soils, geomorphology, pollen studies, fire ecology, settlement and subsistence patterns, local ecology, paleofauna, cultural chronology, sampling theory, and glaciation patterns among others. The argument can become quite complex, but all parts play a role. Some points in an argument must be repeated as part of the effort to tie the entire suite of arguments together. The inductive approach is implicitly narrative, or at least can appear to be narrative, especially because so many separate topics are summarized and discussed to contribute to a general concluding model. According to

Dincauze (2000:22), "paleoenvironmental studies integrate the historical sciences. Dincauze (2000:21) notes that:

> There are many styles of scientific research. The physical sciences are experimental and quantitative. The natural and historical sciences are more descriptive and qualitative, less suited to controlled experiments. The former enjoy particular prestige for the precision of their methods and results. The latter stumble along dealing as well as possible with the complexities of the world and the intricacies of human perceptions, motives, and interpretations.

There is no body of work for this part of North America that covers the research performed in this dissertation. Paleoenvironmental reconstruction has turned to pollen analyses and the few, largely unpublished, macrobotanical studies for the southern Alberta region to derive some general ideas about what the vegetation may have been like during the late Pleistocene and early Holocene. Both types of studies represent vegetation inventories from pond environments and pollen studies can also contain pollen that has blown from areas hundreds of kilometers away. Thus, the recovered remains do not represent specific archaeological site locales where trees and flowers grew and people came to live. Work on sites in the Plains that have examined macrobotanical remains and that have focussed on features in sites, such as hearths and pits, do reveal cultural relationships that existed between the occupants of the site and some aspects of the local environment. There are no studies that have used bulk sediment sampling as a general approach to examine the entire site grid, as opposed to just the parts of the site that have clear cultural associations like hearths and pits. Asch and Asch Sidell (1988) have recommended that at the very least, a series of column samples should be extracted from sites. Their work at the Napoleon Hollow site in the American Midwest has been particularly innovative in its examination of the relationship between botanical remains and stratigraphy based on quantification of remains from a 280 cm profile (Asch and Asch Sidell 1988).

There has been no 'model' study to draw upon for the present study. There is not doubt that the next study of this nature could present an entirely different approach to the macrobotanical problem, building upon the foundation established in the present work.

Meaning of Materials Recovered and Context

The materials recovered from flotation analysis of the bulk sediment samples from the Tuscany site include bone, microdebitage, charred botanical remains, insects and mollusks. It should be noted that some analysis was begun with the mollusks but was discontinued due to a lack of time. The only remains that were analyzed were the charred botanical remains. These remains represent a biased sample because the charring that occurred at some time during the past resulted in the preservation of only these remains. How these charred remains were produced and how they came to be deposited within the sediment matrix, represent questions that will be addressed through various analyses.

The majority of the seeds recovered from the flotation analysis are seeds that have hard seed coats that are nut-like. The bulk of the seeds recovered are from a *Juniperus* species (juniper) and *Arctostaphylus uva-ursi* (kinnikinnick; bearberry). Both possess hard seed coats. It is suggested here that the resistant qualities of the seed coats have led to their preservation via charring and also to their continued survival within the paleosols through time. Both plants produce seeds that are contained in berries that can be useful to people. One of the purposes of the spatial analysis has been to determine whether cultural and/or natural processes have governed the production and/or distribution of the charred botanical remains. The spatial analysis specifically addresses only the buried paleosols within the site grid and the charred botanical remains contained therein.

"The laws describing general regularities in formation processes are known as c-transforms (for cultural) and n-transforms (for noncultural or environmental)" (Schiffer 1987:22). Two sets of data are represented in the present analysis. One set of data is related to the occupation of the site by people during the past and includes artifacts whose deposition falls under the laws governing c-transforms. Very little specific analysis of the cultural material from the Tuscany site has been carried out to date. The faunal remains have been identified to species level where possible. The lithic debitage has not been analyzed. However, both sets of remains have been plotted spatially across the site (only the lithic data are presented here) and demonstrate patterns related to discard processes. Until further analysis is carried out, little else may be said about these artifacts. The other set of data represented in this analysis comprises the charred macrobotanicals, whose deposition falls under the laws governing n-transforms. The examination of the charred macrobotanical remains suggests that their distribution is largely related to natural processes of burning and subsequent deposition.

The seeds recovered are the result of "n-transforms" (for noncultural or environmental) (Schiffer (1987:22). N-transforms fall under the "theories and theoretical systems of other sciences" (e.g. in this dissertation, geology, botany, geography, ecology) (Schiffer 1987:22). If the macrobotanical remains are considered the product of human intervention at the site, then some discussion of discard and use processes in the formation of the site is applicable. This is one of the problems that is addressed, and upon which a conclusion is drawn, at the appropriate place in this dissertation.

Taphonomic Approach

"Taphonomy is the science of the laws of embedding or burial. More completely, it is the study of the transition, in all details, of organics from the biosphere into the lithosphere or geological record." (Lyman 1994:1). Gifford (1981:366) provides a similar definition/description of taphonomy as "...an area of paleontological research that defines, describes, and systematizes the nature and effects of processes that act on organic remains after death." Lyman (1994) notes that archaeologists should be especially concerned with taphonomy because this is what archaeology is primarily about. Lyman's (1994) work is focused on taphonomic problems related to faunal remains in archaeological contexts and this appears to be the case for most of the literature specifically related to understanding taphonomic processes.

It is quite clear that macrobotanical remains also fall under the definitions cited here. In fact, there are sources (e.g. Hally 1981; Mikisicek 1987; Minnis 1981; Pearsall 1988, 2000), that actually do address problems of taphonomy that are related to macrobotanical remains, however, this is not explicitly stated in the discussions. This dissertation represents a taphonomic study of a charred macrobotanical assemblage. It is a study of the processes of formation of the depositional record and the reasons (as best as can be determined) why the assemblage has preserved and the processes that may have had influences on it through time.

PROBLEMS THIS DISSERTATION WILL ADDRESS

A number of ideas have already been presented related to problems that will be discussed throughout the dissertation. The first problem relates to the initial hypothesis and questions about the changes in vegetation through time. While there may be an issue with the oldest dates derived for the Tuscany site and the interpretation of the stratigraphy, it has been established that the oldest sediments are at least ~10,000 radiocarbon years BP in age. Thus, the chronology for the section of paleosols that are the focus of this dissertation begins at that time and ends at the event that deposited the Mazama tephra, which dates to 6730 ± 40 C^{14} years BP (Hallet *et al.* 1997). This represents the period of interest for which the general vegetation cover of the site is reconstructed.

A relatively large assemblage of charred macrobotanical remains was recovered with the flotation analysis. In total 67,074 charred plant fragments were recovered, of which 1477 are charred seeds. One of the most difficult problems with this material, besides identifying the seeds and structures, was identifying how closely linked the relationship was between the charred botanical remains and the cultural occupation of the site. This was examined in great depth and the conclusion may not be what some would expect. However, it is a carefully examined problem and the conclusion was not arrived at without very serious consideration.

As noted above, comparison with the local ecology is imperative in studies such as these so that it is clear that an ancient seed rain is indeed the object of inquiry. Local and regional ecology are problems addressed here and the information derived provides a tentative vegetation analogy for the Tuscany site during the past.

A Lusk cultural component, related to the Plains/ Mountain complex is associated with part of the charred macrobotanical assemblage. Its relative position within the site stratigraphy is critically assessed and has been dated to approximately 7800 radiocarbon years BP (specific dates will be thoroughly discussed below). There are problems related to general Plain's chronological taxonomy and the placement of the Tuscany site within it. These will be discussed. The vegetation reconstruction has important implications for the models of settlement and subsistence patterns that have been developed for the Foothills and Northwestern Plains regions.

An extensive discussion of general trends in the vegetation changes observed during the late Pleistocene and early Holocene in the eastern flanks of the Rocky Mountains, the Foothills, and southeastern Alberta is presented through summaries of pollen and macrobotanical research. These provide a comparative vegetation inventory that also provides a test for the fit of the local inventory from the Tuscany site. Some of the problems associated with pollen studies have to be considered in this general assessment. Nevertheless, corroborating evidence between the two sets of information is derived and confirms the general trends.

SITE AND DATA USED

During the summers of 1995, 1996 and 1997, the Department of Archaeology at the University of Calgary conducted archaeological excavation of the Tuscany habitation site (EgPn-377), located in northwest Calgary (Figure 1) at UTM coordinates 11PGU938654 and legal description: NW¼, Section 4, Township 25, Range 2, West of 5th Meridian. Excavations at the site were completed by the fall of 1997. The excavation grid covered an area of 94 square meters (only 91 units were excavated completely). The actual areal extent of the site is estimated to have been approximately 20 meters by 20 meters. The size of the site is estimated as the entire site was not completely excavated when the Tuscany Excavation Project ended. Thus, only approximately one quarter of the site was excavated.

The Tuscany habitation site (EgPn-377) was located in a fairly large depression in the landscape, often termed swales in Alberta. These features of the landscape represent locations that have a limited areal extent with an outer, roughly circular perimeter at one elevation and an inner circular area of lower or depressed elevation relative to the surrounding land surface. They appear 'bowl-like' within the overall landscape. The Tuscany habitation site depression itself covered an area of approximately 40 meters diameter N-S and 55 meters E-W, thus being more oval in shape. High spots surrounding the depression were located along the west,

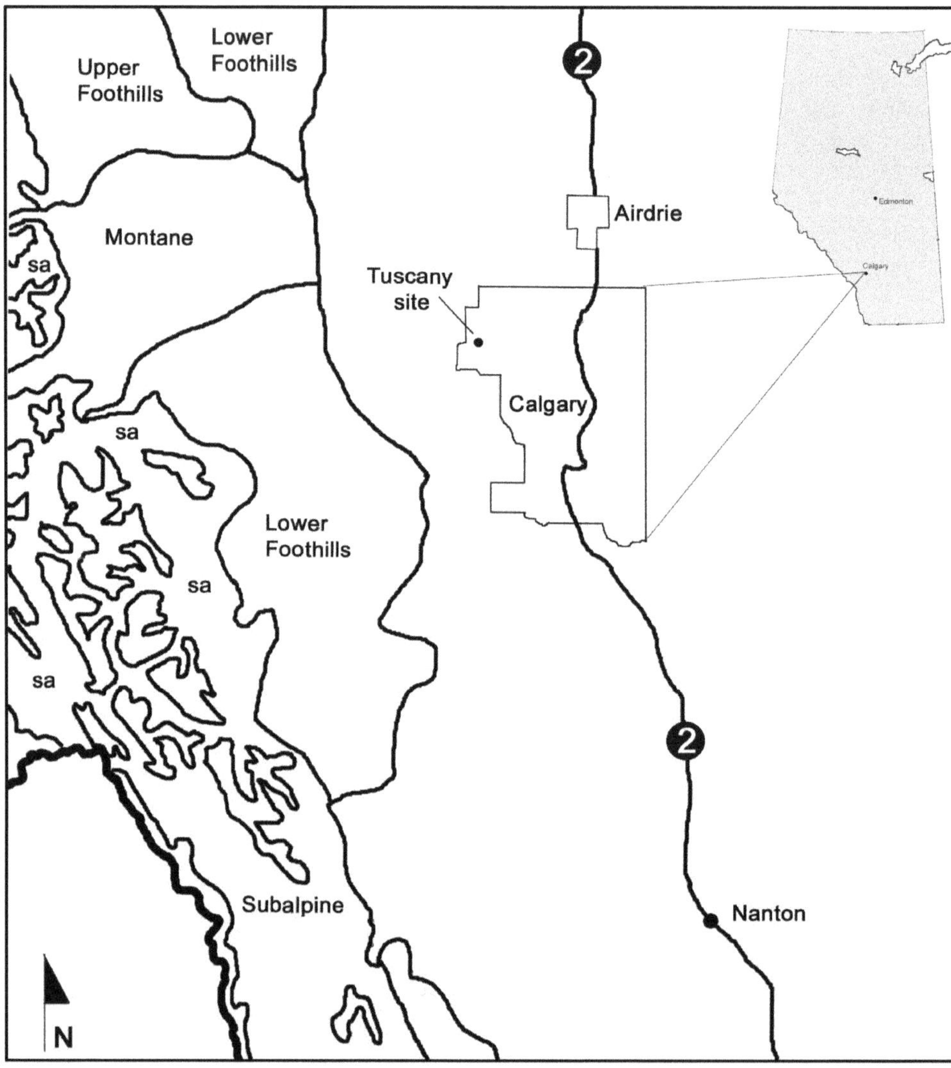

Figure 1: Map showing the location of the Tuscany site within Calgary

north and northeast boundaries (see Figure 2). There was an average of approximately 2 meters height difference between the high ground surrounding the depression and the lower elevation points within the depression itself. The high points or perimeter that surrounded the lower depression area spanned a diameter distance of approximately 100 meters N-S and E-W. While the area does not appear circular in Figure 2, it did appear to be relatively circular in the field. The lowest area within the depression represented a potential area for concealment in the landscape depending upon the direction from which it was approached.

The site was located in an area determined to have once been under the northwest part of Glacial Lake Calgary. This large glacial lake was associated with the last glacial maximum of the Laurentide and Cordilleran ice sheets during the terminal Pleistocene period. The site area was first exposed when Glacial Lake Calgary drained at some time during the terminal Pleistocene. The oldest sediments encountered at the site may thus date to approximately 13,900 calender years ago (Fisher 1999) and are represented by coarse to medium grained sands. All the sediments deposited above them are more recent in age. A series of buried paleosols immediately overlie the Glacial Lake Calgary sands. A very distinct Mazama tephra deposit caps the buried paleosols and provides a well-dated upper boundary at 6730 ± 40 ^{14}C years BP (Hallet *et al.* 1997). The details of the site chronology and its inherent problems will be addressed in detail through the dissertation.

During the first year of fieldwork, thirty-six units were set up within a grid oriented north-south (see Figure 3). Information provided by an archaeological survey assessment report completed by Fedirchuk, McCullough and Associates (FMA), Calgary, indicated the possible presence of a buried tipi ring between 40-50 cm below surface in the central area of the Tuscany depression. Initial expectations for the Tuscany habitation site were focussed on the possible buried tipi ring, a type of site that is not commonly encountered in southern Alberta during archaeological surveys or reconnaissance. Further interest in the Tuscany site also stemmed from its resemblance to the Hawkwood site (Van Dyke and Stewart 1984) (EgPm-179), located in northwest Calgary. The Hawkwood site, excavated in the early 1980's, was contained within two large depressions. The Tuscany

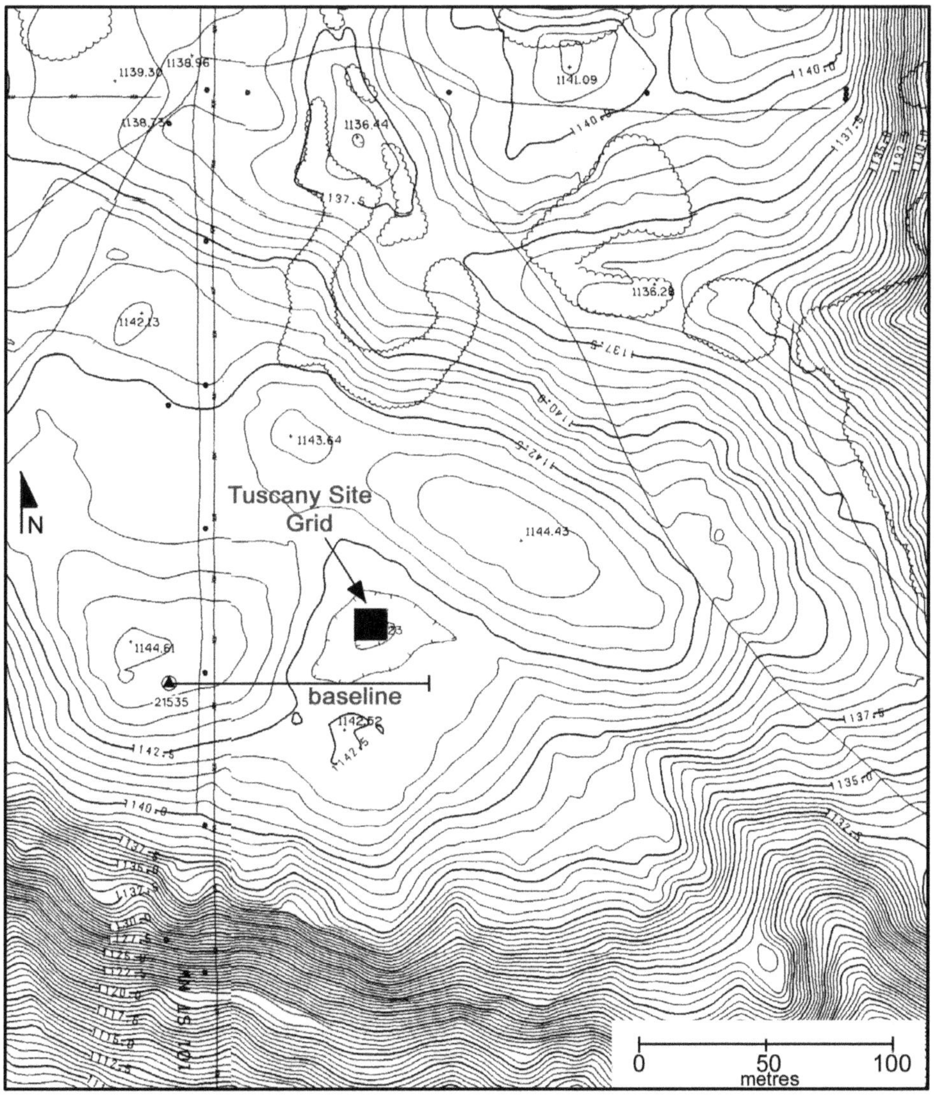

Figure 2: This topographic map identifies the variability in the surface terrain around the Tuscany site. The Bow Valley escarpment begins dropping down to the south approximately 50 meters from the excavation grid. The excavation grid lies within a distinct depression in the landscape, with high spots to the west, north and northeast. The edge of 12 Mile Coulee is visible in the upper right corner of the map, where the contour lines are closely spaced. The 14 x 13 meter grid is drawn to approximate scale (error margin may be 1/2 meter). Source: Adapted from Maps – Sections 4NW, 5NW, 8NW, 9NW, T25.R02.W5 (1:2500), Engineering and Environmental Services Department, © The City of Calgary 1995. With Permission

habitation site consisted of one large depression, described as a swale by FMA. Interest in the Tuscany habitation site was primarily related to the potential comparability of its appearance and materials to those of the earlier, Hawkwood excavation (Fedirchuk, McCullough & Associates, Ltd. 1990).

The uppermost soil of the Tuscany habitation site (EgPn-377), comprising the section from immediately above the Mazama tephra boundary to the surface landscape, was identified as a chernozem (Agriculture Canada Expert Committee on Soil Survey 1987). Chernozems are soils that develop in grassland settings and have very little differentiation within them in terms of characteristic horizons. Archaeologically, such soils are generally excavated in ten or five centimeter arbitrary levels because of the lack of differentiation. The initial approach adopted at the habitation site was excavation using a combination of both five and ten centimeter arbitrary levels. Ten-centimeter levels were used through the upper part of the excavation as part of the exploration strategy to identify the main features quickly. Five-centimeter levels were used later when it was recognized that a finer-grained approach might yield more meaningful information.

During the first year of excavation (1995) the first two levels were each ten centimeters deep and followed the contours of the surface of the unit. The third level varied enormously in depth throughout the units, but averaged ten to fourteen centimeters overall and leveled the unit off to 'tabletop' flatness. Thereafter, levels four through ten

								92	93	94	
								73	74	75	
							32	33	34	35	36
					23	24	25	26	27	28	
					16	17	18	19	20	21	22
76	77	86	60	61	11	12	13	14	15	30	
72	71	87	58	59	6	7	8	9	10		
68	70	88	56	57	1	2	3	4	5		
63	69	89	37	38	39	40	41				
62	67	90	46	47	48	49	50	51			
42	66	64	82	83	84	85	43	44			
52	65	78	79	80	81	53		45			
29	31		91	54	55						

Figure 3: The Tuscany habitation site grid at the end of 1997. All numbered units were excavated. North is the top of the grid

were generally excavated in five-centimeter increments, again with some variability. Levels eleven to twenty-three were excavated in ten-centimeter increments.

Methodology was changed and standardized for the remaining units opened and completed during the final two years (1996, 1997) of excavation. The sod layer was removed as the first level, to a depth of ten centimeters below the highest point measured on the surface. This approach leveled off the units quickly within 1 or 2 levels. The chernozem section was excavated in ten-centimeter, arbitrary levels only; there were no five-centimeter levels in the upper section. The Mazama ash and first two distinguishable palaeosols were excavated as natural levels of variable depth. Finally, the remaining paleosols were excavated in five centimeter, arbitrary levels to sterile, Glacial Lake Calgary sands (see Figure 3). The final level also represented a natural level as excavation of the units ceased when the sand was encountered, revealing the old buried surface of the sand before any materials were deposited overtop. By the end of the 1995 field season it was clear that the site was nearly two meters deep with the possibility of five cultural components at various depths within the sediments.

There were three defined sedimentary sections visible in the site profiles. Please note that the word section is not used in a geological sense here. It is used here to provide a category (in the literal sense of 'section') to delineate parts of the profiles from the Tuscany site as they relate to other types of profile analyses that have been carried out. As will be visible in the discussion on the sediments in Chapter three, several different approaches and descriptions are summarized. Overlapping terminology/ jargon is problematic. The use of the word 'section' here, is intended to stop the use of the same word for different phenomena or defined categories.

The upper sedimentary section was a developed chernozem soil, which was approximately one meter deep and developed in sediments proposed to be aeolian in origin based upon composition and appearance. Unfortunately, no grain size analysis has yet been carried out on any sediments from the site. The upper chernozem section contained three possible cultural components. Immediately underlying the chernozem soil was a sedimentary section identified as Mazama tephra, hence dated ca. 6700 years BP. This provided an important stratigraphic marker for this site (discussed in greater detail below). Immediately underlying the Mazama tephra were a series of paleosols, which contained the lowermost cultural component. The first paleosol of the series exhibited visible differentiation between an A horizon and a B horizon. For this reason, these two soil horizons were excavated as natural levels. Thus, a combination of three methods was finally used to excavate the Tuscany habitation site; arbitrary ten-centimeter levels, arbitrary five-centimeter levels, and natural levels.

As a result of this approach, there is only a very generalized match between units across the levels in the site. This is visible in the profiles that have been created for the north-south gridlines, the northings and eastings, within the grid. Level 17 (Q) in unit 2, for example, is not necessarily at the same general depth as level 17 (Q) in any other unit. Comparability of levels across the site was further complicated by the sloping nature of the depression itself, since levels were removed in a stepwise manner from one unit to the next. This strategy of excavation did present a problem for the spatial analysis and an alternative method was devised to view the data. Measurements had been taken at the top of each level and were easily converted to actual depths in meters above sea level (masl). The levels for each unit were mapped on graph paper to recreate a series of profiles along the northing and easting grid lines. These profiles were used to compare the variable depths and density of materials where botanical and cultural materials were found within the paleosols. The 'stepwise' view of the levels from one unit to the next is clearly visible in the profile drawings created for the spatial analysis discussion of the site.

BULK SEDIMENT SAMPLING STRATEGY: LEVELS OF SAMPLING

Bulk sediment sampling is not a common component of excavation projects taking place in the Plains region. There has been only minor focus on the recovery of plant remains by archaeologists working in the Plains. High rates of perishibility for plant remains result in poor representation in the archaeological record. While there has been an increase in the use of flotation analyses to look more closely at plant remains and small finds (e.g. see Lennstrom and Hastorf 1995), it is not a widely used method in the Alberta Plains. Time and monetary considerations generally govern the extent of sampling schemes within excavations that precede construction projects such as Hawkwood in 1981 and Tuscany in 1995/96/97 (both subdivisions in the City of Calgary).

While the Tuscany habitation site was recovered as a salvage excavation project preceding subdivision construction, it was also conducted as an archaeological field school through the Department of Archaeology at the University of Calgary. One aspect of the research strategy for the Tuscany habitation site included the collection of a series of bulk sediment samples from each unit by level and provided one-percent random samples for flotation analysis. This had been established as a standard procedure for University of Calgary field schools by Jon Driver in 1981 (Driver *et al.* 1982). The bulk sediment, sampling scheme employed for the excavation project was relatively simple. A bulk sediment sample (10 cm x 10 cm x 10 cm; 10 cm x 10 cm x 5 cm) was taken from the southwest corner of each level of each unit provided there was no sediment disturbance. If the usual location showed disturbance another location was chosen for the sample.

In total, 1043 samples from the paleosols located below the Mazama tephra boundary were processed. Remains recovered included charred plant fragments, charred seeds, terrestrial mollusks, insect remains and animal bones, as well as cultural material in the form of small flakes.

SUMMARY OF CHAPTERS

Chapter two provides an extensive summary of the methodology used in the recovery and examination of the macrobotanical materials. Though not a common approach in the northern Plains, bulk sediment sampling is a strategy utilized by many archaeologists around the world as part of their analyses of archaeological sites (e.g. see Pearsall 2000). Once taken, such bulk sediment samples must be further processed by water, by chemical flotation methods, or by a screen sieving method before they will yield any tangible data for analysis. The methodology and equipment used by projects throughout the world to process bulk soil samples are highly variable and lend great complexity to what might first appear to be a simple and straightforward processing procedure (e.g. see Daimant 1979; Lange and Carty 1975; Pearsall 2000; Struever 1968; Watson 1976). Processing of bulk sediment samples from the Tuscany site comprised a large part of this study and methodology requires detailed discussion. Once it was recognized that a substantial amount of macrobotanical remains had been recovered through the flotation processing, methods were developed to examine and identify the materials and will also be discussed.

In Chapter three a fairly detailed summary of the climate, ecology, geology and geomorphology of the Calgary area is presented. This information provides the background to understand how the environment of the site has changed through time. The cultural remains and features from the site are summarized and discussed, providing background for the spatial analysis in Chapter five.

Chapter four provides an extensive description and discussion of the charred seeds recovered from the Tuscany flotation processing. A number of important plant species were identified from the charred seed remains. Each of these plants are examined at length to develop perspectives of the role they played in the environment during the past and whether or not they may have been of importance to the Aboriginal populations that occupied the landscape.

Two plant species, kinnikinnick (*Arctostaphylus uva-ursi*) and juniper (*Juniperus* sp.), were found to be ubiquitous across the area of the site grid through time. Chapter five examines the spatial distribution of these two plants, as well as the other plants found to occur at the site through the time. The spatial distribution of the cultural materials is examined here as well to place it within the paleoenvironmental framework developed for the site. The distribution of the charred plant remains appears to be related to activity around a hearth, in the sense that it has been modified/crushed by trampling. A small, Lusk point related to the Plains/Mountain complex was found in association with the hearth feature and provides a timeframe based on the cultural chronology developed for the greater Plains region of the northwestern United States and Canada.

Chapter six provides a synthesis developed from the information summarized in the previous four chapters along with other lines of evidence. Together, these are used to derive ideas about the evolution of the paleoenvironment through the late Pleistocene and early Holocene. As mentioned, the Lusk projectile point places the site within a chronological framework. The implications of this framework are contrasted with the information derived from other dated sites with similar sediment stratigraphy in the Calgary area, along with other dated sites along the eastern flanks of the Rocky Mountains. The implications for Aboriginal populations living within this paleo-landscape are explored along with the evidence derived thus far on known Plains/Mountain complex lifeways.

An extensive summary of a number of pollen and macrobotanical studies is provided in Chapter seven. The studies included in the discussion are located in southern Alberta, eastern British Columbia, northwest Montana, and southwest Saskatchewan. One of the problems for paleoenvironmental research in the Plains region in Alberta is the lack of lakes deep enough to provide sediment cores for pollen analysis that extend back to the terminal Pleistocene period. This has meant that ideas about environmental change through time have relied on pollen analyses from regions like the mountains, foothills, and boreal forest, all peripheral to the plains. More recently, a number of sites have been located that have provided sediments containing well-preserved macrobotanical fossil remains that have been identifiable, providing valuable information. One problem with these sites, however, is that they have often been waterlogged through time and provide a local aquatic inventory of the paleoenvironment. The charred botanical remains recovered from the Tuscany sediments are from non-aquatic plants and provide an important inventory of plants from a very different ecological niche. This niche appears to fit in between the paleoenvironments described to the east and west of the site.

Chapter eight summarizes all the information available from the sediments excavated at the Tuscany habitation site. A series of sites in the greater Calgary region are summarized and contrasted with the Tuscany site. Chapter nine summarizes the conclusions of the dissertation.

Chapter 2 METHODOLOGY

SEDIMENT SAMPLING AND STANDARDIZATION

Technically, the samples that are referred to as "bulk soil samples" are really "bulk sediment samples". The term "soil" is firmly entrenched in the literature (e.g. see Pearsall 2000) that discusses sampling procedures related to flotation processing and analysis. However, the term "sediment" will be used here to identify the type of samples collected for this research project.

It is apparent that archaeologists and paleoethnobotanists have been focusing their attention on very specific features within archaeological excavations—with the intent of deriving as much data as possible on subsistence and activity patterns. Sediment samples have been taken from features like hearths, pits for storage and/or refuse, burials, and middens. In an ideal world, the best excavation methodology would be to sieve or use some flotation method on all material from an excavation to recover every cultural artifact possible. This is impossible. No excavation has the time, money or human-power to carry out such a procedure. The next best solution is probability sampling where some portion of the known universe (in this case, several cubic meters of ground) is sampled to derive an unbiased estimate of what is probably contained within it.

Pearsall (2000:66) recommends a "blanket sampling" strategy for flotation in which researchers "...collect soil for flotation from all excavation contexts... ...it is an easy strategy to carry out in the field." A blanket sampling strategy may recover remains in unexpected locations—places where remains would not normally be expected. However, Pearsall (2000:67) also qualifies her recommendations for implementing blanket sampling. The integrity of features must be maintained – hearths should be sampled separately from the surrounding house floors, pits from middens, floors from wall trenches, and so on.

Lennstrom and Hastorf (1992:1995) discuss bulk sediment sampling design from a paleoethnobotanical perspective, providing concrete examples from their excavations in Pancán, Peru. Studies such as these provide relevant examples of bulk sediment sampling designs and the pitfalls that can be avoided.

Lennstrom and Hastorf (1992) discuss two methods to take sediment samples: bulk sediment and scatter samples. According to Lennstrom and Hastorf (1992:206) bulk sediment sampling removes sediment "...from a single location, of contiguous matrix, from a single context." Scatter samples consist of "...several small "pinches" of sediment collected throughout a given provenience, level, feature, or context" (Lennstrom and Hastorf 1992:206).

Pearsall (2000:69) describes "...three commonly used techniques for taking flotation or fine-sieve samples: "pinch" or composite sampling, column sampling, and point sampling." Pearsall's (2000) pinch/composite sampling is equivalent to Lennstrom's and Hastorf's (1992) scatter sampling, but with the additional recommendation to collect standard sediment volumes. Pearsall's (2000) point sampling is more or less equivalent to Lennstrom's and Hastorf's (1992) bulk sediment sampling. Pearsall (2000) recommends the use of narrow balks as long columns, which can be bagged, labeled and written up at once. Point samples are small and from specific areas like the same corners of a grid of fifteen, 50 x 50 cm squares, or specific points like inside or under ceramic vessels (Pearsall 2000:71)). In all cases, sediment volumes should be recorded, according to Pearsall (2000).

The sediment samples collected at the Tuscany habitation site are a combination of Pearsall's (2000) column and point sampling techniques and fall quite well under Lennstrom's and Hastorf's (1992) definition. Sediment samples were taken as consistently as possible, from the southwest corner within each level of each unit in the excavation grid, much like point samples according to Pearsall's (2000) recommendation. Thus, theoretically, they should have represented a column of dirt. In actual practice, they became what might be called "point-bulk-sediment-samples". It transpired that the degree of disturbance to the site sediments by Richardson ground squirrels (*Spermophilus richardsonii*) was quite high. It was simply impossible to remove a bulk sediment sample consistently from the same location from one level to the next. In some cases the bulk sediment samples became thoroughly randomized in their locations from one level to the next through the depth of the unit.

Initial expectations for the possibility of encountering features, pits or hearths during excavation was low. After a review of the literature, the procedures of Lennstrom and Hastorf (1995) were contemplated. In their excavations at Pancán, Peru, they undertook a program of sediment sample collection from features and their adjacent locations. Samples from above and below the features, pits or hearths were taken along with the samples from within them. This procedure provided extremely interesting results when comparisons were made of the macrobotanical remains found between and among adjacent areas.

Only one possible hearth feature was found in the lower palaeosols (Unit 39 Level S) at the Tuscany habitation site. The feature was very difficult to differentiate from the surrounding sediment matrix as the excavator came down upon it. Thus, no sample could be collected from above the hearth in the manner recommended by Lennstrom and Hastorf (1995). However, the remainder of the feature was intact and was collected in its entirety. Sediment from underneath the feature was collected as well (this feature will be discussed further, below).

The method for collection of the sediment samples was very straightforward. The bulk sediment sample represented only 1% of the total volume of sediment per single level, whether or not it was an arbitrary ten or five centimeter, level. Thus, they were 10 cm x 10 cm x 10 cm or 10 cm x 10 cm x 5 cm in final, trimmed dimensions. This translates into final volumes of 1000 ml and 500 ml, respectively. In fact, most of the sediment samples were much larger once measured in the laboratory, averaging somewhere around 1700 ml in volume for the ten-centimeter levels and 800 ml for the five centimeter levels. The size of the samples collected from the habitation site were the decision of the project director. A recommendation by Pearsall (2000) is to consistently collect the same sizes of samples from their various contexts. This makes sense in that each archaeological site contains its own unique types of sediments and artifacts. The sediments of the Tuscany habitation site were interesting. Though they contained a large percentage of sand of aeolian origin, there was also a certain percentage of clay in them. The one-liter and half-liter samples collected at the habitation site appear small compared to other sites discussed in the literature (e.g. see Lennstrom and Hastorf 1992; Schock 1971; Pearsall 2000). Since they are not collected often in Alberta Plains archaeology the optimal sample size for this area has not been established as yet. However, one-liter samples appear to have been adequate for this project and would be recommended as the minimal size collected for future research.

The mechanics of extraction of the bulk sediment samples was relatively simple. The location of the sample within each level was recorded on the level map in the unit notebook. The general size and type of sediment sample (bulk sediment sample vs. feature sample, etc.) was recorded under the "samples" heading in the unit notebook—for each individual level. The sediment block was trimmed to its required final dimensions, sliced out and deposited into a ziploc bag. The unit information was written on each sample bag with a permanent felt marker. As well, a bulk sediment sample card was filled out, inserted into a smaller sealable ziploc bag and dropped into the sediment sample bag before it was finally sealed. The extra sample card served as a backup for information processing in cases where the 'permanent' writing on the ziploc bags failed. The cards were also saved in the laboratory where the measured sediment volume was recorded when the sample was prepared for flotation processing.

The excavation methodology undertaken at the Tuscany site had important ramifications for the data recovered from the flotation process. Arbitrary 10 cm, 5 cm, and natural layers were used to excavate the site. It appears that the natural layers were not clearly seen by excavators and there is little correlation between them from one unit to the next, as will be visible in the profiles that will be presented. The result of this excavation method is a set of levels for each unit in the grid that is, essentially, not comparable to any other unit at a one to one scale. Thus, the entire site cannot be considered as one large core with multiple comparable samples or samples that could be amalgamated to yield the kind of data appropriate for a computer generated pollen diagram. Unfortunately, few of the levels match; their data cannot be amalgamated in any way. Each unit is an entity unto itself.

An alternative method was developed to view the trends in the data. A series of wall profiles were created for each northing and easting in the site grid. The elevations of each level in each unit were calculated and plotted on graph paper. These were then scanned and redrawn on the computer. A total of 22 profiles were created, however, only a selection will be presented in the spatial analysis.

There is one final problem resulting from the sampling methodology that proved critical to the entire spatial discussion. When the bulk sediment samples were processed, all the sediment collected for each sample was measured and used. Thus, each sample comprised a random volume of sediment in milliliters. These random volumes had to be corrected in order to make the recovered material from the flotation samples comparable statistically. Each sample was corrected to represent 1000 ml of processed sediment. Thus, a calculation was made for each set of recovered botanical remains from each sample so that it represented the number of items that would have been recovered had the sample actually been 1000 ml in volume. For example, a 1600 ml sample with

6 seeds became a 1000 ml sample with 4 seeds and a 600 ml sample with 8 seeds became a 1000 ml sample with 13 seeds. All the spatial projections are based on the corrected values. Thus, the number of seeds represented on the density and distribution maps, and in the profiles, is not the actual number of items recovered through the flotation analysis; they are probability projections. The total number of charred plant remains recovered via flotation for the site grid was 67,074, of which 1477 (2.2%) were charred seeds. With volume correction, the total number of charred plant fragments was 80,476, of which 1630 (2.025%) were charred seeds. This represents a loss of 0.175% of the total number of charred seeds by using the volume correction approach. The sediment samples represent 1% of the volume of the site. The charred seed total (1477 seeds) of 2.2%, thus represents only 0.022% of the entire site volume and the corrected volume sample total constitutes a 0.020% representation of the site, a difference of 0.002%. These are not significant percentage differences.

The volume correction approach was initially intended for the charred plant fragments, which are represented in far greater magnitude than the charred seeds. The corrected value (80,476) for the charred fragments is larger by a factor of 19% compared to the original figure (67,074). In the spatial analysis, this difference in total number of pieces becomes more significant if we are looking for general patterns of depositional distribution, which was the case. However, it was determined that the volume correction approach (everything to 1000 ml) to the data had to be consistently applied and the charred seed data was corrected as well despite the much smaller frequencies represented. Since the frequencies of the charred seeds are very small relative to the total number of charred fragments recovered and the samples represent only 1% of the site, actual numbers should not be focussed upon too critically. The identities of the seeds as representatives of plants that grew in the paleoenvironment are far more important than are the numerical values from these statistically small samples.

FLOTATION PROCESSING OF BULK SEDIMENT SAMPLES

Introduction

Archaeological research in palaeoethnobotany or palaeobotany relies very heavily on the analyses of macroremains—the larger botanical materials that are visible to the naked eye and identifiable under low magnifications (Pearsall 2000:11). These remains are recovered from archaeological contexts, usually via some method of sediment sampling collection. Pearsall (2000:12) does note that the recovery of small material remains in situ may be considered more advantageous by some excavators, but actually introduces a serious number of biases into the data. The major bias introduced immediately, stems from the collection of larger and more visible remains like large seeds or charcoal, while smaller items like *Chenopodium* seeds are overlooked (Pearsall 2000:12).

To some measure, bulk sediment sample collection overcomes the recovery biases that are a product of the differential visibility of remains contained within the sediment matrix. However, once taken, bulk samples, or any other type of sediment samples, require researchers to give thoughtful consideration as to the methodology used to process them since that methodology determines what analyses may be carried out on the small material remains derived. Sediment samples can be processed a number of ways, including water flotation, chemical flotation, water sieving, and dry sieving.

Dry sieving did, in fact, take place at the Tuscany habitation site. However, it was dry sieving at the ¼ inch scale where all the sediment from the excavation was processed for larger finds that may have been missed in excavation including items like small bones, flakes and tools. This sieving scale was far too large for the kinds of macrobotanical remains found in the sediments in this particular site.

Dry sieving as a method of processing sediment samples, however, has proven to be problematic in macrobotanical studies because the abrasive action inherent in sieving tends to be destructive of plant remains, especially charcoal (Pearsall 2000). The point of any sieving method is to separate out all the different size classes of botanical and other remains from the sediment matrix that contains them. Paleoethnobotanists have found that flotation techniques allow the recovery of all classes of small material remains. Sediment texture and structure may be important factors to consider when choosing a method to recover small material remains.

There is a great deal of variety in the construct and purpose of flotation systems. They can be manual or mechanical. Manual systems are labor intensive but usually inexpensive to build and maintain. They generally involve buckets, tubs and barrels to wash the sediment away and recover botanical and other remains left behind in the screens. Pearsall (2000) notes, manual flotation is generally slower and very tiring, thus requiring a good labor pool to complete the work.

Mechanical systems are usually more involved than manual systems. Less labor is required to run them but they can be very expensive, large, awkward and not easily transported, and noisy if they need a motor to run parts of them. Pearsall (2000) provides an excellent review of the history of the development of flotation techniques as well as a review of the various general types of systems that have been designed and used most often by archaeologists and paleoethnobotanists to recover small material remains. Since Pearsall's review is very thorough, it will not be summarized here but is recommended reading for anyone interested in pursuing flotation as a method to derive data for analysis of small material remains at a site.

The Tuscany habitation site grid ultimately grew to include ninety-four excavation units. Bulk sediment samples were taken from ninety-one of those units. Each unit averaged approximately twenty-six levels, of which half were comprised of the paleosol sediments, which were the focus of this research project. Ultimately, two different flotation methods were used to process the bulk sediment samples. Both methods had positive and negative aspects to them as will be discussed.

The First Flotation System of the Tuscany Archaeological Project

The first method utilized to separate small material remains from the Tuscany bulk sediment samples was one that had been utilized for a number of years at other field school sites. Thus initially, there was no need to choose and/or design a system that would be appropriate for processing the bulk sediment samples from the Tuscany project excavations. However, while working with the flotation system several problems were encountered and these require an in-depth discussion.

The first system used by the Tuscany Project appears to be a combination of different systems. A picture of a system of nearly the same design can be found in Pearsall's (2000:26) discussion of froth flotation. The tub itself is a polyethylene construction and holds about 30 gallons of water (a very rough estimate). The tub appears to have been designed for froth flotation, which was intended to overcome the problems of water and chemical flotation with charred botanical materials (Pearsall 2000:26). Froth flotation uses a collector, often kerosene, which is added to the water before flotation (Pearsall 2000:26). The collector coats the charred remains which then make better contact with rising air bubbles and hence get carried to the surface (Pearsall 2000:27). The addition of a frothing agent as well, "...lowers the air-water surface tension so that the air bubbles can cluster on the surface without coalescing" (Pearsall 2000:27). The Tuscany flotation system did not employ the use of either a collector or frothing agent. The original tub was modified to use forced air as an agitator. The air was jetted through holes in copper tubing that had been anchored in the bottom of the tub. Pushing air through the water caused a roiling action, which served to force the lighter fractions present in the sediment samples to float to the surface of the water and wash out into a fine mesh screen.

Tub flotation of the sediment samples from the Tuscany habitation site was a fairly simple procedure that took place in one of the laboratories in the Department of Archaeology at the University of Calgary. The basic equipment included a source of running water, a source of forced air, the modified tub, and two different sized screens (approximately 1.5 mm and .5 mm) to collect the small material remains that washed out from the sediment. Forced air agitated the tub of water. The 1.5 mm screen mesh was clipped to an insert which slid into the interior of the flotation tub to catch the heavy fraction of the material remains in the samples. The sediment samples were dumped into the water and then filtered through the 1.5 mm screen insert. Any pieces larger than the screen size would stay in the insert and most of the remaining sediment matrix would wash through and settle to the bottom of the tank. Running water circulated through the tub along with the forced air and served to wash the light fraction materials down an overflow tube and into the .5 mm screen. Flotation time varied from sample to sample depending on the volume and texture of the sediment. The amount of clay present was a significant determining factor in the length of time required to process a sample. The overall average flotation time was about one hour per sample. After one hour of air and water circulation the tub water would become relatively clear. The screens were removed and the materials caught in them were washed down to one corner using a small latex hose attached to a sink faucet. Two sets of newspapers were labeled and laid out on the lab counter. The screens were banged/tapped out over the newspaper to catch the light (.5 mm) and heavy (1.5 mm) fractions, which were then dried for a number of days before being sorted. The identification and counting of the material remains was completed with the aid of Wild and Zeiss low-powered microscopes at 10-40x.

As mentioned above, the clay content in bulk sediment samples can prove problematic for swift flotation with the tub method. Initially, some samples were left circulating in the flotation tub for nearly four hours and still contained clay clumps that had to be gently broken down by hand. As well, due to the large number of samples waiting to be processed (from a total of 1043) some dried out completely to 'cement-like' consistency by the time they were processed. Their volumes were difficult to measure (large lumps) and had to be estimated in some cases. They also had to be prepped/softened before they could be processed.

The solution was to deflocculate all of the samples in advance of running time with Tri-sodium Dodecahydrate Phosphate (TSP). One-gallon ice-cream pails with lids were utilized for this purpose. The measured sediment sample was placed in a pail. A measured amount of TSP, 10% of the sample volume, was sprinkled over the sample. The pail was then filled with warm water. The sample was stirred, capped, and left overnight to soak. Despite this pretreatment of the sediment samples, they still often took up to two hours to run through the flotation tub. Nevertheless, samples were processed in this manner from August, 1995 until January 1998, by which time approximately 550 samples were completed.

Problems with the tub flotation system

Initially, the flotation tub was simply an empty polyethylene container. The forced air system and hose to fill the tub and agitate/circulate water were separate items that were hung over the edge. After a few months of use, the tub was modified so that the water and air systems were attached/anchored through holes in the bottom of the tub. The tub was set on top of a board with wheels for ease of mobility and the old plastic drain tube was replaced with a motorcycle tire tube made of more flexible rubber. A number of minor adjustments made the whole system much easier for students to handle. Excepting the problems with the clay content of the sediments, the flotation processing carried on continuously and slowly.

The flotation tub was large and at least thirty gallons of water were required to fill it for operation. The water was continuously circulated for one hour per sample and thus many more gallons ran down the drain during that time. The amount of water used was never a contentious issue because the system was operated on University of Calgary property. However, the tub did take a long time to fill and drain and students began to run two to three samples with the same tub of water. This led to a question of contamination between samples. Though the water appeared relatively clear after an hour of circulation, small rootlets and other materials were visible floating in the water. Contamination between light fractions probably did take place. However, contamination also took place between samples due to the age of the flotation tub itself. The tub was acquired during the 1970's and had been well used by 1995/97. The plastic was quite scratched and gouged. It did not matter how carefully it was scrubbed and flushed out—small rootlets and organic fragments continued to cling to the plastic, only to wash off into the next, fresh tub of water. Thus, the question arose whether small seeds were being washed from one sample to the next. The implication was that there were two separate potential sources of contamination of the light fractions of all the samples processed in this manner.

Another problem requiring consideration was the heavy fraction screen. The mesh was standard window screen that can be purchased at any hardware store. When the sediment samples were dumped into the flotation tank, the heavy parts settled quickly to the bottom. Pieces smaller than the mesh size either floated out as the light fraction with the water and air circulation or, if they were saturated from their soaking in the TSP, sank to the bottom and fell through the mesh. There were *Chenopodium* seeds present in some samples and the question began to arise whether small seeds of similar sizes (1 mm to .5 mm) might be dropping through the mesh to be flushed down the drain. A sample of sediment from the bottom of the flotation tub was saved and dried for examination. While no seeds were found in this material it became apparent that quite a large amount of the charred plant material present in the sediment matrix was falling through the larger mesh and being flushed down the drain. Realistically, most of the remains were too small to begin to quantify, but the question of possible loss of information remained.

The overflow tube, which carried the light fraction materials down to the fine screen, was another part of the tub flotation that may be considered mildly problematic. The water circulated and overflowed gently into a catching trough in the front part of the tub. Sediment always accumulated in the trough and had to be rinsed down with each sample change. From time to time the rinsing process was neglected and this led to probable contamination of samples from those previously run. The long overflow tube itself required rinsing after each sample and if neglected was probably yet another source of contamination from one sample to the next.

The screens for the tub system were large and unwieldy. They were awkward to handle and very tiring to work with. Sharp angles and corners caused numerous cuts and abrasions.

In summary, there were five main reasons to consider adopting an alternate methodology: 1) water wastage; 2) slow processing times; 3) four possible sources of contamination of the light fractions; 4) both the light and heavy fraction screens were large and awkward to handle, making the work tiring and unnecessarily difficult; and 5) loss of material through the heavy fraction mesh. When the Tuscany project ended in the fall of 1997 several hundred bulk sediment samples remained to be processed. The average number of samples processed during an eight-hour day was between four and six. Washing the screens out, filling and draining the tub and other daily laboratory chores contributed to this low rate of processing. At that rate, the task of flotation was estimated to require at least five months or more of concentrated work, even if that was the *only task* done each day—not a realistic expectation.

The Alternate Method: Bucket Flotation

During the fall of 1997 I met L. Scott-Cummings at the Third Biannual Rocky Mountain Conference, held in Bozeman, Montana. During a discussion on flotation methodology and the dilemma of processing the Tuscany sediment samples she recommended adopting a bucket flotation method to speed up the processing time. The instructions were available on the World Wide Web at an address with America Online (AOL.com). This methodology is outlined below.

Tools:

1 round 3-gallon bucket.

Long gloves (optional), which you will want if your water is very cold.

A long handled stirring spoon – tall enough to reach from the top to the bottom of the bucket (optional).

.25 mm mesh – you may place this mesh inside a sieve or attach it to a support. I use an old geological sieve that I have taken the sieving mesh out of. I attach the .25 mm mesh using a hose clamp to the bottom portion of the sieve, just below the heavy ring. This lasts for several seasons. I purchase a polypropylene .25 mm mesh. You could substitute chiffon as long as the holes are fine. [We use the same .25 mm sieve for each sample floated during the day – our light fraction does not dry in the sieve.]

1 mm mesh sieve

Gel capsules containing 50 parched poppy seeds for control. (I buy gel capsules at a health food store.)

Instructions:

1. Fill bucket with water about 2/3 full.
2. Measure dirt – I usually float about 1-2 liters at a time. Add contents of 1 gel cap to dirt.

3. Pour (don't dump) 1-2 liters of dirt into bucket.
4. Stir to create a vortex. You may use your hands (with or without long gloves) or a long stirring spoon, etc.
5. Allow vortex to slow, but not stop. Pour liquid through .25 mm mesh (or chiffon). Rinse light fraction lightly with running water (a hose, at the sink, or from a bottle).
6. Add water until bucket is about 2/3 full.
7. Repeat the stirring and pouring until the water is clear. I usually complete at least 3 cycles of fill-stir-pour. Sometimes it takes 6 or 7 (or more cycles until all floated remains are poured off).
8. When the supernatent is clear, let the light fraction dry in the chiffon or empty the .25 mm mesh onto a newspaper (or other convenient drying tool) and label it.
9. Pour the heavy fraction through a 1 mm mesh sieve, then label and dry the remaining heavy fraction (we use newpaper when we're in the lab).
10. If you have more dirt to float, start with more water and float another 1-2 liters.

The advantage to this method is that by swirling and creating a vortex, then pouring before the vortex has stopped you are capturing many of the "near floatables", not just the seeds and vegetation bits that float to the top.

You must practice to develop a technique where the pouring is fast enough to capture the botanic remains in the light fraction, but slow enough to keep from pouring out the fine portion of the heavy fraction and having a very messy light fraction to examine. Watch for dark specks, (which should be charcoal) that travel just along the top of the heavy fraction and are visible as you are almost finished pouring off the liquid. You should be able to pour most of these out, while leaving the heavy fraction behind.

As with most methods, this is easier to demonstrate then to write up. For those of you who add poppy seeds (or any other control), you should be getting over 90 % recovery – this is a closed system until you screen the heavy fraction. I recommend parching the poppy seeds for controls, since fresh seeds tend to sink and become mixed with your heavy fraction (and lost).

Obviously, since the bucket is rinsed out between samples, there is no possibility for contamination.

Instructions dated: 11-19-1997.

This method of flotation is extremely simple and inexpensive. It is also very portable and does not necessarily need to be carried out indoors or at a sink. Most indoor plumbing would be unable to accommodate the high volume of sediments that would flush through the pipes. Some type of sediment trap would be required. One great advantage of the method is that much less water is required per sample to complete the flotation process. The screens can be smaller and easier to handle, thus reducing early fatigue. Because it is a closed system there is no contamination of light or heavy fractions if the bucket is rinsed out very thoroughly each time a sample is run.

There were, however, two aspects of the tub flotation that were desirable: the circulation generated in the tub by running water and its agitation by forced air. There was no way to contain a forced air system in a small, two-gallon bucket. It was possible though, to circulate the water using a piece of latex tubing as a small, flexible hose attached to the faucet of a sink. An alternate method of bucket flotation was developed to process the Tuscany bulk soil samples (as outlined).

Tools:

1 small (2 gallon/8 or so liters) bucket

1 rubber cake spatula (the thicker version) with a new wooden handle

1-250 micrometer geologic sieve (U.S.A. Standard testing sieve No. 60)

1-1.00 millimeter geologic sieve (U.S.A. Standard testing sieve No. 18)

wood blocks – set the geology sieves on top of them in the sink for good water flow with no back-up.

50 centimeters of latex tubing (1/2 inch or 1 centimeter in diameter). Attach tubing to faucet with screw-on adapter – available as plumbing hardware.

A fine camelhair oil painting/model painting brush.

A fine-tipped dental pick.

A stack of newspapers.

Sticky labels and a waterproof pen to label samples.

Gelatin capsules. These were purchased from a pharmacy in lots of 1000 (size 0) for the best price but they are available at health food stores in 100 lots as well, at greater expense. Size 00 gelatin capsules were only available in 100 lots at greater expense.

Gelatin capsules containing 50 parched poppy seeds for control.

Instructions:

All bulk soil samples in this case were pre-soaked overnight with Tri-Sodium Phosphate and water in 1 gallon plastic ice-cream pails. This procedure is outlined above.

1. Set the 250 microns sieve on top of the blocks in the bottom of the sink.
2. Open a pail and gently stir the sample with the rubber spatula. The high clay content of the Tuscany soils made them quite sticky and pre-soaking and stirring was required. Other sites may not need this step in the procedure.
3. Gently pour the sediment mixture into the bucket. Add the 50 poppy seeds.
4. Rinse the ice-cream pail out over the bucket to remove the complete sample. Run water into the bucket to fill it about 2/3 full. At this point, the latex tubing works very well. The end of the tubing can be pinched gently to give pressure to the water. This spray causes the sample to froth because of the TSP content. The froth collects many of the floatable materials present in the samples. This action circulates the water in the bucket.
5. Tilt the bucket over gently until it begins to pour off through the 250 micron sieve in the sink. The latex tubing should be in one hand and directed

into the bucket so that water circulation continues. The floatable materials flow over the edge of the bucket with the froth created from the pressure spraying. This spray is not strong enough to cause any breakdown of items in the sediment samples.

6. As the water becomes clearer, the tube can be dropped and the bucket tipped over until all the water runs out through the sieve. This process can be repeated several times. A gentle flow of water from the tubing can be used to liberate floatable materials, if the bucket is tipped over to a 20 degree angle (or less) over the sieve. This procedure takes 5 to 25 minutes depending on the sample texture.
7. Change sieves. Pour the remaining sediment through the 1 mm sieve as quickly as necessary. Rinse the bucket out thoroughly.
8. Use the latex tubing to rinse the material in the geology sieves to one side. Gentle spraying from the underside of the sieves, held tilted, moves the material quickly.
9. Bang out the sieves onto labeled newspapers. Fold papers and set aside to dry. The smaller, brass geology sieves should be banged out straight up and down, never on their edges. Otherwise, they will twist/warp and will not be stackable for other types of sieving tasks.
10. The small paintbrush is useful to remove items clinging to the sieves. The dental pick can be used to remove small items stuck in the 1 mm sieve mesh.

As previously mentioned the average bulk sediment sample sizes were 1700 milliliters for 10 centimeter arbitrary levels and 800 milliliters for 5 centimeter arbitrary levels. These volumes were quite manageable with the equipment recommended here. During the early stages of using this new methodology it was already apparent that the recovery of items from the samples was actually better than that of the tub flotation procedure. This was clearly demonstrated by the recovery rates for very small, round, black items which appear to be the sporangium of a type of root fungus. The sporangium, tentatively identified as *Cenococcum* sp., averaged less than one millimeter in size and was often smaller than .5 millimeters. If these remains were recovered in the tub flotation procedure, they could be found numbering up to as many as eight in the light fraction. However, when recovered in the bucket flotation procedure samples, they numbered in the tens and twenties, and in some cases as high as the seventies. The recovery rate of these sporangia illustrates the significant difference between the efficiencies of the two methodologies.

Probably the most important aspect of changing methodologies halfway through the project was in the amount of time that it took to finally finish processing the remaining bulk sediment samples. The time required for prior processing has been outlined already, above. The last 493 samples processed by bucket flotation took approximately five weeks to complete. These bulk sediment samples were from 1997, the final summer of excavation of the Tuscany habitation site. This task took a fraction of the time it had taken to process the previous 550 samples using the tub flotation method. Bucket flotation is tedious and tiring because it requires complete attention and no other tasks can be done at the same time. However, two to four samples can be processed per hour, which doubles/quadruples the number of samples processed over the same amount of time by tub flotation. Quality of information is an extremely important consideration of any methodology but the time required to acquire it must be taken into account as well. In the case of flotation analyses it is highly desirable to process samples quickly and efficiently. Sorting and analysis of the materials recovered are the areas where time and attention should be invested.

Experimental Water Sieving: Alternative Methods to Derive Samples

During processing of the sediment samples with the tub flotation method, it became increasingly apparent that large amounts of materials were being flushed out of the bottom of the flotation tub and down the drain. It became important to know what macro-remains were passing through the bottom screen insert from which the heavy flotation fractions were being derived. Before settling on use of the bucket method of flotation, a number of other methods of washing bulk sediment samples were explored. While none of these methods were used to process the bulk sediment samples from the Tuscany habitation site, they will be summarized here. Perhaps information from this experimental exercise may be useful to future studies.

Other methods of separating bulk sediment samples include dry sieving and water sieving. Water sieving is a method that has been used in Mediterranean sites (e.g. Daimant 1979; Limp 1974) as an alternative to tub flotation methods. Experimental work comparing the two separation methods, via run times and amounts of recovered materials, were conducted by Limp (1974). According to Limp (1974:340) the water separation technique employed and described (the 'French' method) was two to three times as fast as the flotation method. However, that flotation method is not clearly described, making it difficult to assess the significance of these results. The method derived a heavy and light fraction from a series of sieves, somewhat similar to results from the Tuscany flotation methods.

There is no equipment or sieving system at the University of Calgary equivalent to that described by Limp (1974). However, soil samples are often dry-sieved for soil studies such as grain size analysis. This is carried out with a series of geological screens/sieves, which come in various mesh sizes. It was determined that such sieves would suffice for experimental water-sieving work with some of the bulk sediment samples from the Tuscany site. Five screen sizes were chosen to sieve the samples by size class: 1.19 mm, .85 mm, .625 mm, .425 mm, and .3 mm.

Three methods were developed for testing. A step by step description of each of these methods is provided. For the first and second test methods only samples deflocculated with TSP were sieved through the screens. The third sample was not pretreated with TSP

The first water sieving test

1. Prepped a bulk sediment sample by deflocculation with TSP.
2. Stacked the geology screens in the sink (on wooden blocks for drainage).
3. Stirrred the deflocculated sample thoroughly. Poured sediment mixture into the top screen slowly. Rinsed it through with tap water. At a certain point the bottom screen (.3 mm) began to plug up with silt, which quickly escalated out of control. The screens were really too small to run 1800-2000 ml. volume sample through them. There was simply too much material to be caught in the lower three screens. It became apparent that the screens were silting up badly when water started seeping out between the seams of the screens. Once this started happening, the screens had to be unstacked and each one sprayed under high pressure using the latex hose to break through the silts and clays so the screens could drain. This method did not work well with a large sample containing large amounts of clay and silt. The time required to run the sample was 45-50 minutes, making it roughly equivalent to tub flotation.

The second water sieving test

1. Prepped a bulk sediment sample by deflocculation with TSP.
2. Stacked the geology screens in the sink (on wooden blocks for drainage).
3. Stirred the deflocculated sample thoroughly. Poured a portion of the sediment into a 500 ml beaker and set the beaker on the top screen of the stack then ran water into the beaker so that it filled and overflowed. The overflowing water carried the sediments out and they poured gently through the screen. Topped up the beaker with more sample when it became clear. At a certain point the beaker became full of the heavier sands that could not be carried out by the overflowing water. This was rinsed slowly through the screens. Again, the screens began to silt up and would not drain. The screens had to be unstacked and drained with high-pressure spray.

The third method of water sieving

1. Measured the volume of a sediment sample
2. Subsampled the volume by 10% and placed the subsample in a 500 ml beaker.
3. Stacked the geology screens in the sink (on wooden blocks for drainage). Set the beaker on the top screen. Ran water into the beaker to circulate the sample till it overflowed gently. This subsample was not deflocculated with TSP, which made an incredible difference in the way the sediment broke up in the beaker. The palaeosols from the Tuscany habitation site did have a distinct clay portion, which was enough to make sieving without the TSP pre-treatment problematic. The clay clumped in the bottom of the beaker and when poured out onto the screen it continued to remain clumped. The clumps were difficult to break up under pressure spray from the latex hose. This method of processing these sediment samples was not recommended at all.

Summary discussion on experimental water sieving

One of the most important points discovered with this series of experiments was recognition of the difference that clay content can make to flotation of sediment samples. It became apparent that the only viable method to process the Tuscany palaeosols quickly and efficiently involved pre-treatment of every single sample with Tri-Sodium Dodecahydrate Phosphate and water at least one day in advance of running the samples through some sort of flotation device(s).

In retrospect, it is obvious that the geological sieving screens were really too small for processing the larger sediment samples. Larger sieving screens constructed like the geological sieves would make the first method a viable alternative to water flotation. Smaller sediment samples or subsamples from larger samples would make this method feasible as well. The primary problem with the first experimental method was the large size of the sample and the silt build-up in the sieve with the smallest mesh size.

One of the interesting differences also observed between flotation and water sieving was the cleaner appearance of the charred plant remains from the sieving process. The charred plant remains from the flotation processing always appeared a bit dirtier or muddier and hence not as outstanding. The charred remains from water sieving were darker, obviously charred, and showed up better as a result. This difference in appearance showed up most clearly under the low-powered microscope. Why there was such a noticeable difference in appearance was not completely clear. Possibly it was related to the way that water washed over the material once it was in the right sized sieve. Not only were the materials cleaner overall, but they also appeared less battered by the rinsing aspects of water sieving.

Conclusions on Water Flotation and Water Sieving

In retrospect, the method of bucket flotation that was finally adopted to finish processing the Tuscany bulk sediment samples was a cross between flotation and water sieving. The light fraction was acquired by a flotation procedure. The TSP functioned as a frothing agent to some measure and dispersed the clays, which made the samples easier to wash through the screens. The heavy fraction was technically acquired via a water sieving process and generally contained the greatest amount of material in terms of size and volume of recoverable materials. In fact, the light fractions often contained a few grams of fine-grained sand and silt, which had to be sorted through. However, the increased recovery of smaller items like the small spore bodies and small seeds indicates that the methodology was more rigorous overall.

Bucket flotation proved to be a superior method in terms of cost, time, and final product. It is not generally advisable to change methodologies halfway through a project, however, in this case better data was the final result of this strategy.

Sorting of the Light and Heavy Flotation Fractions

Sorting the light and heavy flotation fractions is probably one of the most important stages in the whole procedure of processing bulk sediment samples. The equipment required is quite simple and includes a low-powered microscope, two glass petrie dishes, tweezers that can pick up small and fragile items, a very fine oil paint brush, a thick oil paint brush (cat's tongue recommended), a biology probe, a small water container (a slide film canister will do), a large piece of glossy newsprint paper, a can of anti-static spray (for clothes), some kind of labeling system, and a very fine-tipped pen. In this project the final method of containment of items culled from the matrixes was gelatin capsules, 2 x 2 inch ziplocs and 3 x 4 inch ziplocs.

In the very beginning of the project (1995) the petrie dishes used for the sorting under the microscope were the small, disposable plastic type. These dishes were problematic for this task as they built up a static charge very quickly. Sometimes the very small botanicals and small shells in the samples would spontaneously repel away from the tweezers when static charges built up. Most often these items were never recovered again. Glass petrie dishes were purchased instead. Glass dishes build up static charges as well—not as quickly as the plastic dishes did but the effect was exactly the same. Sometimes a very interesting item would launch itself into the 'ether' never to be seen again! There was a solution and hence the recommendation to purchase anti-static clothing spray. Oddly enough, this type of a spray eliminates the static build-up problem with the petrie dishes. It needs to be sprayed on generously and then buffed. The anti-static-effect lasts for several days and when it tapers off, one can simply wash the petrie dish and then re-apply the spray.

The choice of tweezers for sorting can be critical to the task as well. Several types of snail shells were found in the palaeosol samples and were extremely fragile in most instances. They were very diagnostic, providing information about the environment at the time that the sediments and palaeosols were forming. Bulky tweezers that required a firm grip were not appropriate for delicate sorting—in fact they destroyed data. There are fine-tipped tweezers available for work in biology (dissection) and a variety of tweezers are available in local pharmacies that can be purchased at lower cost.

Each site will have different kinds of diagnostic materials present in the sediment samples. In this particular site the diagnostic botanicals and some of the non-botanical items were small and fragile. Nearly all of the botanical material was charred, or appeared charred within the palaeosol bulk sediment samples. Several thousand, gelatin capsules were used to contain these fragile items and have proven quite adequate for the job.

The methodology for sorting the flotation samples was quite straightforward and will be briefly summarized here.

1. Spread out the sheet of glossy newsprint on a counter.
2. Set a petrie dish in the center of the paper.
3. Selected the dried, light and heavy fractions of a sample.
4. Carefully unfold and open one of the fractions. Newspapers were used to dry and store the flotation samples. Once dried out, they could become stiff and unwieldy. Smaller newspapers, if available, are recommended. Most of the time, some rootlets or lumps of clay would stick to the paper. A probe was useful to pick these items off. Once everything was loose, the paper could be tilted and everything would slide down into the petrie dish. Material that fell out of the dish was retrievable in the same manner from the glossy paper underneath. In this way, loss from samples was kept to a minimum.
5. People developed their own method of sorting through samples under the low-powered microscope. In the case of the Tuscany materials there were often several hundred pieces of charred plant fragments to pick up. A miniature dustpan was made with a piece of index card. The charred plant fragments were swept up with the large paintbrush and tipped into a gelatin capsule. This proved much faster than picking each item up one at a time. The very fine paintbrush was useful to pick up very fragile pieces. Once the brush was dipped in water and squeezed off, most small items stuck to it and were easily deposited into a gelatin capsule. Each item saved was stored in a gelatin capsule. Each capsule was saved in a 2 x 2 inch ziploc bag that had been properly labeled. All 2 x 2 bags for each level were kept together in a 3 x 4 inch ziploc bag. All the ziploc bags for each level of each unit were hung on a binder ring. Thus any sample is immediately accessible to view at any time.
6. A recording sheet was kept for each level that had a sample requiring sorting. The light and heavy fractions were differentiated. A simple, tally system accounted for diagnostic items found in either fraction.

Parts of the methodology adopted for the palaeobotanical aspects of the Tuscany project were derived from Pearsall (2000). However, there is only so much information that can be gleaned from literature sources. Ultimately, it was found that the best way to derive methodology was to work with the materials. It became immediately apparent which ideas worked well and those that did not. Innovation and flexibility were two characteristics required for this work.

IDENTIFICATION OF THE SEEDS

All 1043 flotation samples acquired from the paleosols of the Tuscany habitation site were sorted under a low-powered binocular microscope between magnifications of 10x and 40x. Initially, a simple separation was made which differentiated lithics, bones, snails, botanicals and other items, from one another. Once the samples were completely sorted they were further separated into two collections, one containing only the botanicals and snails, and the other containing the remaining items.

There were several different types of charred seeds found in the collection. Many samples had more than five seeds, some had up to seventeen with several different types. The high quantities of individual items were problematic. A pictorial record of the collection had the potential to become very expensive with the purchase of film and developing costs. Additionally, one of the problems with taking pictures with a regular camera is that poor quality pictures are not apparent until they are developed. The collection has to be handled more than once to get a useful final product.

A new methodology was adopted which was intended to cut a number of steps out of the recording procedure. Each level was examined individually, including the light and heavy fraction. The charred/uncharred seeds were taken out of the gelatin capsules and arranged on piece of graph paper that had been taped to the viewing plate of the low-powered microscope. The graph paper provided a scale in millimeters for the seeds. This was useful when the comparative collection was used to identify the seed types as seed size is a critical part of identification at the species level.

A digital camera was used to take pictures of the seeds through one of the oculars of the low-powered microscope. The camera was a Sony MVC-FD73 Digital Mavica with a Quick Access FD Drive (2x) and a 10x optical zoom option. The recorded files were JPEG (640 x 480). Each floppy disk contained an average of 30 recorded images. A total of 21 floppy disks were used to inventory a substantial part of the charred seed collection. The recorded images were immediately viewable in a playback mode and if judged inadequate, were deleted right away and re-taken. Close-up shots were best taken at the lowest magnification of the microscope and the 10x zoom on the camera was used to pull the image in closer to the size required. The digital images can be opened immediately on a computer and edited for printing. For the purposes of this study the images were printed on laser-quality bond paper at 600 x 600 dpi. The quality of the images was adequate for a comparative pictorial inventory. Seeds considered diagnostic were also inventoried by unit and level using an index card system. A line drawing of each new type of seed was completed with a clear background scale for comparative identifications.

The comparative collection used for identification of the charred seeds is maintained by Dr. Alwynne Beaudoin and is presently located at the Provincial Museum of Alberta, Edmonton. The seed collection was a side-project for a number of years and museum botanists added seeds collected during field trips. Dr. Beaudoin has added material to the collection as well. During the summer, 2000, a seed collection donated by Agriculture Canada was inventoried and catalogued, which boosted the total assemblage to approximately 1600 items.

Identification was a slow process. Developing familiarity with the charred specimens from the Tuscany samples was the first step and was accomplished through the line drawings and the pictorial inventory developed. Familiarity with the different modern seed types was partially accomplished through the use of Montgomery's (1977) work with seeds from eastern Canada. Finally, simply looking at different seeds from the comparative collection under the microscope helped develop a knowledge base of the different seed types. A pictorial inventory was also made from the comparative collection using the digital camera. This proved to be a powerful tool. Color pictures were taken of various seeds from plant species that can be found in the present Alberta environment. These picture files were stored on a laptop and eventually on a CD-ROM disk, which can be used for identification in other projects.

As items were identified a smaller inventory was established for work with an Environmental Scanning Electron Microscope (ESEM) available for research at the Foothills Hospital Medical School, Calgary. Several important charred seeds and charred plant remain structures were mounted on stubs. Higher magnification enables the identification of things like the pore structure of wood that can differentiate deciduous wood from coniferous. One of the attractive features of the ESEM is that it is environment friendly. Scanning Electron Microscopes (SEMs) require specimens to be sputtered with a gold film to survive in the vacuum chamber environment. This step is not necessary with the ESEM and other than mounting, samples are unaffected by the process of examination. The picture quality from this equipment is superior to any other methodology. The digital format picture files are easy to access and edit on personal computers.

POLLEN VERSUS MACROBOTANICAL REMAINS

One of the means of deriving information about extant vegetated environments through time is via palynological studies. As noted by Moore, Webb and Collinson (1991:1), pollen grains of both recent and ancient age can be used in studies that endeavor to trace vegetational history via individual plant species or vegetation communities. Birks and Birks (1980) note that "...pollen analysis is the principal technique used to reconstruct Quaternary environments." Pollen diagrams are interpreted in terms of flora, vegetation, and environment, and the information provided can be compared for different areas (Birks and Birks 1980). The following discussion will provide background information to illustrate the differences between pollen studies and the macrobotanical research and methodology undertaken with this project.

Most palynological research is carried out on cores that have been acquired from lake and/or pond sediments which are expected to contain a record of local and regional pollen accumulation that extends back in time to the early formation of these bodies of water. In the studies conducted in western Canada, various methods are used to acquire sediment cores for pollen analyses. Some methods mentioned in the literature include a modified Livingston piston sampler (White 1987), a portable percussion coring system designed for operation from an ice surface (Reasoner and Hickman 1989), a Hiller peat sampler (MacDonald 1982), and power-assisted coring techniques (Vance, *et al.* 1993)

Once cores are acquired, they can be preserved in a frozen state to better maintain their integrity (e.g. Reasoner and Hickman 1989). The main objective of coring should be to acquire a sample that includes the most complete section through the deposit targeted for research (Birks and Birks 1980:38). Cores are subsampled after assessment by researchers so as to derive maximum information from them and include sampling schemes based on core length/depth and visible stratigraphic components. In southern Alberta subsamples are described as being processed by various methodologies including the standard methods of Faegri and Iversen (1975), for example. Processed samples are most often mounted on microscopic slides for the next stage of analysis.

Identification and counting of pollen grains is the basic method employed by the different analyses reviewed here. According to Birks and Birks (1980:165), the number of pollen grains counted in a study depends on the problem being investigated. The count should be high enough that it represents a random sample of the pollen grains present, otherwise the count achieved will not be reproducible (Birks and Birks 1980:165). The total number of pollen grains counted is very important to derive realistic representative numbers of individual taxa present in the samples. As noted by Pearsall (2000), counts of total pollen is a common approach, which can be adjusted depending on the variety of taxa present. For a smaller number of taxa, e.g. 20, counts of 150-200 grains are adequate and 200-grain counts yield 75-85% accuracy for the more common species present (Pearsall 2000). To derive representations of taxa that are more rare in the environment, the counts must be increased, which translates into a greater investment of time.

The type of count is important to the final calculation of representation. There are relative counts and absolute counts. Relative/percentage count proportions must total to 1, which means that when one taxa increases another *must* decrease even if the absolute amount of the other taxa has not actually changed (Pearsall 2000). In absolute pollen frequencies there are no dependant relationships to the amounts of other taxa as they are calculated by counting "...the number of pollen grains per unit quantity (by weight or volume) of deposit" (Faegri and Iversen 1964:40).

Counts of each taxon identified are included in a final total referred to as the pollen sum. Taxa included in the pollen sum can vary. A number of the studies reviewed during the present study indicated clearly what is included in the pollen sum, others did not. Some studies included pteridophytes, others did not. All pollen studies examined here subdivided the pollen core into pollen zones. Zones are defined by Birks and Birks (1980:168) as "...a body of sediment with a consistent and homogeneous fossil pollen and spore content that is distinguished from adjacent sediment bodies by differences in the kind and frequencies of its contained fossil pollen and spores." Assemblage zones are generally named for the most dominant taxa present in the counts.

The pollen diagrams examined for this dissertation presented percentage pollen sum data from the study sites. Pollen percentage data has been critically examined over a number of decades. It has been recognized that this data is open to misinterpretation depending on the method used to calculate percentage values of taxa against the other taxa identified in the analysis (e.g. see Davis 1963; Faegri and Iversen 1964; Pearsall 2000). The important point made here is that one must understand the methods used to count and display pollen data if one is to comprehend the patterns occurring.

Perhaps the most important information that can be drawn from pollen analyses is the identification of the different species or genera present in the landscape of each study location. They do present lists of overstory and understory combinations. Some of the plants listed are distant or regionally distributed as their pollen is airborne and carried to the pond or lake location. Other plants such as the aquatic taxa are obviously local flora growing in and around the ponds or lakes studied.

Pine pollen has to be critically evaluated when it shows up in samples from sites in the mountains, foothills and plains because it is produced prolifically and can be blown hundreds of kilometers. Thus, when pine pollen is tallied in amounts of 15-20% in samples it is most likely that there were no pine trees in the area, or if there were they were few in number. However, when pine pollen counts in samples reach proportions of 50-85% it is very likely that pine trees are at least in close proximity to the lake/pond being studied, an assumption that is being accepted here.

Sediment cores taken from lakes and ponds can provide excellent pollen records because of the quality of pollen preservation in such anaerobic environments. One of the problems for southern Alberta, part of the northern Great Plains region, is the lack of lakes and ponds from which good quality cores can be extracted. Similar problems appear to exist in the adjacent Great Plains region of the United States (Barnosky *et al.* 1987). Unfortunately, many of the pollen studies from the Plains region do not have the time depth of other pollen studies. Many of the bodies of water available for study were formed later in the Holocene and their sediment accumulation simply does not extend to the late Pleistocene-early Holocene timeframe.

However, even if there were more lakes with accessible sediments available there are problems associated with pollen studies due to: 1) long-distance dispersal of pollen, 2) over-representation of prolific anemophilous taxa like conifers, 3) poor taxonomic control due to identification of pollen grains to only the family or genus level, 4) representative bias in samples due to poor preservation of the pollen of some plant species (e.g. *Populus*), low natural production of pollen, and entomophily (pollination relying on insects) (Vance *et al.* 1995; Yansa 1998). Another problem noted by Beaudoin (1993), is the coarse time scale inherent in pollen studies because of the way cores are sampled. Samples themselves can span decades and the intervals between them through the core can span several centuries.

Pollen may be well preserved in many cases but its exact geographic origin is essentially unknown. Recovered pollen may be derived from a large, undetermined region. Vance *et al.* (1993:84) point out that present-day ecotone sites do yield good evidence of climate change through time because of their location between ecological and climatic zones. Thus, movements of ecological boundary zones and associated vegetation changes can be roughly determined and are informative (Vance *et al.* 1993:84). However, these authors recommend that pollen analyses be supplemented with data from other proxy indicators to develop more detailed reconstructions of paleoclimatic events (Vance *et al.* 1993:84).

A number of recent studies of Plains paleoenvironments in Alberta and Saskatchewan have utilized macrobotanical analyses and methodology to acquire data for study (e.g. Beaudoin 1996; Beaudoin and Halsey 1989; Beaudoin and Wright 1993; Steinhauser 1991; Vance 1991; Yansa 1998; Yansa and Basinger 1999).

Postglacial ice-related features called kettles often formed small ephemeral ponds, which have proven to be repositories of plant and mollusk remains that preserved in the anaerobic environment of pond sediments. These remains are providing important data on the local flora of sites that can be compared to pollen studies. Pollen analyses reveal a more regional perspective of plants present in landscapes whereas macrofossil analyses yield a very local perspective of the vegetated landscape.

According to Yansa (1998), there are countless prairie potholes in the hummocky moraine areas on the northern Great Plains and they may be filled with well preserved stratigraphy and macrofossil remains. Studies such as Yansa's (1992) investigation of one of these prairie potholes in southwestern Saskatchewan, show one new direction for paleoenvironmental reconstruction. Intense study of sites at different locations on the landscape could lead to a much more detailed paleoenvironmental reconstruction of western Canada during the late Quaternary once the data from the local scale macrobotanical studies are combined with the more broad-scale information acquired from pollen studies.

CHARRED PLANT FRAGMENTS

In the Tuscany flotation project the macrobotanical remains constituted the largest part of the materials recovered. Nearly all the macrobotanical remains were charred. There were two main categories of macrobotanical remains recovered: charred plant fragments and charred seeds/seed fragments.

The criterion for sorting the charred plant remains out of flotation samples was fragment size. If a piece of charred plant could be more readily picked up with fine tweezers, it was counted and recorded on the flotation tally sheets. In general, pieces less than half a millimeter (<.5 mm) were not picked out, but the remainder of the sample was saved in a separate bag which was then stored with the diagnostic items that were recovered. Thus, everything was saved in case there is ever a need to re-examine the samples for something that may have been missed with the initial sort. An estimate was made of the number of smaller pieces left behind in the remainder of the sample. However, these numbers were not used in the database created. They will be useful for possible future work.

No attempt was made to try and recover the smaller remains by flotation using a heavy liquid, like sodium polytungstate or zinc chloride (see Pearsall 2000 for descriptions of heavy flotation methodology and available chemicals) or a seed blower, which is a chemical free method to separate samples (Pearsall 2000). There were too many samples and not enough time to attempt such recovery. The decision was made to ignore the very small charred plant remains in all cases/samples. However, there is the possibility that the smaller particles are representative of herbaceous vegetation like grasses and the broad leaves of plants. This idea has the potential to be investigated at a later date because all the sample remains have been kept.

All charred plant remains were quantified by a simple tally of the number of recovered pieces in each sample. The charred plant fragments were tallied for each level of each unit in the site. The results of this simple tally indicated that there were definite trends in the data that repeated from one unit to the next. There were definite points of high accumulations of charred material through the profiles.

In fact, the methodology for evaluating charcoal accumulation rates in lacustrine sediment studies is similar in approach. For example, in Morrison's (1994) study of regional fire history in south India, sums of charcoal particles per cubic centimeter and sums of total area of charcoal per cubic centimeter for each sample are charted against the stratigraphic layers in the study core. These simple summation tallies show peak totals that can be associated with historical and archaeological events where significant amounts of vegetation burning took place. The data correlate well with a pollen analysis that shows more open vegetation existing in the landscape after burning episodes (Morrison 1994).

A brief summary of work by Conrad and Koeppen (1972) at the Brewster site in Iowa discusses charcoal specimens recovered through flotation analysis. In this study, the charcoal specimens were identified to the species level, or the genus level if species was not possible. The total number of specimens was then summed by weight to the nearest 0.1 grams. These sums were then compared against one another to determine the relative amounts of each species or genus of tree or shrub wood found in the site through the depth of the excavation. As amounts of the different tree and shrub species increased or decreased, speculations could be made about climatic changes taking place or clearing for agriculture by human populations in the area (Conrad and Koeppen 1972).

These two studies utilized different summation approaches for charcoal data and produced results that could be interpreted in a meaningful way. The simple tallying of the charred plant fragments from the Tuscany flotation samples yielded changing totals for each level in a unit. It might be considered inconsistent because each person sorting samples decided what was included in the final count. However, there is a point where the fragments of charred plant are too small to pick up using tweezers and that was fairly consistent from one sorter to the next. An objective evaluation of the methodology leads to the conclusion that a weight summation of the charred plant fragments from the Tuscany samples would probably have yielded results similar to the simple counts. There was a fairly consistent size range to the charred plant fragments recovered from the flotation samples suggesting that each sorter picked out similar sized fragments. This can be verified so some degree based on the storage method used to prevent crushing of these samples. All the charred specimens were contained in size 0 gelatine capsules. There were very few pieces that did not fit into the capsules, which again suggests that at least the upper size range of charred plant fragments was consistent between sorters.

Charcoal Research

The charred plant fragments are comprised of small, charred pieces that are still very woody in appearance and in structure at higher magnification. The fragments appear to be some type of wood that has not quite burned to the point of becoming charcoal—they are not amorphous. They still have visible structures at high magnification that can be used to differentiate them as conifer (soft wood) or deciduous (hard wood) woods. This topic will be addressed in greater detail, below.

Charcoal is an amorphous inorganic carbon compound derived through the incomplete combustion of plant tissues. The formation of charcoal from wood is the result of an exothermic reaction where the combustion is supported through the production of the combustible gases, carbon monoxide and methane, and maintained by low levels of oxygen (Mac Donald *et al.* 1991:53; Patterson *et al.* 1987:4). The reaction takes place between 280-500°C, however, at 500-600°C the charcoal is converted to carbon monoxide through a process of glowing combustion at the glowing surface (MacDonald *et al.* 1991:54). Charcoal is very chemically resilient and survives the processing procedure used to derive pollen from sediment samples for pollen analysis. Point count estimates or direct measurements are made of charcoal fragments on pollen microscope slides to calculate charcoal accumulation rates through time in sediment cores (MacDonald *et al.* 1991; Patterson *et al.* 1987).

Fossil charcoal has been used by paleoecologists to examine the importance of fire regimes, such as their influence on plant populations, hydrology, and nutrient budgets (Clark 1988:81). A method now frequently used to examine the history of vegetation and fire regimes in an area is through detailed analyses of charcoal from lake sediments (e.g. MacDonald *et al.* 1991; Clark 1988; Swain 1978). Quantitative analysis of microscopic charcoal from lacustrine deposits has archaeological applications as well, potentially answering questions related to human settlement and land use changes through time (e.g. see Morrison 1994).

Charred plant fragments recovered by flotation were ubiquitous in the Tuscany site sediments. They were found in varying quantities in most of the excavated levels of the 91 excavated units in the grid. The primary difference between the earlier studies and methods just discussed, and the present study, is related to the size classes of charred materials recovered for investigation. As noted by MacDonald *et al.* (1991), the size of the charcoal fragments that are generally counted on microscope slides are between 50 μm^2 and 10,000 μm^2 or even larger. The size of the charred plant fragments recovered from the Tuscany flotation varies from very small particles, 250-350 µm in maximum size, to particles as much as 8-10 millimeters in length and width, and sometimes larger. The piece in Figure 15 is approximately 4.5 mm across and is representative of the average size of the larger pieces that were sorted out of flotation samples.

Size of charcoal particles generated from fire events appears to be an extremely important topic in studies using a microscopic approach. While there have been a number of studies utilizing the microscopic approach, there is now some question about whether there is any correlation between the known fire activity around lakes and the amount of measurable microscopic charcoal that is incorporated into the lake-bottom sediments in a manner similar to a pollen rain. According to MacDonald *et al.* (1991), more recent studies indicate that in some cases, peak fire years appear to have no influence on the amount of microscopic charcoal that is deposited in nearby lakes (e.g. see Swain 1973, 1978; Clark 1988b). This pattern was confirmed by their study conducted in Wood Buffalo Park, in northern Alberta.

The problems appear to be related to taphonomic processes which Patterson *et al.* (1987:4) define as "...the processes acting upon an object from the point at which it is produced to the point where it is finally sampled." In the case of microscopic charcoal studies, the original amount produced, modified by dispersal, deposition, and preservation, coupled with sampling, preparation and counting methods, all affect the final results (Patterson *et al.* 1987:4). In Clark's (1988a) examination of problems with charcoal analyses the focus is on transport of materials via the air and by water. Both pose serious problems for good distribution and deposition of charcoal. It is noted (Clark 1988a:68) that, "paleoecologists have adopted working hypotheses that govern their interpretations of stratigraphic charcoal data... ...it is presumed that "large" fragments remain closer to the source than do "small" fragments, which are spread evenly over a wider area." This hypothesis is attributed to Patterson *et al.* (1987) who provide a lengthy examination of the problems.

Morrison (1994:679) notes that, "... the dynamics of charcoal transport are much more complex...", especially when the principles of physics are applied as in J. S. Clark's (1988a) study, summarized by Morrison. Wind transport has different effects on different size classes of particles. Wind can pick up the smaller fractions and move them around by suspension, sometimes to great distance from the source. Wind can also move larger particles by saltation along a surface, where one particle strikes another, knocks it into the air, and when it falls it may strike another into motion, effectively moving many particles downwind. Larger particles are moved by traction, which is simply movement along the surface.

At the Tuscany site the two means of charcoal transport that appear to be evidenced by the size classes of charred plant fragments recovered are saltation and traction. Transportation by airborne suspension does not appear to be a major contributing factor to the deposition patterns of the charred plant fragments.

Given the size range of charred plant pieces recovered from the Tuscany flotation samples, there are of course, questions about their deposition within the sediments at the site. A spatial analysis has been completed for the charred plant remains through three dimensions; across the site in planview and down through the site by depth. These ideas will be explored further, below, but the point here is that it is clear from the analysis that an extraordinary amount of charred plant remains has been deposited throughout the paleosols below the Mazama tephra.

It is noted by Patterson *et al.* (1987:19) that in contrast to pollen research there does not appear to have been much research exploring problems related to charcoal deposition. This appears to be true and there is little research on the scale of the present study. Patterson *et al.* (1987:19) ask the question: "Are soils more useful for providing an indication of local fires, or does the soil-charcoal relationship suffer from the shortcomings evident in the soil-pollen relationship?"

While the question is posed from a paleoecological perspective, it can still be answered here. It is very clear that the soil and charred plant fragment relationship at the Tuscany habitation site has not been particularly problematic. The survival of the charred plant remains through time has been very good. Further, the charred plant remains survive the processing method very well. The charred seeds and seed fragments have also survived well. Examination of both sets of data will demonstrate the usefulness of the present study, especially in the realm of paleoenvironmental information.

Chapter 3
CALGARY SETTING, IMMEDIATE SITE SETTING AND SITE CULTURAL FEATURES

CLIMATE

Climate Classification of the Calgary Area

According to Klivokiotis and Thomson (1986:2), a number of factors control the climate of Calgary including latitude, general atmospheric circulation, geographic location, and relief features of the Rocky Mountains. Elevation and local topography are additional smaller scale factors. Latitude position controls the amount of daylight and sunlight angles and there is a seasonal range in the availability of solar radiation, which is instrumental in the great temperature differences between summer and winter. According to Klivokiotis and Thomson (1986:4) the climate of Calgary is classified as continental. This is indicative of greatly contrasting temperatures between daytime and nighttime and also between winter and summer. Location on the eastern side of the Rocky Mountains limits potential moderating effects of warm, moist air from the Pacific. However, warm westerly winds will occasionally draw mild air down to lower elevations, leading to warmer temperatures and melting snow. This phenomenon is known as the "chinook" and greatly ameliorates Calgary winters.

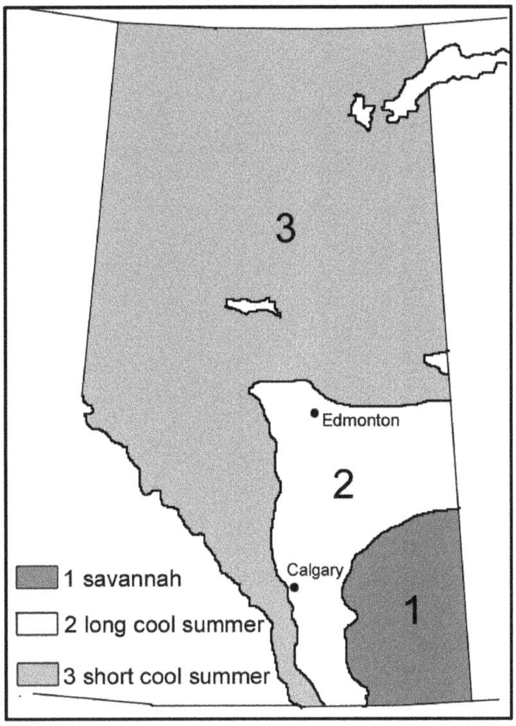

Figure 4: Koeppen's climatic zones (using data for 1931-1960). Source: adapted from Longley (1967:66)

Longley (1967:67) notes that most of Alberta is located in the cool, temperate zone of the classification scheme devised by German scholar, W. Koeppen. In this classification, the Calgary area is divided into two sub-zones, one with a short, cool summer and the other with a longer, cool summer. The latter has a summer with a mean temperature of over 10 degrees Celsius for four months. In Figure 4, Calgary can be found within zone 2 experiencing long cool summers. Calgary is also very close to the border of zone 3 where short cool summers are the normal pattern.

The southeast corner of the province (east and south from Drumheller) is classified as a savannah climate because of deficient and cyclical rainfall. The boundaries of the three zones match up fairly well with the Boreal Forest, Aspen Parkland and Grassland boundaries in the Ecoregion classification developed during the 1980s and 1990s (to be discussed below).

In MacMillan's (1986) summary, the climate of Calgary is characterized by temperatures that are warm in summer and relatively cool in winter, though moderated by Chinook winds. MacMillan (1986:16) notes, the Koeppen classification defines Calgary as having "...a Humid Microthermal climate that is cold with humid winters and cool summers". Chinook winds cause variability that is difficult to characterize.

Table 1: Biophysical and Climatic Characteristics of Ecoregions surrounding Calgary

Ecoregion	No.	Reference Vegetation	Reference Soil	Dominant Climatic Regime	Temperature (C°) Mean Summer (May-Aug)	Month of Maximum Precipitation	Total Precip. (Mm) Summer	Temperature (C°) Mean Winter (Nov-Feb)	Total Precip. (Mm) Winter	Total Annual Precip (Mm)
Fescue Grass	3	Rough Fescue-Parry Oatgrass	Black Chernozem	Prairie	13.8	June	214	-3.9	84	445
Aspen Parkland	4	Aspen and Rough Fescue Grassland	Dark Brown and Black Chernozem	Prairie-Boreal	14.4	July	259	-8.7	53	412
Montane	5	Lodgepole Pine and Douglas-fir (Succession to White Spruce and Douglas-fir)	Eutric Brunisol	Cordilleran-Prairie	11.9	June and August	210	-5.5	93	440
Subalpine	6	Lodgepole Pine (Succession to Engelmann Spruce and Subalpine Fir)	Eutric Brunisol	Cordilleran	9.4	July	263	-8.9	146	568
Alpine	7	Heather	Brunisol	Cordilleran	**7.6	July	**206	--	--	--
Lower Boreal-Cordilleran	9	Aspen-Balsam Poplar-Lodgepole Pine (Succession to White and Black Spruce and Balsam Fir)	Grey Luvisol	Boreal-Cordilleran	12.8	July	295	-7.8	60	464

** estimated values

Source: (Adapted from Map – Ecoregions of Alberta, W. L. Strong Ecological Land Surveys Ltd. 1992b)

Wilson (1983) outlines a Plains climate that is the result of three air masses: the Mild Pacific from the west, the Arctic from the north and the Tropical Maritime from the south. In the Calgary Plains area, Mild Pacific air dominates in summer and winter. Western Canada's weather is noted to be cyclonic but physiography, especially the Rocky Mountains, influences the whole pattern. Thus, the Pacific air mass loses its moisture by the time it reaches the eastern, downslope to the Plains. As well, orientation of mountain passes causes prevailing wind directions to be from the northwest in Calgary.

ECOLOGY

Alberta Energy and Natural Resources initiated a project, in 1979, to map the ecoregions or major climatic zones of Alberta. First, the province was divided into three main ecoprovinces including the Grassland Ecoprovince, the Cordilleran Ecoprovince and the Boreal Ecoprovince. These were further subdivided and thirteen large ecoregions were identified based on climate as expressed by the vegetation (W. L. Strong, Ecological Land Surveys Ltd. 1992a:1) (Table 1). It is noted that these large ecosystems are hard to recognize and map, primarily because the variation of both vegetation and climate within an ecoregion can be almost as great as that between the ecoregions themselves (W. L. Strong Ecological Land Surveys Ltd. 1992a). However, individual ecoregions are identifiable by the repetition of vegetation/site condition sequences and distinctive ecological relationships within the region. Information for the ecoregions discussed here is summarized from W. L. Strong, Ecological Land Surveys Ltd (1992a).

These ecoregions have been further subdivided into ecodistricts based on physiographical and/or geological patterns that characterize the landscape (W. L. Strong Ecological Land Surveys Ltd. 1992a:5) (Figure 5 and Table 2).

The categories in Table 2 are fairly self-explanatory except for "unit", which requires discussion. "Unit" is an alphanumeric code comprised of the ecoregion number and a letter code that describes the physiographic component identified for the ecodistrict.

Two of the ecoregions have been delineated for the immediate Calgary area. The Fescue Grasslands ecoregion is found on the east half of the city and the Aspen Parkland ecoregion is found on the western half. The dividing line between the two ecoregions runs north-south approximately halfway through the city. The Tuscany habitation site is located near the eastern border of the Aspen Parkland ecoregion. Vegetation at this location is actually transitional, being comprised of species dominant in both ecoregions.

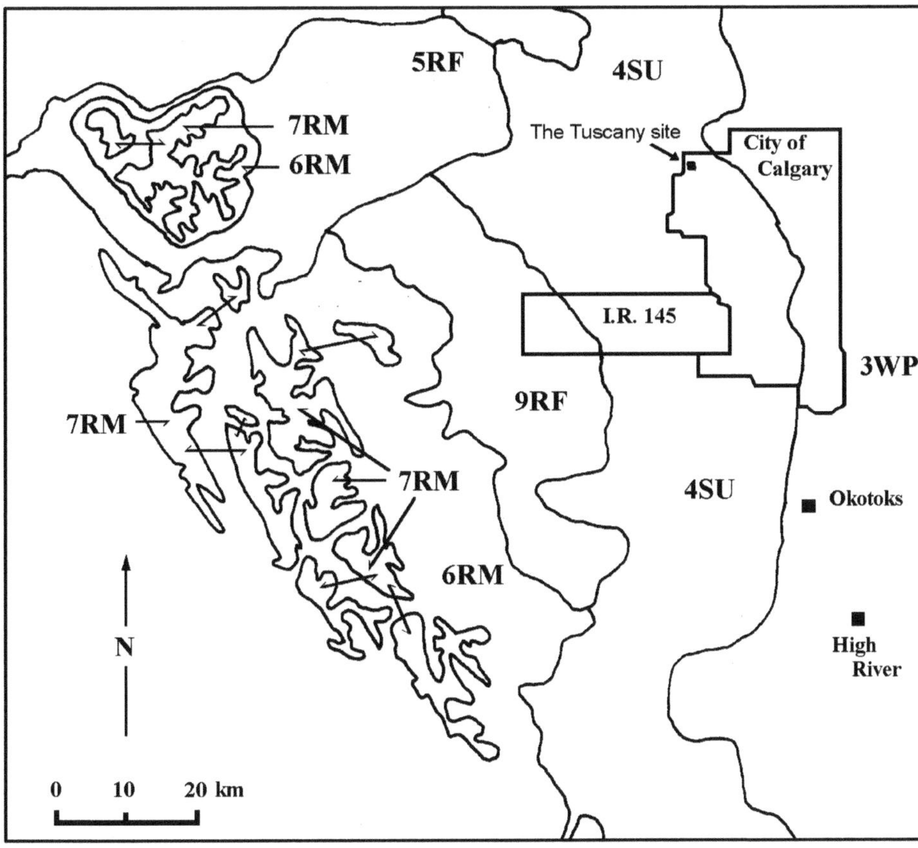

Figure 5: The Ecodistricts surrounding the greater Calgary region. See Table 2 in Chapter 3 for the descriptions of the alph-numeric codes

Table 2: Physiography and biophysical characteristics of Ecodistricts in the Calgary region

Physiographic Component	Unit	Landform/ Parent Materials	% Slope	Soil Great Groups	Moisture Regime	Dominant Vegetation
Southern Alberta Uplands	3SU	Rolling morainal plain	6-9	Black Chernozems	Submesic	Grasslands
Western Alberta Plain	3WP	Level lacustrine plain	0-5	Black Chernozems	Submesic	Cleared
South Saskatchewan River Valley	3SV	Steep slopes and terraces of Bow River valley	31-45	Eutric Brunisols	Submesic	Grasslands
Southern Alberta Uplands	4SU	Undulating morainal plain	6-9	Black Chernozems	Submesic	Cleared
Rocky Mountain Foothills	5RF	Ridged Foothills topography	31-45	Black Chernozems/ Eutric Brunisols	Submesic	Grasslands and coniferous forests
Rocky Mountains	6RM	Steep saprolite slopes	16-45	Eutric Brunisols	Mesic	Lodgepole pine and Engelmann spruce-Subalpine fir forests
Rocky Mountains	7RM	High mountain peaks	+71	Not soil	Subxeric	Absent or low-growing herbs and shrubs
Rocky Mountain Foothills	9RF	Ridged topography with morainal surfaces	10-45	Gray Luvisols	Mesic	Mixedwood forests

Source: (Adapted from W. L. Strong Ecological Land Surveys Ltd. 1992a: 59-61)

Fescue Grass Ecoregion

The Fescue Grass Ecoregion in the east half of Calgary is a grassland, however, Strong (W. L. Strong, Ecological Land Surveys Ltd. 1992a:25) notes that its climate is not typical of grasslands due to two related factors: increasing elevation and location near the Rocky Mountains. As a result, the area experiences cooler summers and warmer winters compared to the Dry Mixed Grass and Mixed Grass ecoregions to the east. Increased elevation of the Fescue Grass Ecoregion results in increased precipitation for the area. Importantly, the precipitation and temperature characteristics of the ecoregion parallel those of cordilleran climates, rather than grassland.

Annual precipitation for the Fescue Grasslands is 445 mm with 214 mm (mean of range 135 mm – 345 mm) falling during the summer on average and 84 mm during the winter. The characteristic that really differentiates these grasslands from the others is the mean summer temperature of 13.8°C, which is approximately 1.5°C lower than the Mixed Grasslands. There are fewer growing days and freezing temperatures occur in the late summer and early fall. However, winter temperatures are warmer than other Alberta ecoregions, because of the influence of chinooks during the course of the season.

The ecological conditions of the Fescue Grasslands are important to the background discussion because they are so close to the Tuscany habitation site, which today contains a mixture of Plains and Boreal vegetation. Six ecosystems are common to the Fescue Grasslands area. The reference ecosystem is dominated by Rough Fescue-Parry Oat Grass, which grows well on moderately well-drained Black Chernozems in dry/warm conditions. Also associated with these conditions are two ecosystems where Needle-Wheat Grass or Buckbrush are the dominant vegetation. In the transition ecosystems that are wet and cooler, the soils change from Chernozems to Brunisols and Regosols and the vegetation is dominated by trees and shrubs in the form of Saskatoon, Aspen, Willows and Poplars. These three sets of trees/shrubs characterize the last three of the six ecosystems common to the Fescue Grasslands. Some of the common low-growing herbs found in the Fescue Grasslands include everlastings (*Antennaria* spp.), sticky purple geranium (*Geranium viscosissimum*), lupines (*Lupinus* spp.), golden bean (*Thermopsis rhombifolia*), pasture sage (*Artemisia frigida*), common yarrow (*Achillea millefolium*), northern bedstraw (*Gallium boreale*), and fleabanes (*Erigeron* spp.). The Black Chernozem soils are characteristically shallow (Ah horizons of less than 30 cm) but of high organic content.

Aspen Parkland Ecoregion

The Aspen Parkland Ecoregion, as defined by W. L. Strong, Ecological Land Surveys Ltd. (1992a), begins approximately halfway (east-west) within the boundaries of the city of Calgary. This ecoregion is defined by "...the co-occurrence of Chernozemic soils (Black and Dark Brown) and a mixture of native grassland and deciduous forest plant communities in the form of a parkland" (W. L. Strong, Ecological Land Surveys Ltd. 1992a:27).

The Aspen Parkland Ecoregion is a transitional zone between the boreal and grassland ecoregions and is thus a mixture of the two areas in terms of climate and vegetation. As well, it is situated in the mid-Alberta storm track, which greatly influences the climate, differentiating it from the ecoregions surrounding it. The ecoregion receives more precipitation overall than grassland ecoregions, but as well, more precipitation falls during the month of July rather than June. Summer precipitation ranges from 234 mm to 323 mm, with a median value of 259 mm. Winter precipitation has a median value of 53 mm, similar to the grasslands. Total annual precipitation averages 412 mm. Summer temperatures average 14.4°C, which is actually warmer than the Fescue Grassland to the east. Winter temperatures, with a median of –8.7°C, are the coldest within the Grassland Ecoprovince. Importantly, for this discussion, the Aspen Parkland that occurs west of the Fescue Grassland in the Calgary area is probably influenced more by the interactions between Grassland and cordilleran climates than by Grassland and Boreal climates. This influence is strongest during winter months and is probably due to location or proximity to the foothills and Rocky Mountain Ranges, clearly seen on Figure 5.

Three physiognomically distinct types of vegetation distinguish the Aspen Parkland from surrounding ecoregions. Stands of aspen (*Populus tremuloides*), which are actually clone clumps, are found surrounded by grassy landscapes in the southern part of the ecoregion. Moisture conditions, slope and aspect are important factors enabling the growth of these stands of trees, which comprise approximately 15% of the vegetation cover of the ecoregion. Grass and shrub communities dominate the remaining landscape vegetation.

The dominant grasses of the reference site in this scheme include rough fescue (*Festuca scabrella*), bluebunch fescue (*Festuca idahoensis*), June grass (*Koeleria macrantha*), and needle grasses (*Stipa* spp.). Low-growing herbs include old man's whiskers (*Geum triflorum*), sticky purple geranium (*Geranium viscosissimum*), and northern bedstraw (*Galium boreale*).

While the grassland areas are found in the drier parts of the Aspen Parkland, shrub communities are found in places where moisture accumulates and remains into the summer, such as north-facing slopes and ravines. The more common shrubby vegetation found in these areas include saskatoon (*Amelanchier alnifolia*), prickly rose (*Rosa acicularis*), buckbrush (*Symphoricarpus occidentalis*), snowberry (*Symphoricarpus albus*), and silverberry/wolf willow (*Elaeagnus commutata*). Dark Brown and Black Chernozem soils are typically found in grassland and shrub communities.

Studies indicate that the Aspen Parkland ecoregion was probably dominated far more by grassland vegetation prior to the settlement and homesteading period following Contact times. Since pioneer settlement of these areas, fire frequencies have been reduced greatly, allowing aspen and shrubs to expand southward and eastward into grassland landscapes. Another explanation or hypothesis for aspen expansion centers on the diet and behavior of bison. Campbell *et al.* (1994), propose that aspen trees were browsed upon, wallowed around, trampled and

Table 3: Relative abundance of tree species per ecoregion

Ecoregion	No.	Aspen	Balsam Poplar	Cottonwoods	Black Spruce	Engelmann Spruce	White Spruce	Black Pine	Limber Pine	Lodgepole Pine	White Spruce	Balsam Fir	Subalpine Fir	Douglas Fir	Alpine Larch	Tamarack	Paper Birch
Prairie Ecoprovince																	
Fescue Grass	3	R	R	O	–	–	R	–	–	–	–	–	–	–	–	–	–
Aspen Parkland	4	D	O	O	R	–	O	–	–	–	–	–	–	–	–	–	–
Cordilleran Ecoprovince																	
Montane	5	C	O	O	–	–	C	–	O	D	–	–	–	C	–	–	–
Subalpine	6	O	O	–	R	C-D	R	–	R	D	R	–	C	R	R	–	–
Alpine	7	–	–	–	–	R	–	–	R	–	R	–	–	–	R	–	–
Boreal Ecoprovince																	
Lower Boreal Cordilleran	9	D	C	–	C	–	C	–	–	D	–	O	–	–	–	O	O

D Dominant, a prominent species in terms of cover and influence on the environment.
C Common, an abundant species but secondary to dominant species.
O Occasional, a species that occurs infrequently.
R Rare, a species that occurs very infrequently.
– Does not occur.
Source: (Adapted from W. L. Strong, Ecological Land Surveys Ltd.1992:32)

toppled by bison in the past and the demise of the bison herds in western Canada coincides with the initial expansion of the aspen parklands.

These are important observations. Likely, both the lack of fire and the decimation of the bison herds allowed the aspen to expand into the grasslands over the past 120 years. It is apparent that the landscape can change in appearance quite quickly with modification to just one factor. The expansion of the Aspen Parkland has taken place over a period of approximately 120 years. These observations of rapidly changing vegetation landscapes have important implications for both the archaeological and paleoecological records and the examination of changes that take place through time.

Discussion

This classification scheme reveals an interesting and fairly complex landscape for the greater Calgary region. The influences of modern human activity are greatest in the more immediate areas around the city of Calgary. Clearly, the city's urban sprawl has caused great change in the landscape, to the point that it will eventually become unrecognizable in terms of classification schemes such as those outlined here. The introduction of foreign, ornamental tree species, shrubs, herbs, flowers and weeds has already changed the local vegetation landscape greatly. Added to this effect is the tremendous growth of natural stands of aspen and spruce in the Bow River valley and its tributaries since the turn of the last century. Archival photos show a river valley with far less arboreal vegetation in it than is found today.

Tree Cover in the Calgary Region

W. L. Strong, Ecological Land Surveys Ltd. (1992a) provides a summary of the relative abundance of 16 tree species that are found in the ecoprovinces and ecoregions in Alberta. Table 3 provides a summary of this information for those ecoregions peripheral to the greater Calgary area. It is apparent that there are definitely far fewer trees in the area immediately surrounding Calgary. Of those trees present in the landscape, the deciduous trees dominate with Aspen being the most prominent in cover and in its influence upon the local environments. The difference between the areas east of Calgary, with few trees in the landscape, and the area west, with more tree cover, is also apparent.

Nose Hill Park

The seven habitats identified for Nose Hill have parallels in the areas surrounding the Tuscany habitation site. The information presented here thus provides background relevant to the local Tuscany landscape. Nose Hill Park is a very distinctive landmark in the city of Calgary, rising 122 meters above the Bow River valley in the northwest quadrant. It is underlain by the Porcupine Hills Formation, which has been covered subsequently by glacial till at the end of the Pleistocene glaciation. Information for this discussion is summarized from ***Nose Hill*** (Hallworth 1988).

Today, the topography of Nose Hill is quite varied and provides a series of habitats that are distinguishable from one another by vegetation: Grasslands, South-facing Slopes, Ravines, Shrub Communities, Aspen Groves, Disturbed Areas, and Spring Areas.

The Grasslands of Nose Hill are identified as representative of the "mixed prairie". These grasslands once covered much of the southern Alberta landscape except the southeastern corner. The grass species found on Nose Hill include blue gramma grass (*Bouteloua gracilis*), June grass (*Koeleria macrantha*), spear grass (*Stipa comata*) and sweet-grass (*Hierachloe odorata*). This habitat is not what would be considered a true prairie, where the height of the grasses may be in the range of 1.5 to 2.5 meters. Grasslands of the Calgary area contain only medium-height grasses.

South-facing slope habitats are also covered with mixed prairie grasses of medium height, along with some shrubs and low-growing herbs including prairie crocus (*Anenome patens*), golden bean (*Thermopsis rhombifolia*), yellow umbrella plant (*Eriogonum flavum*) and moss phlox (*Phlox hoodii).*

Shrub habitats are found in areas where more moisture accumulates than that required for aspen groves. Thus, shrub communities occur more often on ravine slopes, which are shaded during midday. Shrubs found on Nose Hill include buckbrush (*Symphoricarpos occidentalis*), silverberry/wolf willow (*Elaeagnus commutata*), saskatoon (*Amelanchier alnifolia*), and chokecherry (*Prunus virginiana*).

Aspen Grove habitats are found in areas where there is good moisture and are a feature characteristic of Nose Hill Park. The groves are produced by clone growth of aspen, which results in a round shape with older and taller trees in the center and younger ones on the peripheries. Some low-growing herbs found in the understory include yellow pea vine (*Lathyrus ochroleucus*), American vetch (*Vicia americana*), star-flowered Solomon's seal (*Smilacina stellata*), western Canada violet (*Viola canadensis*), and buckbrush.

Ravine habitats have formed along the slope edges of Nose Hill where snowmelt erosion has cut into the landscape. Small local habitats occupy these slopes including Grasslands, Shrub, and Aspens. These unique locations yield different moisture regimes, slope aspects and sun and wind exposure. These areas experience differing flowering dates because of these factors.

The Ravine habitats also have some springs in them. Spring habitats are found at the head of ravines.

Disturbed areas include gravel pits, abandoned fields and overgrazed and overused places in the landscape. Abandoned fields have been invaded by weedy plant species including Canada thistle (*Cirsium arvense*), blue bur (*Lappula echinata*), Russian pigweed (*Axyris amaranthoides*), and stinkweed (*Thlaspi arvense*).

Coulee Environments

The Tuscany habitation site is not a coulee site. It is, however, situated on the periphery of "Twelve Mile Coulee", which is a long channel oriented north-south and extending to the northwest edge of the city limits. There is a small stream that runs through the bottom of the coulee from a source well away from the Tuscany site. While the stream flows quite well during spring runoff, it often dries out by the middle of summer, leaving small evaporating pools which could be a source of water for people living on the terrace above, if they were willing to transport it somehow. The stream is partially responsible for the formation of the coulee, which has a typical, deep "v" shape from its erosional origins. The coulee cuts through the first terrace above the Bow River valley where the Tuscany site is located. The coulee ends where it opens and drops down to the level of the floodplain of the Bow River. The slopes of the coulee, immediately to the east of the Tuscany site, are fairly steep and present a rigorous climb for the unfit. There is easier access to the coulee about 1 to 1½ kilometers north of the site.

The coulee presents a unique microhabitat in terms of the moisture regime and plants that grow in the bottomland and along the slopes. Coulees have formed all over the plains region of southern Alberta. They are very unique and rich micro-environments, with diverse vegetation (Willock 1990). Twelve Mile coulee has a rather twisted course that yields slopes facing several different directions. South-facing slopes tend to be drier and have grasses and low-growing herbs as cover. North-facing slopes retain much more moisture and have larger trees and shrubs with accompanying understory as vegetation cover. There is thus, great diversity in the number of plant species found in the coulee. The results were good for hunting-gathering populations that were living in the area in the past. At various times, depending on moisture conditions and climate, the vegetation could have been similar to present conditions, which would mean some berries like saskatoons and chokecherries may have been available for harvesting. Diverse vegetation would also have attracted animals, perhaps making hunting easier. Environmental conditions are an important factor that must be considered in interpreting a human occupation of a landscape.

SOILS AND VEGETATION OF THE TUSCANY HABITATION SITE

The Tuscany habitation site partially fits into the soil and vegetation descriptions provided for both Fescue Grassland and Aspen Parkland. The modern topsoil at the site was a chernozem and was about one meter thick in the bottom of the swale. The chernozem developed in sediments that were aeolian in origin, fairly sandy and moderately well-drained. While there were no trees present on site, there were trembling aspen (*Populus tremuloides*) and balsam poplar (*Populus balsamifera*) in the vicinity, approximately 200 and 300 meters from the swale respectively. The swale contained low shrubs in the form of a large patch of buckbrush (*Symphoricarpus occidentalis*) and small clumps of shrubby cinquefoil (*Potentilla fruiticosa*). Also dominant in the landscape but not present in the swale itself, were large patches of silverberry/wolf willow (*Elaeagnus commutata*). Another common shrub to the area was soapberry/buffaloberry (*Shepherdia canadensis*) although it was also not present in the swale.

Low-growing common herbs included everlastings (*Antennaria* spp.), golden bean (*Thermopsis rhombifolia*), pasture sage (*Artemisia frigida*), common yarrow (*Achillea millefolium*), northern bedstraw (*Galium boreale*), and fleabanes (*Erigeron* spp.). Other herbs present in the landscape included nodding onion (*Allium cernuum*), prairie onion (*Allium textile*), a camas species (*Zigadenus* spp.), crocus (*Anemone patens*), scarlet gaura (*Gaura coccinea*), prairie sage (*Artemesia ludoviciana*), yellow umbrella plant (*Eriogonum flavum*), gaillardia (*Gaillardia aristata*), old man's whiskers (*Geum triflorum*), wild blue flax (*Linum lewisii*), purple prairie clover (*Petalostemon purpureum*), and star-flowered Solomon's seal (*Smilacina stellata*).

The environment at the Tuscany habitation site was more complicated than it first appeared. Initially, the site appeared to be a fairly undifferentiated prairie landscape. However, the plants growing on the site and in the surrounding landscape were diverse. Further complicating the diversity and availability of different plant species, was the presence of Twelve Mile Coulee, just a few hundred meters to the east of the site. Table 4 includes many of the plants found in the surrounding landscape near the Tuscany habitation site. All of the plant species listed were observed within 1½ kilometers of the Tuscany site during the three-year duration of the excavation project. The plants are listed in alphabetical order by scientific name, which is the easiest form of referencing for comparative literature or for viewing references in available plant guidebooks.

Table 4: Plants observed in the Tuscany habitation site and surrounding area

Plant No.	Scientific Name	Common Name
1	*Achillea millefolium*	Common yarrow
2	*Actaea rubra*	Baneberry
3	*Allium cernuum*	Nodding onion
4	*Allium textile*	Prairie onion
5	*Alnus incana*	Speckled alder
6	*Amelanchier alnifolia*	Saskatoon
7	*Anemone canadensis*	Canada anemone
8	*Anemone multifida*	Cut-leaved anemone
9	*Anemone patens*	Prairie crocus
10	*Antennaria nitida*	Small-flowered everlasting (pussy toes)
11	*Apocynum androsaemifolium*	Spreading dogbane
12	*Arctostaphylos uva-ursi*	Kinnikinnick; bearberry
13	*Artemesia frigida*	Pasture sage
14	*Artemesia ludoviciana*	Prairie sage
15	*Astragalus crassicarpus*	Ground plum
16	*Astragalus missouriensis*	Missouri milkvetch
17	*Betula spp.*	Birch
18	*Campanula rotundifolia*	Harebell; common bluebell
19	*Castilleja breviflora*	Indian paint-brush
20	*Castilleja spp.*	Indian paint-brush
21	*Chenopodium album*	Lamb's quarters
22	*Cirsium arvense*	Canada thistle
23	*Comandra umbellata*	Pale bastard toad-flax
24	*Cornus stolonifera*	Red osier; dogwood
25	*Cynoglossum officinale*	Hound's tongue
26	*Disporum trachycarpum*	Fairy bells
27	*Dodecatheon conjugens*	Prairie shooting star
28	*Elaeagnus communtata*	Silver berry; wolf willow

Plant No.	Scientific Name	Common Name
29	*Equisetum spp.*	Common horsetail
30	*Erigeron spp.*	Fleabane species
31	*Eriogonum flavum*	Yellow umbrella plant
32	*Fragaria glauca*	Wild strawberry
33	*Fritillaria pudica*	Yellow-bell
34	*Gaillardia aristata*	Gaillardia
35	*Galium boreale*	Northern bedstraw
36	*Gaura coccinea*	Scarlet gaura
37	*Geranium viscosissimum*	Sticky purple geranium
38	*Geum allepicum*	Yellow avens
39	*Geum triflorum*	Old man's whiskers
40	*Hackelia floribunda*	Stickseed
41	*Hedysarum spp.*	Hedysarum
42	*Heuchera richardsonii*	Alum root
43	*Hierachloe odorata*	Sweet grass
44	*Hordeum jubatum*	Foxtail barley
45	*Hyoscyamus niger*	Black henbane
46	*Juniperus communis*	Ground juniper
47	*Juniperus horizontalis*	Creeping juniper
48	*Lilium philadelphicum*	Western wood lily
49	*Linum lewisii*	Wild blue flax
50	*Lithospermum ruderale*	Yellow puccoon
51	*Lomatium triternatum*	Prairie parsley
52	*Lonicera glaucescens*	Twining honeysuckle
53	*Lycoperdon spp.*	Puffballs
54	*Matricaria matricarioides*	Pineapple-weed
55	*Mentha arvensis*	Wild mint
56	*Monarda fistulosa*	Wild bergamot
57	*Oxytropis macounii*	Early yellow loco-weed
58	*Oxytropis spendens*	Showy loco-weed
59	*Paronychia sessiliflora*	Low whitlowwort
60	*Penstemon nitidus*	Smooth blue beard-tongue
61	*Petalostemon purpureum*	Purple prairie clover
62	*Picea glauca*	White spruce
63	*Populus balsamifera*	Balsam poplar
64	*Populus tremuloides*	Trembling aspen; aspen poplar
65	*Potentilla anserina*	Silver-weed
66	*Potentilla concinna*	Early cinquefoil
67	*Potentilla fruiticosa*	Shrubby cinquefoil
68	*Prunus virginiana*	Choke cherry
69	*Pseudotsuga menziesii*	Douglas fir
70	*Ribes oxyacanthoides*	Wild gooseberry
71	*Rosa acicularis*	Prickly rose
72	*Rubus spp.*	Wild red raspberry
73	*Salix spp.*	Willow
74	*Shepherdia canadensis*	Canadian buffalo-berry
75	*Sisyrinchium montanum*	Blue-eyed grass

Plant No.	Scientific Name	Common Name
76	*Smilacina stellata*	Star-flowered solomon's seal
77	*Solidago gigantea*	Tall smooth goldenrod
78	*Solidago missouriensis*	Low goldenrod
79	*Symphoricarpos occidentalis*	Buckbrush
80	*Thalictrum venulosum*	Veiny meadow rue
81	*Thalspi arvense*	Penny cress; stinkweed
82	*Thermopsis rhombifolia*	Golden bean
83	*Tragopogon dubius*	Goatsbeard
84	*Viola rugulosa*	Western Canadian violet
85	*Zigadenus veneosus*	Death camas
86	*Zizia aptera*	Meadow parsnip

This list is not complete. In Nose Hill Park (discussed above), which lies approximately 6 kms northeast of the Tuscany habitation site, over 200 flowering plants have been identified from very similar habitats to those found in the Tuscany habitation site area. Thus, the plant list for Tuscany provides only the more visible and/or more abundant plant species found in the area. Plants that are rare in the area are not documented in this overview.

Interestingly, of the 86 plants listed, there are only 10 weedy species. Of these, four are weedy species native to North America, including stickseed (*Hackelia floribunda*), foxtail barley (*Hordeum jubatum*), pineapple-weed (*Matricaria matricarioides*), and possibly the death camas (*Zigadenus veneosus*). Lamb's quarters (*Chenopodium album*), is noted as an annual introduced from Europe, however, it is also possibly native to North America. It was first reported in Canada in 1858 (Royer and Dickinson 1999:154). The other five weedy species are introduced from Europe and include Canada thistle (*Cirsium arvense*), hound's tongue (*Cynoglossum officinale*), black henbane (*Hyoscyamus niger*), penny cress (*Thlaspi arvense*) and goatsbeard (*Tragopogon dubius*).

FAUNA OF THE CALGARY REGION TODAY

Though the Calgary area has been part of the modern industrial milieu for a relatively short span of time, there have been great changes in the species of wildlife that can now be encountered in the landscape. Most notable in its absence is the bison (*Bison bison*). Though not completely exterminated, bison are now found only in the National Parks and on game ranches in Alberta. One other large mammal that has been completely exterminated from the Plains region is the plains grizzly (*Ursus arctos*). A number of mammals have been sighted in Nose Hill Park and are described in the Nose Hill guide (Hallworth 1988). Those animals listed include American Badger (*Taxidea taxus*), coyote (*Canis latrans*), northern plains red fox (*Vulpes fulva regalis*), lynx (*Lynx canadensis canadensis*), mule deer (*Odocoileus hemionus*), white-tailed deer (*Odocoileus virginianus*), white-tailed hare (*Lepus townsendii campanius*), varying hare/snowshoe hare (*Lepus americanus americanus*), muskrat (*Onadatra zibethicus cinnamoninus*), porcupine (*Erethizon dorsatum epixanthum*), Richardson pocket gopher (*Thomomys talpoides talpoides*), Richardson ground squirrel (*Spermophilus richardsonii richardsonii*), little brown bat (*Myotis lucifugus*), least weasel (*Mustela nivalis rixosa*), and Prairie long-tailed weasel (*Mustela frenata longicauda*).

Some animals come into the city suburbs via the natural, old pathways along the Bow River, Elbow River and Fish Creek. Black bears (*Ursus americanus*), periodically have to be captured and relocated. Moose (*Alces alces*) occasionally wander into the city as well. There are numerous beaver (*Castor canadensis*) in Fish Creek Park, in the southeast part of the city. Nelson (1969:225-226) describes a "beaver landscape" in the Bow Valley in early Historic times, circa 1750-1885, in which moose roamed widely, taking advantage of the favorable environment created by beaver ponds.

GEOLOGY/TOPOGRAPHY

Drainage

Calgary (51° N latitude and 114° W longitude) is located approximately 75 kms east of the Rocky Mountains at the confluence of the Bow and Elbow Rivers. One of the dominant landscape features of Calgary is the valley occupied by the Bow River and its tributaries. The Bow River is part of the Saskatchewan River Basin, which ultimately discharges into Hudson Bay. The river originates from a lake in the Rocky Mountains northwest of Banff, Alberta (Wilson 1983:32). The Bow River enters northwest Calgary at an elevation of 1130 m and exits in the southeast quarter at 950 m elevation (MacMillan 1987:2). The river valley is in general, relatively flat and deep and can be up to 4 km in width (Wilson 1983:32). Approximately 15,000 years of valley formation history can be reconstructed through the series of terraces and fills, fans and slopewash aprons in the landscape topography today (Wilson 1983:36).

Physiography

According to interpretations by Moran (1986, Figure 3), most of Calgary is underlain by the Porcupine Hills Formation. This formation is comprised of a "series of crossbedded sandstone and calcareous bentonitic shale" layers and was "deposited in [a] continental fluvial environment" (Moran 1986: Figure 3). The sandstones are found to be well cemented and form cliffs along the river valley sides (Moran 1986: Figure 3). The Bow Valley is steep-walled and approximately 60 meters deep in Calgary (Wilson 1986). The Tuscany site (EgPn-377) is located on a large undulating glacial lake plain overlooking the Bow River Valley. In this area the valley is more gently defined by the uplands which flank what was once a glacial lake plain (Wilson 1981:38). Indeed, the lowermost sediments encountered in the Tuscany excavation were Glacial Lake Calgary sands.

Several prominent landforms are found to the north and west of the city center. Nose Hill, on the north side of the Bow River, is the northernmost landform. It has an elevation of approximately 1220 meters and stands about 122 meters above the plain, oriented in a NW-SE direction. Broadcast Hill is located on the south side of the Bow River and Cairn Hill lies 3-4 kms further south, on the south side of the Elbow River. Both of these landforms reach elevations of approximately 1220 meters (Yudcovitch 1967:11).

The Porcupine Hills Formation is the youngest bedrock (Tertiary) outcropping in the Calgary area and overlies the Paskapoo Formation (Tertiary and Cretaceous). It is the Porcupine Hills Formation that underlies the first terrace on which the Tuscany habitation site is located, on the north side of the Bow River valley. It also underlies the Broadcast Hill Upland area on the south side of the Bow River, opposite the Tuscany site. The glacial tills found throughout the Calgary area contain sand grains and angular fragments which are the products of weathering of all of these formations.

According to Moran's (1986:NTS82-0/1) summary of the surface materials of the Calgary urban area, the Tuscany habitation site (EgPn-377) is located in an area where the sediments are sandy gravel overlying silt. The lithogenesis or origin of the material is fluvial channel sediment overlying lacustrine, offshore sediment. The first two meters of the soil and sediment profile at the Tuscany habitation site were determined to be aeolian in origin. These sediments were lying on top of fine-medium textured sands (1 meter deep) and below these were massive clay deposits of unknown thickness. Very fine bedding lines with .5 cm intervals were visible in the sand deposits, revealing their lacustrine origins. These bedding lines had a varve-like appearance, though there were no darker, organic lines in them. The bedded sands consisted of alternating layers of coarser-grained sands 2 mm thick overlying finer-grained sands that were 3 mm thick. The sands do appear to have been deposited in a slow-moving water environment, which may be deltaic. They appear to be bottomset beds, which are "...thin horizontal beds of finer material deposited in advance of the foreset beds" (Marsh 1987:419).

GLACIAL LAKE CALGARY

A previous discussion has noted the ongoing debate on the precise movements, locations and timing of glacial events during the late Pleistocene. The details of this debate will not be addressed here. The primary interest of this discussion is the timing of events related to the existence of Glacial Lake Calgary.

A recent discussion of this glacial event provides new dates for consideration though some problems are noted. Work by Jackson *et al.* (1997) on cosmogenic ^{36}Cl dating of the Foothills erratics train examined the problem of the maximum extensions of the late Wisconsin, Cordilleran and Laurentide ice advance and coalescence. Fisher (1999) has noted that the dating of the seven erratics in southern Alberta yielded an average age of 13,900 years ago (calendar years) with parameter values of uncertainty of approximately 10%.

However, more recent work by Jackson *et al.* (1999) on the dating of the erratics train in southern Alberta has yielded a different set of cosmogenic ^{36}Cl dates that are slightly older on average. Jackson *et al.* (1999) noted that they prefer the dates derived for a zero erosion age on the erratics. Oddly, there are differences between the dates for erratics of Cordilleran origin versus those of Canadian Shield origin. Cordilleran erratics average 17 ka ^{36}Cl years BP with no snow cover correction and 17.8 ka ^{36}Cl years BP with a snow cover correction. Canadian Shield erratics average 14.4 ka ^{36}Cl years BP with no snow cover correction and 15.2 ka ^{36}Cl years BP with a snow cover correction. Reasons for these differences in the ages derived for these erratics are not understood as yet (Jackson *et al.* 1999).

Fisher's (1999) examination of the first erratics study by Jackson *et al.* (1997), proposes that if we accept the average of 13.9 ka ^{36}Cl years BP it provides a maximum age for the extent of the Laurentide ice from the east. As well, the timing of the ice extensions provides the most likely scenario for the existence of Glacial Lake Calgary and a maximum age for the lake, according to Fisher (1999). The more recent set of dates on the erratics does not essentially change Fisher's (1999) argument, it just pushes the timing of events a bit further back from 13.9 ^{36}Cl years BP to a range of 14.4 to 17.8 ^{36}Cl years BP.

One clarification is required for the relationship of cosmogenic ^{36}Cl dates from glacial erratics to ^{14}C radiocarbon dates derived from wood and bone sources. The cosmogenic ^{36}Cl dates are equated to calendar years and had to be extrapolated backwards to find the equivalent ^{14}C date. Thus 13,900 ^{36}Cl years are equivalent to 11,900 ^{14}C years uncalibrated, according to Fisher (1999) who references the CALIB 3.0 ^{14}C age calibration program for derivation of the converted date. Since the CALIB 3.0 program does not convert ^{36}Cl dates to radiocarbon equivalents this will not be attempted for the new set of dates derived here. Further, since there are apparent discrepancies with the new set of chlorine-36 dates, those problems should be resolved before anymore use is made of them. For the purposes of this dissertation Fisher's (1999) derived date of 13,900 ^{36}Cl years BP will be used with the understanding that there are inherent problems associated with these chlorine-36 dates from glacial erratics.

Dating problems aside, this establishes a potential timeframe boundary for the Glacial Lake Calgary sediments and their relationship to the Tuscany paleosol sediments. The initial topography of the site was probably formed during the existence of the glacial lake and was also subsequently subjected to erosional processes as the lake drained at some time following the last glacial maximum. The sediments that eventually developed into the series of Tuscany paleosols were deposited over time on top of the glacial lake sands. They must be more recent in age.

THE POSTGLACIAL DEPOSITS AND CULTURAL CHRONOLOGY

The two meters of sediments excavated at the Tuscany habitation site which were overlying the glacial sands have been interpreted as aeolian origin. They were unstratified, non-sorted materials composed mostly of fine-medium sands along with some silt and clay sized particles. The materials appear to have blown into the site over time.

There were three distinct sections identified in the excavation profiles at the Tuscany site. The top meter of sediments had accumulated through time and developed into an unstratified chernozem that was clearly identifiable in the walls of the excavation. The second section was identified as a Mazama tephra (mentioned above) deposit of varying thickness throughout the site. Underlying the Mazama tephra was the third section comprised of a series of paleosols. These represent a series of developing A and B horizons that were buried before pedogenesis was very advanced.

The Mazama tephra represents the final capping boundary for the paleosols at the Tuscany site. A brief discussion about Mazama tephra is necessary to elucidate the significance of its presence in this site. As noted by Hallet *et al.* (1997:1202), tephra layers of late Pleistocene and Holocene ages that have been found in stratigraphic contexts in sediments throughout western Canada and the northwestern United States provide important chronological markers for a variety of researchers. Mount Mazama, located in Crater Lake, Oregon, was the site of a massive eruption event previously dated to 6845 ± 50 ^{14}C years BP (Bacon 1983) which has provided one of those important chronological marker horizons. During this significant eruption, tephra was carried up to 2000 km north and east from the source vent, spreading over an area more than 1,000,000 km^2 in western North America (Hallet *et al.* 1997:1202). Dating this catastrophic event has been problematic. Fossil plant material has provided most of the datable samples in the form of charred wood, wood, twigs, peat and even pollen (Hallet *et al.* 1997). The problem with most of the material is that it cannot be determined to specifically date the eruption event. Charcoal from the inside of a tree would yield a date that could be hundreds of years older than the actual eruption. Peat can have aquatic mosses in it that contain old carbon. Study of *Drepanocladus crassicostatus*, a bryophyte, has revealed that modern samples contained less than 85% modern ^{14}C (MacDonald *et al.* 1987:839). The ^{14}C deficient carbon appears to have been taken up by the living plants. A replacement process did not incorporate it after the plants died (MacDonald *et al.* 1987:839). Hallet *et al.* (1997) addressed this complex problem by choosing to date only charcoal and twig remains recovered from Mazama tephra from air-fall depositional contexts. A weighted mean age of 6730 ± 40 ^{14}C years BP (7470-7620 cal years BP) was derived from four dates acquired from material that fulfilled the criteria stipulated (6720 ± 70 TO-5192; 6760 ± 70 TO-5196; 6560 ± 115 I-3809; 6880 ± 70 W-4255) (Hallet *et al.* 1997:1207).

At the Tuscany site the Mazama tephra horizon represents the end of deposition of the paleosols and dates to 6730 ± 40 years BP. This event marked the uppermost point from which soil samples were processed for this dissertation. Thus, the focus for this research includes only the study of the paleosols, which date between 10,000 and 6,700 years BP.

As noted above the dating of Glacial Lake Calgary sediments continues to be a question of some debate. The amount of total ice coverage at the end of the Pleistocene and the coalescence of eastern and western sheets in what is now the Calgary region, remains unresolved (e.g. Campbell and Campbell 1997; Dyke and Prest 1987; Fisher 1999; Jackson 1997, 1999; Ronaghan and Beaudoin 1988; Rutter 1980; Vickers 1986). It does appear clear that the cultural material remains resulting from human occupation of the landscape were deposited in stratigraphic contexts following the retreat of the ice sheets and the subsequent deposition of other sediments followed by the formation of soil horizons over time along with the colonization of the region by vegetation.

The following discussion provides a very brief summary of the regional cultural chronology intended as a frame of reference to place it within the geological sequence of events outlined above. As Wright (1995:42) notes, there is limited evidence of Paleo-Indian culture in the Plains and Foothills regions of western Canada except in Alberta and adjacent British Columbia. The Sibbald

Creek site (Gryba 1983), 72 kms directly west of Calgary, contained an occupation level dating to approximately 11,000 years ago. At the Vermilion Lakes site, located in the Bow Valley corridor 50 kilometers northwest of Sibbald Creek, a cultural feature yielded a series of dates, which averaged 10,626 years BP (Wright 1995:44). The Lake Minnewanka site is located on the shores of the reservoir, Lake Minnewanka. This site is located a few kilometers southeast of the Vermilion Lakes site and has yielded cultural material that is also at least 10,000 years old (Landals 1998). Comparison of the lithic remains of the Lake Minnewanka site to those of the Vermilion Lakes site remain inconclusive, however, there are some similarities apparent. Moving out on the Plains, we find the Fletcher site located about 50 km SSE of Taber, Alberta. A sampling of probable sedge seeds (*Cyperaceae*) found below a bone bed yielded an AMS radiocarbon date of 9,380 ± 110 years BP (TO-1097) (Vickers and Beaudoin 1989).

The Tuscany habitation site (EgPn-377) appears to fit into the scheme, both geologically and culturally. The site contains sediments identified as a series of paleosols that date after the regional deglaciation sequence, thus fitting into the time frame following the glacial retreat. These paleosols are the focus of this dissertation. A date of 9950 ± 120 ^{14}C years BP (TO-9262) has been derived from 49 mg of charred plant fragments recovered from Unit 34 Level DD from 1139.50 to 1139.44 masl. This location appears to be the deepest, or lowermost level of all the units within the Tuscany site grid. The cultural remains recovered from the site were positioned in the sediments above the Glacial Lake Calgary deposits and can be inferred to be younger in age. A point recovered from Unit 37 Level S at 1140.21 masl has been determined to be Lusk, based on stylistic attributes. Two dates have been derived from materials in association with the projectile point. A date of 7840 ± 100 ^{14}C years BP (TO-9263) has been derived from 27 mg of charred plant fragments recovered from the hearth feature in Unit 37 Level T from 1140.16 to 1140.11 masl. The second date 7610 ± 70 ^{14}C years BP (TO-9264) was derived from a purified bone collagen extraction of 2900 mg from a bison femur recovered in Unit 47 Level T from 1140.19 to 1140.14 masl. These two dates average to 7725 years BP and yield a date range from 7940 to 7540 years BP. While they do not overlap, they do cluster quite closely, though they have been derived from two entirely different materials within the paleosols and associated with the projectile point.

The paleosols are sandwiched between Glacial Lake Calgary sands on the bottom, and Mazama tephra on the top. A fourth date, 6940 ± 70 ^{14}C years BP (TO-9261), for the Tuscany site was derived from a purified bone collagen extraction of 2000 mg from a bison third phalanx recovered in Unit 26 Level S from 1140.22 to 1140.17 masl. This bison bone was located just below the Mazama tephra deposit within the sediments of the site and fits well within the stratigraphic/depositional sequence now available for the paleosols from the suite of four radiocarbon dates.

USING FAUNAL REMAINS TO DERIVE THE COLONIZING VEGETATION

The main thrust of this next discussion is to examine an alternate approach to develop ideas about the very general types of vegetation that migrated into the greater region of southern Alberta after the retreat of the glacial ice at the end of the Pleistocene. Several bones of a number of mammal species that lived at the end of the Pleistocene have been recovered at a number of locations in southern Alberta. A number of these bones have been radiocarbon dated and are identified to the genus or species level. The fact that these mammals were present in the landscape indicates the presence of vegetation on which they may have browsed or grazed before they died. The remains from Cochrane and Calgary have been recovered from gravel pit contexts. The most recently recovered faunal remains dating to the terminal Pleistocene from St. Mary Reservoir, are from floodplain deposits. Associated with the faunal remains at St. Mary Reservoir are a number of animal trackways that have preserved through time.

Two sets of faunal remains exist that can be used to derive some sense of the vegetation that would have existed in the landscapes to support them in terms of browse and habitat. The first set of faunal remains is from extinct megafauna that died off for unknown reasons at the end of the Pleistocene time period. The second set of faunal remains is from archaeological sites and represents materials recovered from contexts associated with human occupation of the landscape.

Table 5 (a and b) organizes information related to animal species that have been verified to have existed in the Calgary regional landscape and southern Alberta during the post glacial period. The first set of dated faunal remains includes megafauna that became extinct between the last glacial maximum and the early Holocene. The second set includes identified faunal remains from dated and undated archaeological sites relevant to this study. The tables compare the type of forage/browse and habitats preferred by the different mammal species present in the landscape. The only problem with using archaeological faunal remains is that they could be transported from a kill location at some distance to a habitation site. This could encompass an area of several square kilometers distance in any direction. Nevertheless, the implications of a faunal inventory are important to consideration of a vegetation inventory in terms of both forage/browse and habitat cover. Animals necessarily have preferential habitats, which yields some indication of possible paleovegetation inventories. An issue with such reconstructions is, of course, whether or not modern species are adequate analogues for looking at preferences of extinct species, but at a general level this is probably a valid approach. Inference of a very general paleovegetation cover is the sole purpose of this entire discussion.

A number of bones from species of extinct megafauna and other mammals have been have been found in

Table 5(a): ^{14}C dated faunal records of the late Pleistocene in the Calgary area and southern Alberta

Location	Date (radiocarbon years BP)	Reference	Fauna	Forage (Wilson 1983; Banfield 1974; Hills *et al.* 1999)
Clarke Pit, Cochrane	11,370 ± 170 (GSC-613 11,100 ± 160 (GSC-989)	Churcher (1968)	*Equus conversidens* (Mex. Half-ass) *Ovis canadensis* (Bighorn sheep) *Bison bison ssp. cf. antiques* *Rangifer tarandus* (caribou)	Grasses predominantly Grasses, forbs, shrubs/trees Grasses predominantly Lichen, mosses, shrubs, grasses
Gallelli Pit, Calgary	11,330 ± 290 (RL-757)	Wilson and Churcher (1978)	*Equus conversidens* *Bison bison ssp. cf. antiquus* *Mammuthus sp.* (mammoth) *Camelops hesternus* (camel)	Grasses predominantly Grasses predominantly Grasses predominantly but also shrubs Grasses predominantly
Saint Mary Reservoir, SW Alberta Dates from bone collagen were corrected for isotopic fractionation	11,130 ± 90 (TO-7693) 11,330 ± 70 (TO-7696) 10,980 ± 80 (TO-7691) 11,350 ± 80 (TO-8792)	Kooyman *et al.* (in press), Hills *et al.* (1999), Kooyman *et al.* (1999)	*Equus conversidens* *Bison bison ssp. cf. antiquus* *Mammuthus sp.* *Camelops hesternus* *Rangifer tarandus* *Bootherium bombifrons* (muskox)	Grasses predominantly Grasses predominantly Grasses predominantly but also shrubs Grasses predominantly Lichen, mosses, shrubs, grasses Shrubs, forbs and grasses
Griffin Pits, Cochrane	10,760 ± 160 (GSC-612)	Churcher (1968; 1975)	*Equus conversidens* *Ovis Canadensis* *Bison bison ssp. cf. antiquus* *Cervus elaphus* (elk)	Grasses predominantly Grasses, forbs, shrubs/trees Grasses predominantly Grasses, shrubs/trees, forbs
Aquitaine Pit, Bow Valley, Calgary	10,200 ± 280 (GSC-3065)	Wilson (1983)	*Bison bison ssp. cf. antiquus* (extinct form of bison)	Grasses predominantly

sediments in the St. Mary Reservoir near Cardston, Alberta, and in gravel pits in Calgary and Cochrane. Importantly, radiocarbon dates have been derived from collagen extracted from some of these bone samples. The large mammals represented were primarily grazers and browsers subsisting on combinations of grasses, forbs, shrubs/trees, lichen and mosses.

The megafauna remains and trackways from the St. Mary Reservoir site have been dated between 11,300 and 11,000 years BP (Kooyman *et al.*: 2001). The site is located in southwestern Alberta. The megafauna present in the location indicate that the landscape was covered primarily by grasses though there must have been shrubby brush and possibly trees along the St. Mary river banks. The presence of caribou indicates the possibility of an open spruce and fir forest nearby with open lichen and moss covered ground. Caribou can be found today from the Mountains and Foothills region and throughout the Boreal forest in Alberta (Pattie and Fisher 1999).

At Cochrane, near Calgary, there are indications that the vegetation must be diversifying into deciduous and conifer types of cover only 20 kms west of the city. The Clarke Pit at Cochrane with two dates, 11,370 ± 170 and 11,100 ± 160 years BP contained the remains of horse and bison, which prefer grass forage, along with bighorn sheep and caribou which prefer lichen, mosses, forbs and shrubs—vegetation found most likely in spruce and fir covered landscapes. These megafauna remains are not associated with human predation and are therefore not likely to be 'out of the natural habitat' at time of death. While bones may be scattered by scavengers they are not likely to be carried very far away from the place where they perished. Remains of megafauna at the Griffin Pits in Cochrane yielded a date of 10,760 ± 160 years BP, which is a few hundred years after the first set of dates. The presence of bighorn sheep and elk indicates a landscape of mixed vegetation with grasses along with forbs, shrubs and trees. Thus, the general vegetation cover between 11,300 and 10,700 years BP for the area around Cochrane appears to have been more tree-covered in general compared to the vegetation of today.

At the Gallelli Pit in Calgary, 20 kilometers to the east of Cochrane, the megafauna present in the landscape at 11,300 years BP appear to be animals that preferred to browse on grasses. The same situation appears to be the case later in time with faunal material that dates to 10,200 years from the Aquitaine Pit. Only *Bison bison* ssp. cf. *antiquus* remains were recovered. These are not definite indications that the vegetation cover in the landscape was covered by grasses. We can say that grasses, as preferred forage, were probably present. The presence of shrubs and trees are not necessarily precluded.

In Table 5(b) the faunal remains from archaeological sites within 5 kilometers of Tuscany have been summarized. The sites listed have dates that fall within the timeframe boundaries between the glacial deposits and the Mazama tephra at Tuscany. At the Hawkwood site only bison occurs in Component 1 with a date of 8250 ± 330 years BP. Both bison and deer occur in Component 2, dated to 7030 ± 210 and 6820 ± 280 years BP. Though bison

Table 5(b): ^{14}C dated/relative-dated faunal records from Holocene sites in the Calgary region and the predominant vegetation associated for browse and habitat

Location	Date (radiocarbon years BP)	Reference	Fauna	**Forage** (Renecker and Hudson 1993) and **Habitat** (Pattie and Fisher (1999)
Hawkwood site (EgPm-179) Calgary	8250 ± 330 (RL-1554) Component 1	Van Dyke and Stewart (1985)	Bison species (not identified)	Grasses predominantly
Hawkwood site (EgPm-179) Calgary	7030 ± 210 (RL-1276) 6820 ± 280 (RL-1277) Component 2	Van Dyke and Stewart (1985)	Bison species (not identified) *Odocoileus* sp. *O. hemionus* (mule deer) *O. virginianus* (white-tailed deer)	Grasses predominantly Shrubs/trees, forbs, grasses Grasses, shrubs/trees, forbs
EgPn-428 Crestmont Estates Calgary	Date rejected Est.~8000 yrs BP	Vivian and Reeves (1999)	*Ursus* sp. (bear) *Canidae* sp. (dog-like) Blood residue Bison *Ovis* sp. (sheep) *Odocoileus* sp. (deer) *Sylvilagus* sp. (rabbit)	Primarily forest Dependant on species Grasses predominantly Grasses, forbs, shrubs/trees Depends on species – see above Shrub cover
Tuscany site (EgPn-377)	9950 ± 120 (TO-9262) 7840 ± 100 (TO-9263) 7610 ± 70 (TO-9264) 6940 ± 70 (TO-9261)	Siegfried and Oetelaar (2000)	*Bison bison* *Cervus elaphus* *Odecoileus* sp. *O. hemionus* (mule deer) *O. virginianus* (white-tailed deer) *Antilocapra americana* (antelope) *Ursus americanus* (black bear) *Felis concolor* (mountain lion) *Lepus* sp. (hare) *Sylvilagus* sp. (cottontail) *Ondatra zibeticus* (muskrat) *Microtus* sp. (vole) *Lemmus* sp. (lemming) *Tetronidae* sp. (grouse) *Anatidae* sp. (duck) Blood residue *Canidae* sp. (dog-like)	Grasses predominantly Grasses, shrubs/trees, forbs Depends on species – see above Shrubs/trees, forbs, grasses Grasses, shrubs/trees, forbs Forbs, shrubs/trees, grasses Primarily forest Wooded rocky areas Dense shrub to forest cover Shrub cover Sloughs, ponds Grassland, shrubs and trees Bogs to spruce woods Deciduous to coniferous cover Sloughs, ponds Dependant on species

prefer grasses as forage there may have been shrubs and trees present in the landscape. The presence of deer indicates that shrubs and trees must have been in the nearby landscapes. Unfortunately, the species of deer is unknown. However, both mule and white-tailed deer eat shrubs/trees, though mule deer prefer them.

The evidence of fauna at the Crestmont Estates site (EgPn-428) is more tentative. Faunal remains of bear and a dog-like species are present. Bison, sheep, deer and rabbit are indicated only by blood residue analysis from lithic tools. Bison and sheep prefer grasses as forage but the remaining species are found in shrubby or treed landscapes. Again, grass preference, as forage does not preclude shrubs or trees being present in the same landscape. The site was dated by collagen extracted from faunal remains, however, the date was rejected as being too recent for the stylistic attributes of the lithic assemblage associated with them. Thus, this site has an estimated date of approximately 8000 years BP.

The faunal remains inventory at the Tuscany habitation site is the broadest of all the sites listed. The cultural component associated with the assemblage has been dated to the 7800 years BP range. The faunal assemblage indicates a landscape covered with mixed vegetation composed of open grassland interspersed with patches of shrubs and trees. This vegetation would probably appear very similar to that found on the slopes of Twelve Mile Coulee today, several hundred meters east of the site location.

The two sections of Table 5 list animals that were present in the southern Alberta landscape following continental glaciation. The first set of larger mammals is not necessarily associated with humans occupying the landscapes except at the St. Mary Reservoir where recent work has associated butchered horse skeletal materials with human manufactured stone tools (Kooyman *et al.* in press). Inferences can be made about the vegetation that must have been present in the landscape as forage and browse in order for the animals to survive. It appears that vegetation 20 kilometers west of Calgary, in the 11,400-10,700 year BP time period, may have been more wooded to support elk populations. Vegetation cover in the Calgary area during the same time period does not appear to have been wooded though there may have been parkland type vegetation present with enough grassland available to support grazing ungulate species. Part of the problem with relying on the presence of fossil mammal remains to define the vegetation cover is the fact that certainly we have not found remains of all the species that were actually present—more species may yet be found.

At 10,200 years BP *Bison bison* ssp. cf. *antiquus* were still grazing in the Bow Valley in what is now Calgary. Following this time period *Bison bison* ssp. *antiquus* appear to have disappeared from the landscape along with some of the other megafauna. Our understanding of this process is still unclear though destabilizing predation models including human entry into the New World have been proposed as possible explanations (e.g. Whitney-Smith 1997). According to Guthrie (1980) there is no discontinuity between any of the bison populations. Diminished body size is a natural adaptive response and the bison of the early Holocene are the ancestors of the bison of today (Guthrie 1980:47). The bison recovered from the Hawkwood site and dated to 8250 years BP is not distinguished as *B. b. antiquus, B. b. occidentalis*, or *B. b. bison* and it is difficult to discern what may have happened through time in such cases. An examination of bison remains distinguished by their metric variability would show the change in sizes, which along with their locations throughout the landscape, could possibly be correlated to changes identified with vegetation through time in the Great Plains region (e.g. Kooyman and Sandgathe 2001; Wilson 1980).

By 8000 years BP a landscape covered with mixed vegetation cover appears to have existed much closer to Calgary than at present. While it is certain that not all the animals represented in the faunal assemblages of the archaeological sites came from the immediate site vicinities, they were probably killed within certain walking distance of a few kilometers from habitation and camp sites depending on the size of the animal killed.

DETAILED DESCRIPTION OF SITE SPECIFIC TOPOGRAPHY

The Tuscany site, i.e., EgPn-377, covered an area of approximately 400 square meters within the confines of a depressed basin immediately north of an escarpment which forms the first terrace above the Bow River floodplain. The site area is located in LSD NW1/4, Section 4, Township 25, Range 2, W of 5th Meridian (11PGU938654). The description of the depression as it appears today and its dimensions has been discussed previously (Chapter 2). As noted, the depression or swale, appears very bowl-like in the present landscape, with relief of 2 meters elevation between the high points of the perimeter and low points in the bottom.

The paleosols that occur in the swale or depression appear to be isolated features resulting from topography. A series of backhoe trenches and exploratory units around the excavation site proved that the paleosols thinned toward the perimeter of the depression edges, becoming more and more ephemeral and difficult to discern with distance from the center of the swale. Thus the paleosols of the depression were limited in their extent. At the deepest point in the center of the depression the paleosol zone from the bottom of the Mazama tephra to the top of the glacial sands was .87 meters thick (Unit 26). Figure 6 represents the Tuscany grid. The unit numbers are in bold while the thickness of the paleosol zone in each unit is in italics and given in meters. It is quite clear from the measurements shown in Figure 6 that there was a thickness gradient within the paleosols. They were thinnest in the southwest section of the grid. The northeast section of the grid from unit 1-5 and up, to unit 75 in the NE corner of the excavation, contained significantly thicker paleosol deposits. Unfortunately, the depression could not be completely excavated during the three years the University of Calgary Field School was in operation.

CULTURAL REMAINS AND FEATURES

There were indications of two possible human occupations of the Tuscany depression area during deposition and formation of the paleosols during the past. Immediately underlying the Mazama tephra, in the second pedogenic unit, a number of large bones were recovered and identified as *Bison b. bison*. One projectile point, identified as Bitteroot, was found in the transition boundary between the top of the most recent paleosol and the Mazama tephra deposit. The association between the projectile point and the bison remains was not conclusive and very few flakes (<5) were found within the uppermost paleosol. The projectile point represents the Mummy Cave Complex of the Early Middle Prehistoric I, according to Vickers (1986). This complex has been found to date between 7630 to 5255 years ago in the northwest Plains region (Frison 1998:161). Because there is only an extremely tenuous link between the projectile point and the faunal remains, nothing conclusive is drawn here regarding a possible occupation of the most recent or uppermost paleosol. Once the site report is written, these problems may be clarified.

In Figure 7, general boundary outlines are shown for the main activity areas representative of lowermost human occupation of the Tuscany depression locale during deposition and formation of the paleosols. These activity areas are represented by identification of a hearth feature and the distribution patterns observed for lithic debitage and the larger faunal remains. These distribution patterns represent the lowermost and oldest occupation of the depression during the past, dating to 7800 years BP. These distribution patterns are discernable on the horizontal plane, across the site. There is also a vertical distribution pattern discernable with the lithic debitage, the larger faunal remains and the charred botanical remains. These distribution patterns will be discussed at length in the following chapter.

The hearth feature area represented a small location of approximately 50 x 50 cm on the east half of Unit 37 in

								92	**93**	**94**	
								.84 **73**	*.77* **74**	*.71* **75**	
							.62 **32**	*.75* **33**	*.77* **34**	*.67* **35**	*.66* **36**
					.64 **23**	*.59* **24**	no book **25**	*.87* **26**	*.70* **27**	*.80* **28**	
					.50 **16**	*.70* **17**	*.75* **18**	*.70* **19**	*.59* **20**	*.69* **21**	*.67* **22**
.40 **76**	*.49* **77**	*.50* **86**	*.58* **60**	*.53* **61**	*.60* **11**	*.67* **12**	*.71* **13**	*.65* **14**	*.65* **15**	*.59* **30**	
.51 **72**	*.48* **71**	*.39* **87**	*.29* **58**	*.52* **59**	*.58* **6**	*.63* **7**	*.54* **8**	*.57* **9**	*.56* **10**		
.41 **68**	*.53* **70**	*.54* **88**	*.54* **56**	*.59* **57**	*.66* **1**	*.61* **2**	*.61* **3**	*.62* **4**	*.43* **5**		
.50 **63**	*.49* **69**	*.51* **89**	*.45* **37**	*.56* **38**	*.57* **39**	*.63* **40**	*.50* **41**				
.60 **62**	*.51* **67**	*.53* **90**	*.51* **46**	*.44* **47**	*.44* **48**	*.49* **49**	*.45* **50**	*.44* **51**			
.47 **42**	*.39* **66**	*.54* **64**	*.47* **82**	*.51* **83**	*.47* **84**	*.45* **85**	*.41* **43**	*.36* **44**			
.37 **52**	*.43* **65**	*.52* **78**	*.43* **79**	*.43* **80**	*.45* **81**	*.39* **53**		*.32* **45**			
.47 **29**	*.48* **31**		*.35* **91**	*.45* **54**	*.30* **55**						

** Numbers in bold = unit numbers; Thickness of Paleosols (meters) = italicized

Figure 6: The Tuscany habitation site grid at the end of 1997. North is the top of the grid

Level S (N521, E526) (1140.16 masl to 1140.11 masl) and Level T (1140.11 masl to 1140.06 masl). The hearth feature itself was 23 centimeters east of a Lusk projectile point, which was also found in Unit 37 Level S. The hearth feature was actually quite small and was not observed during the excavation of Level S. It was observed during the excavation of Level T and was only partially recovered as a result. It appeared as a greyish-white stain within the reddish colored paleosol (designated as Layer J in the wall profiles discussed below). Only the bottom half of the hearth feature was recovered as a flotation sample. The entire hearth sample of 5900 ml of sediment was processed for flotation. No charred seed remains were recovered from flotation of the hearth sediment samples. A total of 687 charred plant fragments were recovered from the entire 5900 ml sample. These results can be compared to the bulk sediment sample that was also collected from the southwest corner of Level T in Unit 37. The bulk sediment sample was 500 ml in size and contained only 70 charred plant fragments and no charred seeds. Each sample total can be recalculated to represent a 1000 ml sample volume in order to compare the rates of recovery. If the hearth sample were actually only 1000 ml, the number of charred plant fragments recovered would equal 116 pieces. If the bulk sediment sample from the southwest corner were actually 1000 ml, the number of charred plant fragments recovered would equal 140 pieces. The rates of recovery of charred plant fragments are not significantly different for these two sample types from Level T of Unit 37. The bulk sediment sample may be considered to provide an unbiased control sample for this case. As a control, it reveals that there appears to be very little differentiation between the type and quantity of materials recovered from the hearth feature itself compared to the surrounding sediments at the same level within Unit 37. Thus, except for the fact that an observable stain appeared in the sediments in close proximity to the projectile point recovered at the bottom of Level S in Unit 37, no other cultural association may be extrapolated from the flotation of the hearth feature.

The charred plant materials recovered from the bulk sediment samples were found to occur in each excavated unit in the site grid. An interesting pattern was identified from the distribution of the total number of recovered pieces tallied for each unit. The summed tallies revealed a circular area measuring between 3 x 4 meters diameter around the hearth where these tallies maintained lower frequencies for the charred plant remains. This pattern can be interpreted to be cultural in the sense that it may represent a trample area around the hearth. This trample area may have been the result of a small hut structure.

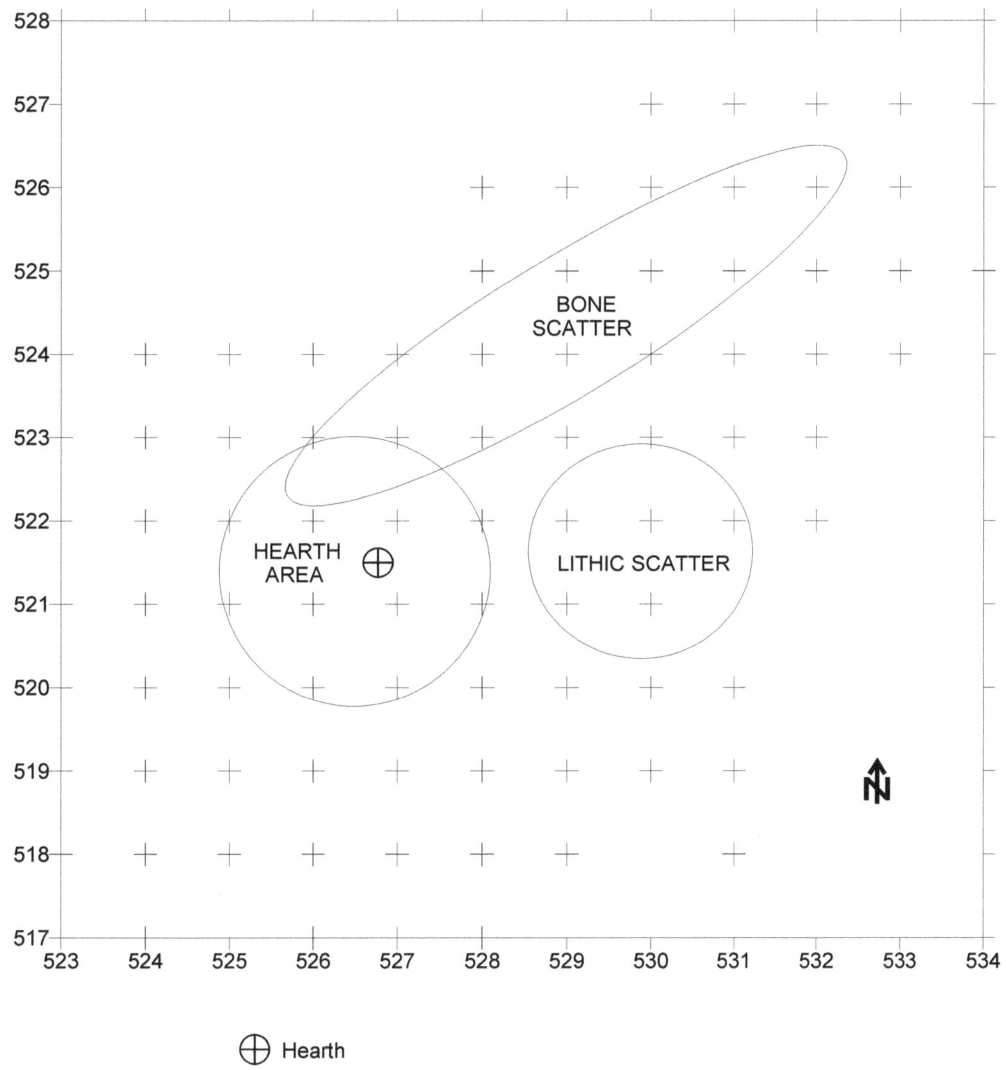

Figure 7: The Tuscany habitation site grid with the general parameters of the three activity areas outlined

There are no direct cultural linkages that can be drawn from any sites in the southern Alberta region. There are two examples that are peripherally relevant to this case because of some similarities of overall site structure. The first relevant case is derived from the Hell Gap site in Wyoming, where the remains of three structures were recorded by Irwin-Williams *et al.* (1973). Three sets of post-holes formed three superimposed circles that were just over two meters in diameter. As noted by Dixon (1999:230), these structures may have been small, skin-covered structures that could have accommodated only a small group of people like a nuclear family. These structures are smaller than the dimensions of the trample zone described for the Tuscany site, however a three or four meter diameter structure is well within the design parameters of a small, skin-covered structure.

Another example of interest to this discussion was found at the Vermilion Lakes site in Banff National Park, Alberta, located approximately 130 km west of Calgary (Fedje *et al.* (1995). Complex stratigraphy contained remains from 12 cultural components that spanned 11 millennia. The older sediments contained a well-stratified section covering the time period from 10,800 to ca. 9000 years BP. In Block 153R2, eight cultural components were aggregated into three groups based on chronological and technological relationships. Group 1 consisted of the components dating between 10,800 – ca. 10,399 years BP and is the occupation of interest to this discussion. The remains of a complex activity area included a semicircular area outlined by postholes, lithic and faunal concentrations, and a surface hearth in the center. The postholes delineate a crescent shape that is noted to possibly have functioned as a windbreak, or there may have been a circular structure. Unfortunately there were no diagnostic artifacts associated with this interesting feature, however, dates on charred wood, a charred seed and bone collagen defined the chronological framework of Group 1. Comparisons were drawn by Fedje *et al.* (1995:92-93) to the nearby, Lake Minnnewanka site (349R), where Clovis lanceolate spear points represent an early cultural complex that would fit the chronology found in the feature area of the Vermilion Lakes site.

The lithic debitage recovered from the excavation of the Tuscany habitation site paleosols presents unequivocal evidence of human occupation of the locale. This human occupation appears to have taken place during the timeframe around 7800 years BP, according to radiocarbon dates derived from bone and charred plant materials associated with the lithic debitage. The raw sums of number of pieces of identifiable flakes have been plotted in a series of profiles created from unit level measurements plotted along the easting grid lines of the site. The lithic debitage presents a unique pattern of distribution quite different from that observed for the botanical materials. The debitage is distributed generally in the upper charred plant concentration layer (to be discussed in Chapter 4) with the highest number of flakes occurring in the same levels as those with the highest number of charred plant fragments. The hearth feature lies within the upper charred plant concentration and the Lusk point was associated with it. The details of the spatial distribution patterns, both horizontal and vertical, will be addressed in the following chapter.

DEFINITION OF THE AMALGAMATED LEVELS BASED ON THE SEDIMENTS AND SOILS

The Tuscany habitation site represents a relatively deep archaeological site. Many archaeological sites (e.g. tipi rings) in the Plains experience low rates of deposition, which leads to shallow stratigraphy containing compressed culture history and a compressed environmental record, if the latter is present at all. The Tuscany site's relatively small area of deep stratigraphy appears to be related to original topography created during glaciation and then subsequent deposition of sediments that retained part of the topography and also contributed to the creation of the depression or swale over the millenia.

The complete stratigraphic profile contained five identifiable sedimentary sections. Within the sedimentary sections, five pedogenic zones were also identified where soil-forming processes had altered the properties of the original sediments. During interpretation of the site wall profiles, a series of letter designations from "A" to "N" were utilized for the breakdown of each individually identified part of the entire profile that could be described. When the wall profiles were interpreted, measured and drafted to scale on graph paper, the aim of the interpretation was to provide a cohesive set of descriptions that applied to each wall profile and thus, comprehensively, to the site as a whole.

The description of the Tuscany site sediments has been broken down into four approaches focusing on the differences in color, texture and means of deposition. Table 6 provides an outline of the relationships between the four sets of descriptions presented here. It should be apparent that there are five sedimentary sections within which the five pedogenic zones are also found. The pedogenic zones each have sets of designated soil horizons. The wall profile layers simply put each unique part of the wall profiles into a logical order from the top of the wall profiles to the bottom, so that they may be described in greater detail. A very general outline of the approximate depths of the major sections observed is also provided.

The sedimentary sections, which simply describe the nature of the sediments themselves, will be discussed first. The first sedimentary section, from the sod surface to an average depth of 1.1 meters in the central area of the excavation grid, consisted of a mixture of fine-medium grained sands and silts with a small percentage of clays of aeolian origin. The interpretation for this material is that it blew in from nearby areas west and northwest of the site, the directions for prevailing winds in the area today.

The Mazama tephra deposits were immediately underlying the upper aeolian materials and comprised the second sedimentary section. This second section has been well dated in western Canada at 6730 ± 40 years BP (Hallet *et al.* 1997) and represents a single volcanic explosion and deposition event originating from Crater Lake, Oregon, in the United States. There was no specific chemical, or mineral analysis carried out on the volcanic tephra deposit in the Tuscany habitation site. However, its color, texture and position within the archaeologically dated stratigraphy of the site, in conjunction with other sites where the tephra has been clearly identified, led to the conclusion that it was indeed, Mazama tephra. (e.g. the Hawkwood site 1985). The thickness of the Mazama tephra deposit was variable through the site due to the undulating nature of the original soil surface. On average it was approximately 10 centimeters thick.

The third sedimentary section consisted of fine to coarse-grained sands, along with silts and a small percentage of fine clays. The texture and color boundary between the Mazama tephra and the lower section of sands, silts and clays was very distinct. In the deepest parts of the site, this third section was 87 cm thick. This section was thickest in the center of the depression and thinned out in all directions as the excavation moved upslope. This section comprised the focus of this dissertation, containing the charred plant materials identified along with the hearth, lithic debitage and bone scatter activity areas. In the deepest areas of the depression, this section extended down to just over 2 meters below the sod surface. Excavation of units in the grid ceased at the bottom of this section.

The fourth sedimentary section was comprised of fine to medium-grained sands. The sands extended to an average depth of 3.6 meters below surface. There was visible bedding in the sand deposits, which also exhibited mottling interpreted to be the result of a significant amount of gleying. The sand deposits may have been the result of the creation of a beach during the existence of Glacial Lake Calgary at the end of the Pleistocene. In the middle of the sand zone there was a white lens containing gastropods possibly of the genus *Stagnicola,* perhaps species *S. arcticus*, which would suggest deposition in a cold body of water (Oetelaar 1997:10).

The final, fifth sedimentary section in the Tuscany site was a very wet and sticky clay deposit that contained

Table 6: Amalgamation of the site sediment interpretations by sedimentary sections, pedogenic zones, soil horizons, and profile layers

Sedimentary Sections	**Pedogenic Zones**	**Soil Horizons**	**Wall Profile Layers**	**Depths Below Surface** (rough approximations)
1. Fine to medium-grained sands and silts with a small percentage of clays – aeolian in origin	1. A developed chernozem below the present-day sod.	A B B/C C	A B C D	0 cm BS ~110 cm BS
2. Mazama tephra – volcanic deposit	Not a pedogenic zone – no soil development	Not a soil	E	~110-120 cm BS
3. fine to coarse-grained sands and silts with a small percentage of clays	2. Possible chernozem with suspended development due to burial by Mazama tephra	A B	F G	~120 cm BS to variable depths within site
	3. Possible brunisol with suspended development	A B	H/I J	Variable within site
	4. Possible brunisol with suspended development	A B	K L	Variable within site
	5. Organic soil with suspended development	A	M	To ~210 cm BS
4. Fine to coarse-grained glacial sands	Not a soil except the upper 10 cm served as the 'C' horizon for zone 5 – remainder exhibited signs of gleying	C (upper 10 cm under the last 'A' horizon constituted a 'C' horizon. Remainder was not a soil	N	~210 – 360 cm BS
5. unconsolidated clay deposit containing some pebbles and some cobbles	Not a soil	Not a soil	Not designated	360 cm BS to unknown depths

small gravel, pebbles and a few small cobbles. This deposit was interpreted as glacial in origin as well and extended from 3.6 meters below surface to unknown depths.

Sedimentary deposition of material takes place in landscapes that continuously evolve through processes such as weathering and successional vegetation regimes. Such processes can serve to change the properties and appearance of the original sedimentary deposits. There were five pedogenic zones identified within the sedimentary sections. Pedogenic zones are identified where pedogenic or soil-forming processes have taken place. These processes have visibly altered the deposited materials in color and texture to the point that horizon boundaries can be delineated within them.

Sedimentary section one (from the sod surface to the Mazama tephra) contained the first pedogenic zone. The first soil-forming unit was the highly visible zone entirely above the Mazama tephra. It included a thick, dark A horizon and underlying thick, pale, calcium enriched B/C horizon, which extended from the surface to 1.1 meters average depth. This pedogenic zone was interpreted to be representative of a well-formed chernozemic soil, commonly associated with soil forming processes in prairie/plains landscapes. These deposits were, however, not of interest to this study as they were located above the Mazama tephra boundary.

The remaining four pedogenic zones (pedogenic zones 2, 3, 4, and 5) were all identified within sedimentary section three. They were located below the Mazama tephra deposits and comprised the focus of this study. These four pedogenic zones represent four buried paleosols consisting of four sets of 'A' and 'B' horizons. These paleosols appear to have experienced suspended development, which would have taken place with burial and subsequent deposition of other sedimentary material overtop.

Pedogenic zone 2 was comprised of the paleosol immediately underlying the Mazama tephra. The 'A' horizon of this paleosol presented an abrupt contact in most units in the excavation. The color and texture change from the tephra deposit to the topmost paleosol was usually quite visible, changing from a pale, cream-colored deposit with the texture of fine flour, to a dark brown deposit of fine sands, silts and clay that was much harder to excavate because of the minor clay content. In excavation Unit 14, the 'A' horizon extended from 111-118 cm below surface (see Figure 8 for description of the profile of excavation Unit 14). The 'A' horizon was underlain by a 10 centimeter, calcium enriched, sandy 'B'

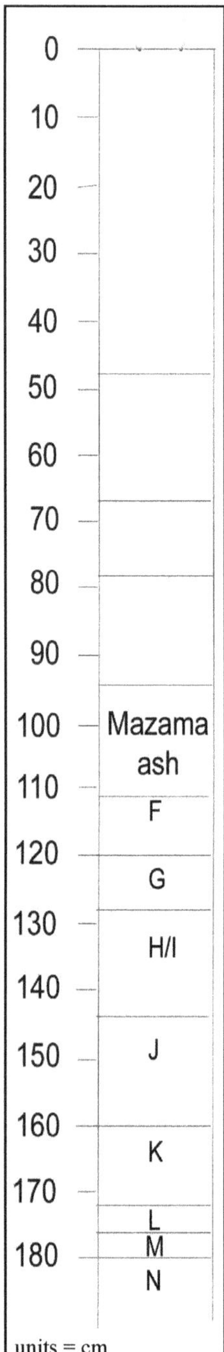

0-10 = Sod layer, surface of unit

10-48 cm BS – 10YR 2/2 very dark brown; texture = silty fine sand, not or slightly cohesive; structure = blocky or columnar; porosity = highly porous; abundant fine vertical roots; no clasts >.5 cm; contact = diffuse; acid reaction = no effervescence.

48-67 cm BS – 10YR 3/2 very dark greyish brown; texture = silty clay, moderately plastic and moderately cohesive; structure = fine granular; porosity = highly porous; moderate abundance of fine roots; no clasts >.5 cm; lower contact = wavy and gradual; acid reaction = major effervescence.

67-78 cm BS – 7.5YR 3/2 dark brown; texture = silty, slightly plastic, slightly cohesive; structure = none; porosity = porous, no clasts >.5 cm; fine roots present; lower contact = abrupt and sloping; acid reaction = violent effervescence.

78-94 cm BS – 10YR 3/4 dark yellowish brown; texture = silty, slightly plastic and slightly cohesive; porosity = porous; structure = amorphous; few, fine vertical roots; no clasts >.5 cm; lower contact = undulating but clear over 2 or 3 cm; acid reaction = violent effervescence.

94-102 cm BS – Darker, top part of Mazama tephra horizon – 10YR 5/4 yellowish brown; texture = silty, slightly cohesive and slightly plastic, falls apart; structure = amorphous; porosity = porous; no clasts >.5 cm; acid reaction = slightly, if at all effervescent, typical of tephra. Layer E

102-111 cm BS – Lighter, lower part of Mazama tephra horizon – 10YR 6/3 pale brown; texture = silty with a little fine clay, slightly plastic and slightly cohesive; structure = none; porosity = porous; lower contact = abrupt. Layer E

111-120 cm BS – 1st A horizon – 10YR 2/2 very dark brown; texture = silty, fine sand, moderately cohesive and moderately plastic but will break up when rolled; structure = fine granular; porosity = porous; none to few roots; no clasts >.5 cm; lower contact = flat and diffuse; acid reaction = no fizz. Layer F

120-128 cm BS – 1st B horizon – 10YR 3/3 dark brown; texture = fine to coarse sand, very slightly plastic, non-cohesive; structure = fine granular; lower contact = wavy and gradual, not diffuse because it takes place over a short distance; acid reaction = slight effervescence. Layer G

128-144 cm BS – 2nd A horizon – 10YR 2/2 very dark brown; texture = clayey silt, not much sand, few grains, quite sticky and cohesive; structure = granular; lower contact = level and diffuse; acid reaction = violent effervescence. Layer H/I

144-161 cm BS – 2nd B horizon – Redder soil – 5YR 2.5/2 dark reddish brown; texture = clayey silt, slightly cohesive and plastic; lower contact = flat and gradual; presence of lots of charcoal. Layer J

161-172 cm BS – 3rd A horizon (proposed) – Darker soil – 5YR 2.5/1 black; texture = clayey silt, slightly cohesive and slightly plastic; lower contact = flat and abrupt; acid reaction = violent effervescence. Layer K

172-176 cm BS – 3rd B horizon (proposed) – Orange sand – 5YR 3/2 dark reddish brown; texture = silty, not cohesive and not plastic at all; lower contact = abrupt and smooth; acid reaction = violent effervescence. Layer L

176-180 cm BS – 4th A horizon (proposed) – Bottom black soil – 5YR 2.5/1 black; texture = silty with a bit of fine sand, slightly cohesive and slightly plastic; lower contact = wavy and diffuse; acid reaction = violent effervescence. Layer M

180- ? cm BS – Glacial Lake Calgary sand; 2.5YR 4/2 dusky red; texture = sand; structure = amorphous; acid reaction = violent effervescence. Bottom of excavation – sterile glacial sediments. Layer N

Figure 8: Profile description of Unit 14 at the Tuscany habitation site EgPn-377

horizon. This 'B' horizon was also quite distinctive in texture and color, though this was much easier to detect near the central area of the excavation. In excavation Unit 14, this horizon extended from 118-128 cm below surface. Thus, the second pedogenic zone extended from 111-128 cm below surface in excavation Unit 14.

Many of the excavation units also contained a number of different elements of bison skeletal material, which projected up from the paleosols into the Mazama tephra layer. Thus, there was in general, little ambiguity identifying the tephra/paleosol boundary. The Mazama tephra and the first 'A' and 'B' horizon of the paleosols were excavated as natural layers due to their distinctive color and texture.

The third pedogenic zone was located immediately below the sandy 'B' horizon. The third pedogenic zone was comprised of a dark 'A' horizon that graded into a 'B' horizon. In excavation Unit 14, this 'A' horizon extended from 128-144 cm below surface. The 'B' horizon then extended from 144-161 cm below surface. Interestingly, the 'B' horizon appeared much redder in hue once it had dried out for a period of time. Significantly, there was a definite increase in the amount of charcoal visible to the naked eye in the 'B' horizon of the third pedogenic zone.

Pedogenic zone four was found immediately underlying pedogenic zone three and consisted of an 'A' and 'B' horizon. In excavation Unit 14, the darker 'A' horizon was observed extending from 161-172 cm below surface and the 'B' horizon extended from 172-176 cm below surface. This 'B' horizon was more orange in color, comparatively (this is an observation, not a Munsell color designation), which distinguished it from the horizons identified above and below.

A final, fifth pedogenic zone was identified immediately below zone four. Pedogenic zone five occurred immediately overtop the Glacial Lake Calgary sands. In excavation Unit 14 the 'A' horizon extended from 176-180 cm below surface. This 'A' horizon may represent the first organic horizon that would have formed over the exposed Glacial Lake Calgary sands following deglaciation. The sediments identified as Glacial Lake Calgary sands represent the 'C' horizon, or parent material, of this fifth pedogenic zone. Distinctive "fingers" of the 'A' horizon were observed, penetrating and/or grading, down into the glacial sands. In some cases, the boundary between the final 'A' horizon and the glacial sands was difficult to discern because of the way they graded together.

The stratigraphy of the units on the periphery of the excavation became difficult to discern with distance from the center of the grid. It became obvious as the excavation progressed that the paleosols comprised a distinctive and complex deposit. This deposit was the result of the landscape morphology itself, patterns of sediment deposition through prevailing winds, local climate, moisture, and processes of pedogenesis combined with influences from the vegetation cover that was best suited to thrive in conditions existing in the area during the early Holocene.

Three partial wall profiles from the Tuscany habitation site (EgPn-377) are presented next. The three profiles are from walls at the extreme edges of the excavation which were all upslope from the center of the depression. The wall profiles reveal the gradual thinning out of all the paleosols as the excavation expanded outward and upslope from the center of the grid.

The descriptions for the sedimentary sections and pedogenic zones described for excavation Unit 14 apply equally to the wall profiles since they describe the same series of changing boundaries. Thus, all the profiles contain sedimentary sections one, two and three, with a small portion of section four (the glacial sands) being exposed in some spots. Pedogenic zones one, two and three are clearly represented as well, while four and five were problematic and are thus discontinuous from one wall profile to the next.

The profile interpretation for excavation Unit 14 was carried out during the 1996 season. Excavation of the site was completed during the summer of 1997 and the final series of interpreted wall profiles were one of the last tasks executed to finalize the field-work. It was apparent that timing was an important factor in the wall profile interpretations. The wall profile interpretation of excavation Unit 14 was completed during the summer in 1996, when it was still quite shaded by the surrounding walls of the unexcavated units. These wall surfaces had not dried out and the profile interpretation was carried out on fairly damp sediments. The combination of shade and damp conditions made color and boundary changes within the sediments harder to discern with certainty. By the end of September 1997, the walls of the excavation had been exposed for many weeks and had dried out. This made a tremendous difference to the visibility of color changes throughout the darker paleosols. A few changes between the first interpretation of excavation Unit 14 and the final interpretation of the wall profiles need to be discussed. The changes actually constitute a refinement of interpretation, as more detail was visible in the dried wall profiles.

The wall profiles are fairly complex drawings, primarily because there was significant disturbance from burrowing Richardson ground squirrels (*Spermophilus richarsonii richardsonii*) throughout the entire excavation grid. The wall profiles were interpreted by a number of people working for the Tuscany Project. During the interpretation, lines were etched into the excavation walls where boundaries were identified by color and texture changes visible in the sediments. These interpreted wall profiles were measured as they were drawn to scale on graph paper. In order to simplify the field interpretation a series of letters have been used to code the different parts of the profiles that were identified as unique by color and texture. Because each layer is separately described with this method of interpretation, the sedimentary sections and pedogenic zones are not apparent. This approach provides precise descriptions of each unique portion of the wall profiles as observed and interpreted.

Figure 9 provides a legend for the breakdown of the designated layers 'A' through 'N', for the three wall profiles that are presented in Figures 10 (east wall), 11 (west wall), and 12 (north wall). The upper section of each wall profile (east, west and north), from the sod surface to the Mazama tephra, includes layers A to D and fits the interpretation just presented above for excavation Unit 14. The upper section of the profile contains the first recognized pedogenic zone 1, which has been identified as a developed chernozem. However, its location above the Mazama tephra (6800 yr BP) boundary excludes it from the present study. The subtle color changes observed within the Mazama tephra in excavation Unit 14, were not apparent in the wall profiles at the extreme edges of the excavation. In some of the outer excavation units the tephra was much thinner, to the point of being discontinuous. The Mazama tephra was presented as a single layer in the extended wall profile drawings.

Layers F and G were fairly clear in all three of the wall profiles. They represent the second pedogenic zone found in the site with F as the first 'A' horizon and G as its 'B' counterpart. Layer H represents the beginning of the third pedogenic unit as an 'A' horizon. It was quite clearly discernable and averaged approximately 10 cm in all profiles. The bottom of layer H is where the first change in the interpretation of the profiles takes place. When the profiles had dried out, a very thin and discontinuous

Layer	Layer Description
A	10YR 2/2 very dark brown; texture = silty fine sand, not or slightly cohesive; structure = blocky or columnar; porosity = highly porous; abundant fine vertical roots; no clasts >.5 cm; contact = diffuse; acid reaction = no effervescence.
B	10YR 3/2 very dark greyish brown; texture = silty clay, moderately plastic and moderately cohesive; structure = fine granular; porosity = highly porous; moderate abundance of fine roots; no clasts >.5 cm; lower contact = wavy and gradual; acid reaction = major effervescence.
C	7.5YR 3/2 dark brown; texture = silty, slightly plastic, slightly cohesive; structure = none; porosity = porous, no clasts > .5 cm; fine roots present; lower contact = abrupt and sloping; acid reaction = violent effervescence.
D	10YR 3/4 dark yellowish brown; texture = silty, slightly plastic and slightly cohesive; porosity = porous; structure = amorphous; few, fine vertical roots; no clasts > .5 cm; lower contact = undulating but clear over 2 or 3 cm; acid reaction = violent effervescence.
E	Darker, top part of Mazama ash layer - 10YR 5/4 yellowish brown; texture = silty, slightly cohesive and slightly plastic, falls apart; structure = amorphous; porosity = porous; no clasts > .5 cm; acid reaction = slightly, if at all effervescent, typical of tephra Lighter, lower part of Mazama ash layer – 10YR 6/3 pale brown; texture = silty with a little fine clay, slightly plastic and slightly cohesive; structure = none; porosity = porous; lower contact = abrupt.
F	1st A horizon – 10YR 2/2 very dark brown; texture = silty, fine sand, moderately cohesive and moderately plastic but will break up when rolled; structure = fine granular; porosity = porous; none to few roots; no clasts >.5 cm; lower contact = flat and diffuse; acid reaction = no fizz.
G	1st B horizon - Sand layer – 10YR 3/3 dark brown; texture = fine to coarse sand, very slightly plastic, non-cohesive; structure = fine granular; lower contact = wavy and gradual, not diffuse because it takes place over a short distance; acid reaction = slight effervescence.
H I	2nd A horizon – 10YR 2/2 very dark brown; texture = clayey silt, not much sand, few grains, quite sticky and cohesive; structure = granular; lower contact = level and diffuse; acid reaction = violent effervescence.
J	2nd B Horizon – Redder layer – 5YR 2.5/2 dark reddish brown; texture = clayey silt, slightly cohesive and plastic; lower contact = flat and gradual; presence of lots of charcoal.
K	3rd A horizon (proposed) – Darker layer – 5YR 2.5/1 black; texture = clayey silt, slightly cohesive and slightly plastic; lower contact = flat and abrupt; acid reaction = violent effervescence.
L	3rd B Horizon (proposed) – Orange sand – 5YR 3/2 dark reddish brown; texture = silty, not cohesive and not plastic at all; lower contact = abrupt and smooth; acid reaction = violent effervescence. This layer contained concretions of amorphous structure – initially they appeared bone-like however it became clear that they had no regular shape at all.
M	4th A horizon (proposed) – Bottom black layer – 5YR 2.5/1 black; texture = silty with a bit of fine sand, slightly cohesive and slightly plastic; lower contact = wavy and diffuse; acid reaction = violent effervescence.
N	Glacial Lake Calgary sand; 2.5YR 4/2 dusky red; texture = sand; structure = amorphous; acid reaction = violent effervescence. Bottom of excavation – sterile glacial sediments.

Figure 9: Legend of the layer descriptions for East, West and North wall profiles

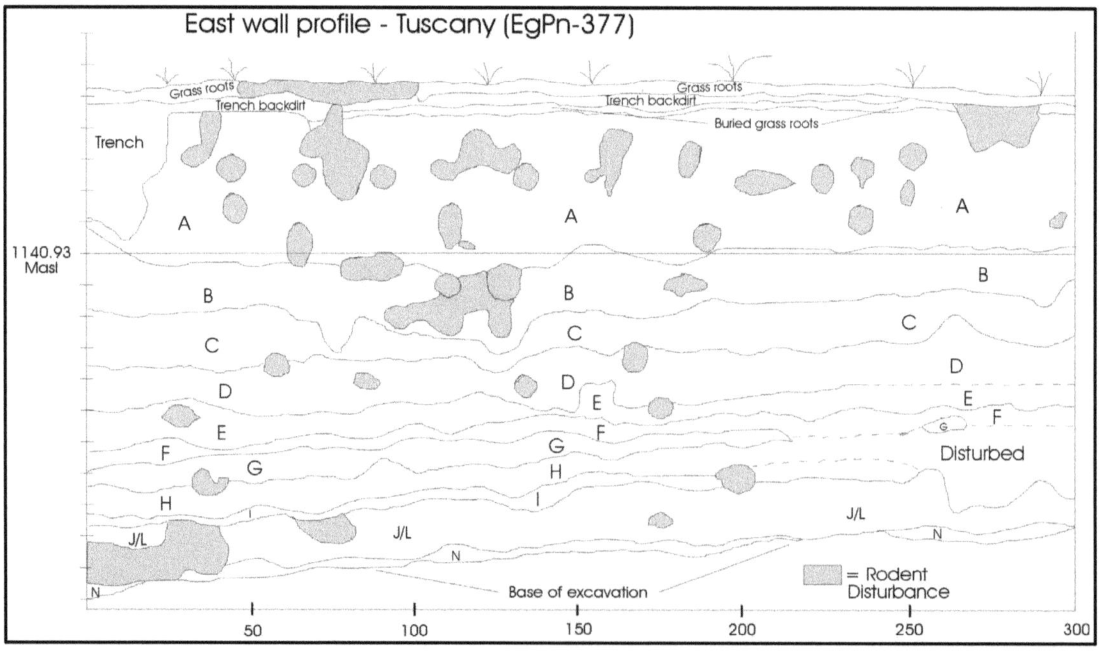

Figure 10: Profile of 3 units along the east wall of the excavation. Bottom scale is in 50 cm increments. Side scale is in 10 cm increments. Disturbed area is probably due to rodent activity in Layer H, however, there was no burrow outline visible

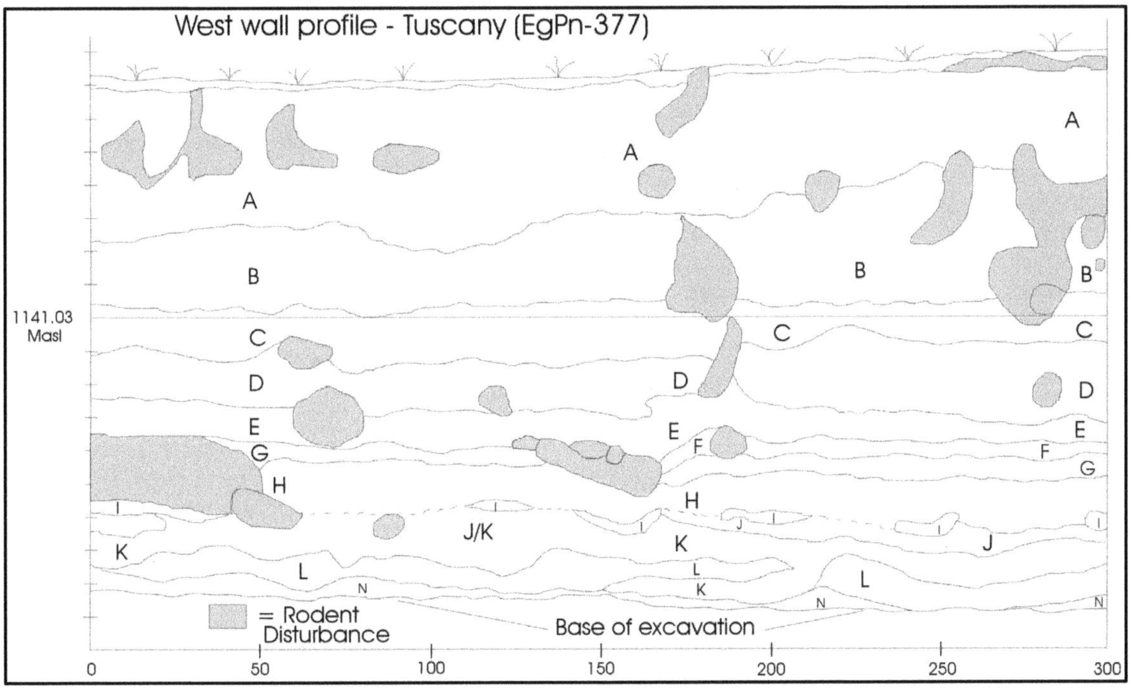

Figure 11: Profile of 3 units along the west wall of the excavation.
Bottom scale is in 50 cm increments. Side scale is in 10 cm increments

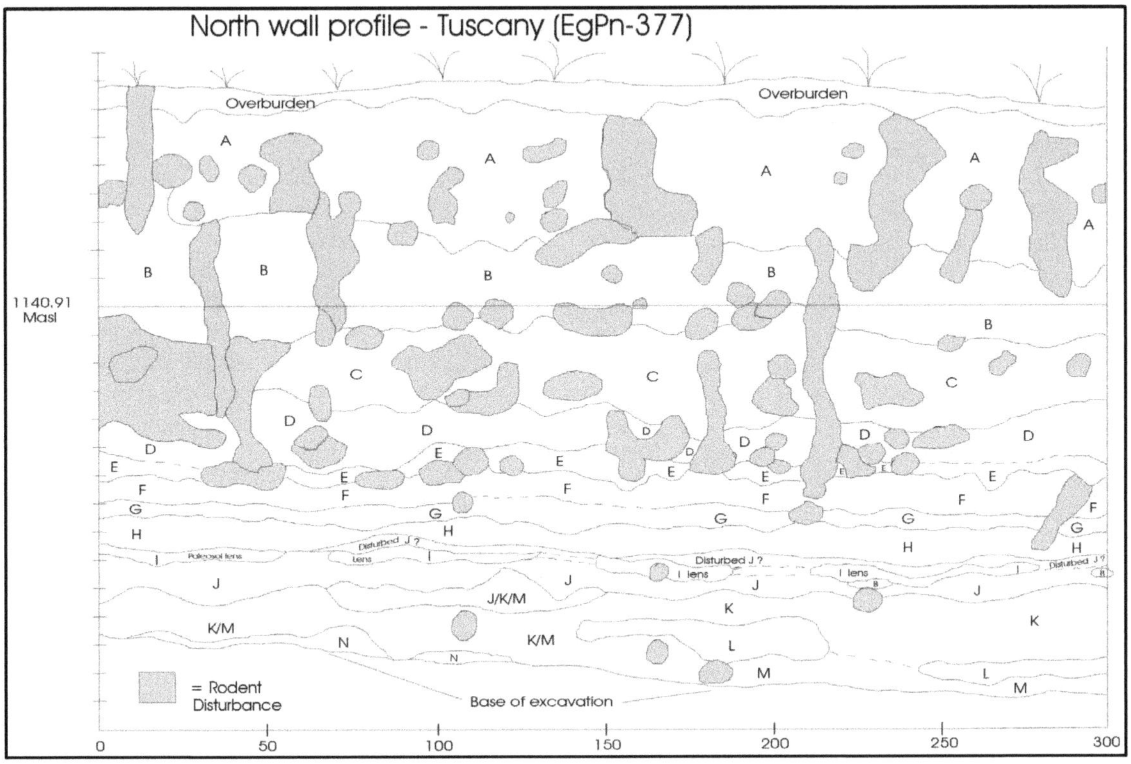

Figure 12: Profile of 3 units along the north wall of the excavation.
Bottom scale is in 50 cm increments. Side scale is in 10 cm increments

paleosol lens (labeled "I") was visible underlying layer H. The combination/relationship of Layers H, I and J is a bit ambiguous. Layer J was noted to be dark reddish brown in color and contained the most visible amounts of charred plant material. Charred plant remains were found throughout the paleosols in amounts that varied with depth. Their relationship to the sediment deposition will be discussed at length, below.

As with Unit 14, Layers K and L could be interpreted as another, or fourth pedogenic unit, with K as the 'A' horizon and L as a 'B' horizon. Layer L is problematic because it was extremely discontinuous throughout the site and contained large numbers of concretions that were odd and amorphous in shape. They dissolved fairly readily in water and were gritty in texture. According to Dr. A. Limbird (pers. Comm. 2001), Department of Geography, University of Calgary, these are most likely $CaCO_3$ accumulations, which have filled in spaces and cracks from plant roots. Thus, they are related to the movement of water through the soil. In the East wall profile (10) Layers J and L were indistinguishable and there was no Layer K. There was also no Layer M above the Glacial Lake Calgary sands either. Layer M may constitute the 'A' horizon of a final, fifth pedogenic unit with the Glacial Lake Calgary sands constituting its 'C' horizon as a parent material. Layer M was extremely dark (5YR 2.5/1 black) and contained charred plant materials within it to the top of the glacial sands. The fact that it is not found in the East wall profile (Figure 10) is related to slope and elevation. Layer M is found initially in the deepest areas of the site and represents the first filling in of the depression/channel scar. While there may have been some deposition in the higher locations, upslope from the center of the depression, it did not survive through time, perhaps being eroded off by wind and water. As the next deposition event took place with Layers K and L forming the 'A' and 'B' horizons, the surface was a bit more level and the infill covered a larger area, slowly appearing to move upslope.

GEOMORPHOLOGY AT THE TUSCANY SITE

It is clear that the sediments of a site can provide an extraordinary amount of information about the history of the site. The pattern of deposition in terms of which deposit came first and what changes there may have been to modify the depositional processes (i.e. subsequent water or aeolian action), can be interpreted by grain size and chemical analyses of the sediments. These types of analysis have not been carried out on the Tuscany sediments. In the meantime, it is still possible to derive the basic history of the formation of the site at a less fine-grained level of analysis. The sediments have been described above. Five sedimentary sections were identified, from the surface down and included, 1) aeolian sands and silts, 2) Mazama tephra, 3) sands and silts, 4) glacial sands, and 5) glacial clays. The clays represent the oldest materials deposited at the site and the aeolian sands and silts of section 1 are the most recent.

The oldest and deepest deposits of clay and sands were identified through exploratory excavation within the site grid. These sediments were sterile in terms of cultural materials and excavation of the units in the grid ended when the fourth sedimentary section was encountered. The final level of each unit was excavated as a natural layer, in the sense that the glacial sands were not dug when they were encountered. The final depths followed the natural stratigraphy of the sand and final elevations were taken for each unit when finished.

The bottom of the Mazama tephra deposit was also excavated as a natural level boundary. When the paleosol deposits were encountered, the excavators were instructed not to dig into them. They were to remove only the remnant Mazama tephra pockets to reveal the original landscape surface as it would have appeared at 6730 ± 40 years BP. The elevation measurements for each unit were recorded at this point as well.

The first two layers of the paleosols (layers F and G in the soil profiles) were also evaluated as being fairly discernable or distinguishable from each other and the stratigraphy below. They were also excavated as natural layers throughout most of the excavation. However, now that the analysis of the site has progressed to the point where the information can be applied, it is apparent that the natural stratigraphy of the first paleosol was not clear to many of the excavators. A series of profiles have been derived from the depth measurements of each unit by level. Once mapped out on graph paper, it is clear that serious discrepancies occurred from one unit to the next, in terms of evaluating where the first 'A' horizon ended (Layer F) and the sandy 'B' horizon began (Layer G). There are good match-ups between units in some cases, but there are obvious errors between other units. There were initial plans to create maps of the changing topography of the grid using the elevation data from the excavation, however, the level of error observed from the paleosol interpretation has negated that option.

There are three surfaces that can be mapped using the Tuscany grid data. The initial prairie surface measurements of each unit appear to be free of large discrepancies. There are some places where the measurements are questionable, but they represent errors in the range of five to twelve centimeters from one side to the other of a unit corner pin. The Mazama tephra-paleosol contact represents the next surface that can be mapped using excavation elevation measurements. The paleosol-glacial sand contact is the final surface that can be mapped using the bottom measurements taken for each unit as it was completed.

Five elevation measurements were taken at the top of each level, and included each of the four corners and the approximate center of each unit. Since most of the bulk soil samples from the units were taken from the southwest corner, the elevation measurements from that corner of each unit were used to create a series of topographical maps of the excavation grid. The three mapped surfaces are presented next with an interpretation of how the landscape has changed.

Figures 13 and 14 present digital elevation model (DEM) views of the changing surface trends through time at the Tuscany habitation site. All maps have been produced using the Surfer 7 software program. Wireframe maps are three-dimensional representations of topography using X, Y and Z data from points on a grid. In the case of Figure 13, the X, Y, and Z coordinates for the southwest corners of all 91 units in the 11 x 11 meter grid have been used to generate the DEM. Due to small-scale variability in the

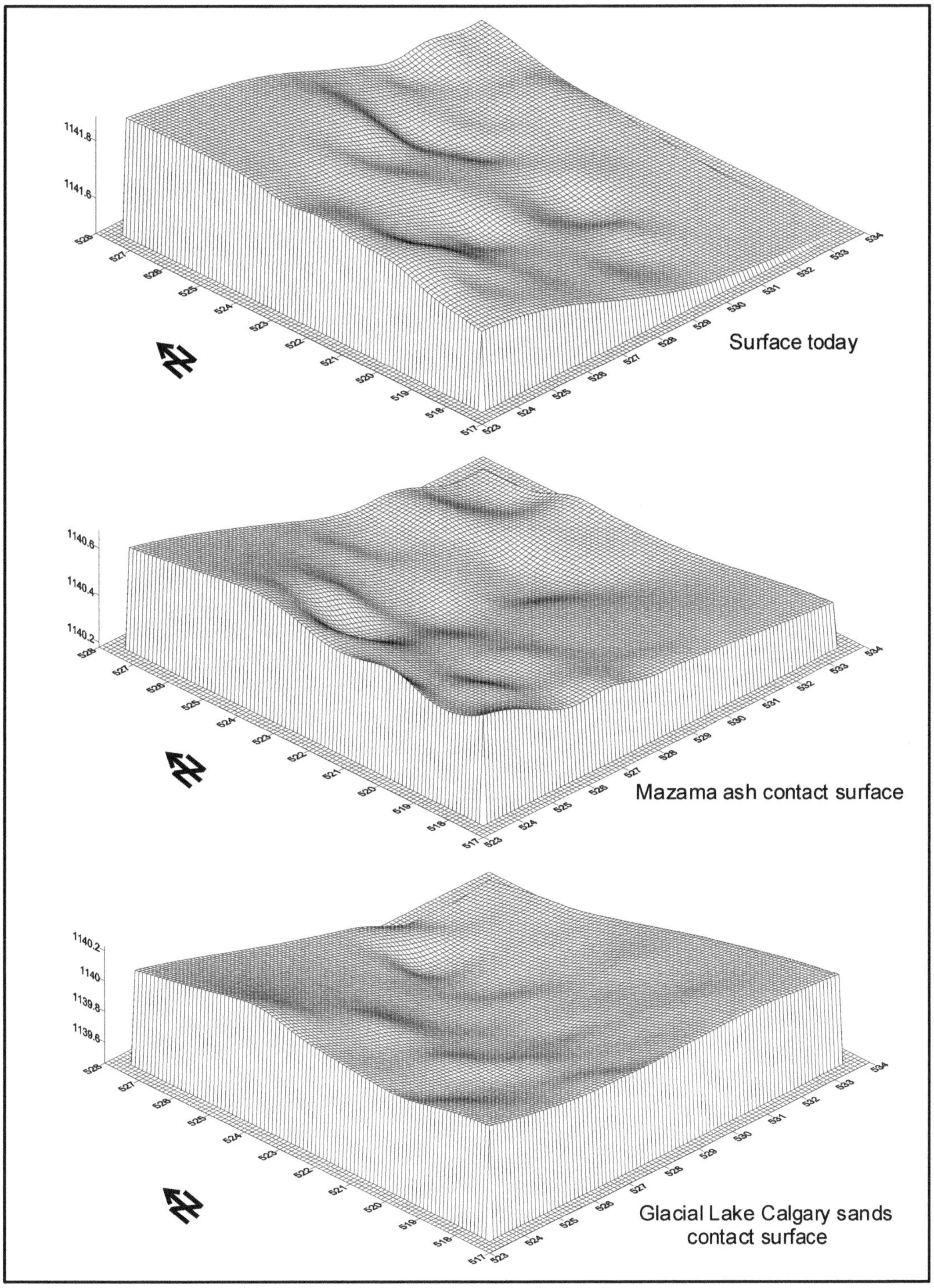

Figure 13. Wireframe DEM grid maps showing the trends of surface change through time. The top view is the surface as it is found today. The middle view is the surface of the depression as it appeared when the Mazama ash fell. The bottom view is the surface of the depression as it appeared after Glacial Lake Calgary drained and vegetation began to colonize the landscape

original grid data, the matrix smoothing function has been used to eliminate noise that was visible on the original maps. There are nodes on the grid with no elevation data. These are the unexcavated units. Surfer 7 interpolates the data that is present and provides a projection that expresses the trends suggested in the data and completes the map of the entire grid for each case presented.

Some interesting trends are revealed in the DEMs. Today, the Tuscany site grid is clearly situated in a depression. Figure 2 presents a topographic map of the area where it is quite clear that the depression is surrounded by a circular perimeter of rising ground. In the wireframe map depicting the modern surface of the site grid we have a view which shows only a portion of the depression—the part that was actually excavated. Though there are data

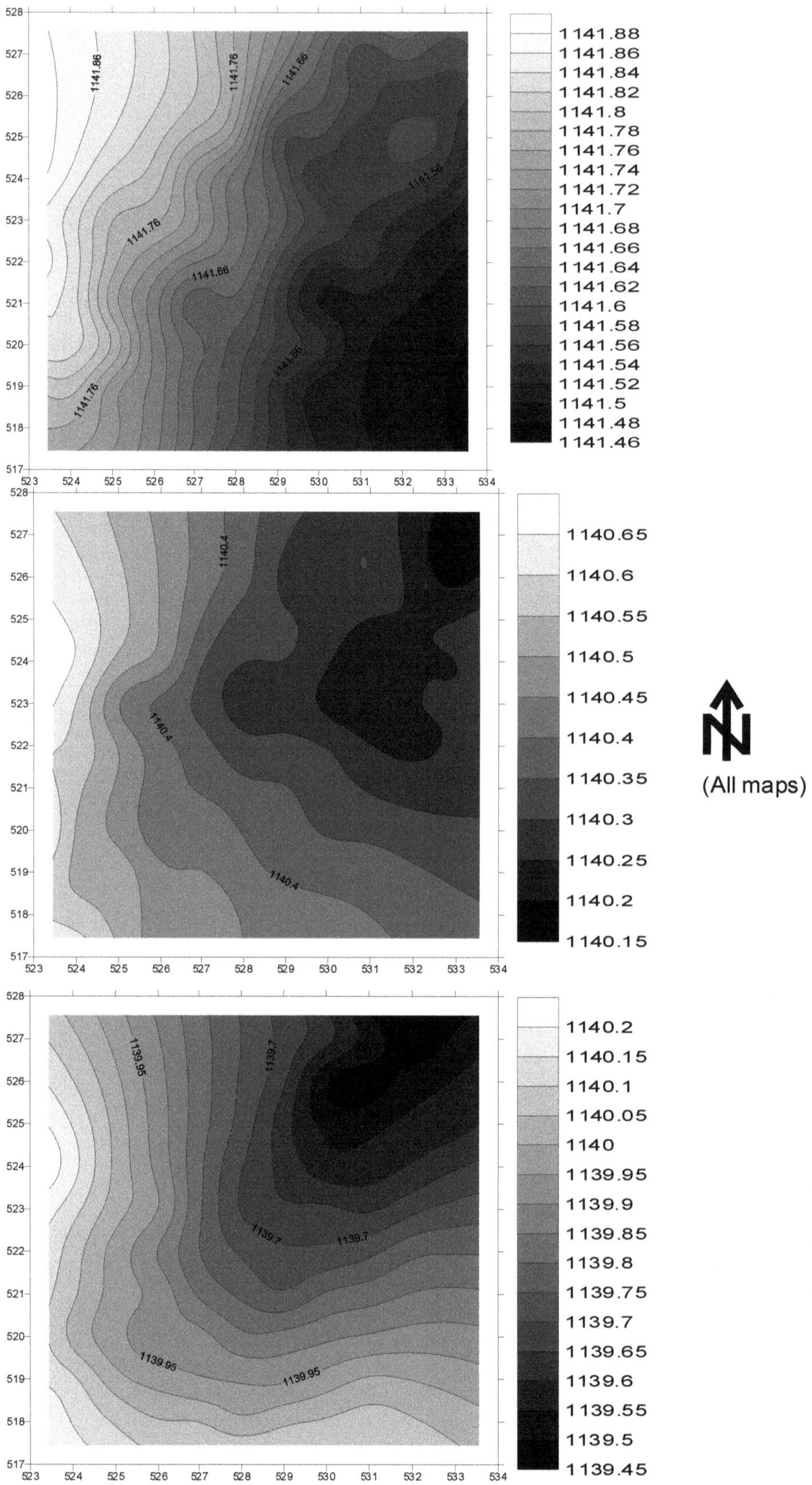

Figure 14: Three DEM landscape surfaces are represented here. The top map is an interpolated view of the land surface of the Tuscany site grid before any excavation took place. The middle map is an interpolated view of the land surface from grid measurements which shows what the surface of the site looked like when the Mazama ash fell. The bottom map is the is an interpolated view of the land surface from grid measurements which shows what the surface was like immediately after the glacial ice retreated at the end of the Pleistocene ~ 14,000 years ago

missing (the unexcavated units) from the northwest and southeast sections of the 11 x 11 meter grid, Surfer 7 has interpolated a very realistic view of the surface trends as they actually appear. The ground did indeed rise to the northwest and it dropped down toward the southeast. The argument can be made that the map projections are fairly reliable, even though there are missing data in the form of unexcavated units within the grid.

If we accept the 'best fit' projections of Surfer 7, we can see the change in topography that has taken place since the retreat of the continental ice sheets at the end of the Pleistocene. Figure 13 presents the wireframe map projections and Figure 14 presents the contour map projections of the Tuscany grid. The wireframe projection for the Glacial Lake Calgary sands contact represents the surface topography of the site upon which the sediments that subsequently developed into the paleosols were deposited. There appears to be a channel scar oriented from the southwest corner to the northeast corner of the grid. Project work by students which examined the sediment and stratigraphy, in all four directions around the grid, indicated the possibility of a V-shaped channel oriented north-south (Oetelaar and Zaychuk 1997). It has been proposed that the channel scar may have been part of the surface drainage pattern when the Bow River was at a much higher level than today. Oetelaar and Zaychuk (1997) have suggested that there was a former streambed of the Bow River located north of the Tuscany site and the channel scar through the grid drained into it.

A series of peripheral units dug at 10, 20 and 30 meters, away from the grid in the four cardinal directions, demonstrated that the paleosols were a discontinuous feature of the landscape. In the unit located 30 meters east of the grid, the reddish layer identified as Layer J in the wall profiles (above), was found only 50 centimeters below the sod surface. This same reddish Layer J was found at depths of 120 centimeters below the sod surface in the deep, central areas within the excavation grid (Oetelaar and Zaychuk 1997). The reddish Layer J was difficult to discern in the profiles of the peripheral units, but it was clear that it was sloped upwards, away from the center of the depression. At 30 meters away from the grid, east and south, the reddish Layer J was clearly thinning out. Layer J was not visible in the peripheral units 10 meters north of the grid nor 10 meters west. Thus, the original paleosol surface appears to have had a finite border to its occurrence in the landscape. This indicates the possibility that the paleosols and associated vegetation were a local phenomenon. This distribution pattern for the paleosols may also be due to erosion processes that removed everything but the materials in the sheltered depression.

As time passed, sediments slowly accumulated in the depression partially filling in the deeper part of the channel scar or depression, which had a leveling effect on the local landscape. By the time of the Mazama ash fall at 6730 ± 40 years BP, the surface appears to have leveled out to some extent. However, there is indication that the slope upward to the northwest had begun to form prior to the Mazama volcanic event. The deeper part of the area that became the Tuscany site is still deepest in the northeast at the Mazama contact surface (see Figures 13 and 14). The pattern of deposition is probably related to colluvial processes and wind, where material is washed downhill to the lowest spots of the land surface during storms as well as being blown in continuously with prevailing winds.

When we compare the wireframe and contour maps depicting the modern surface of today to the projection maps depicting the glacial sands and Mazama tephra surfaces, it is apparent that the topography has changed significantly since the early Holocene. The deepest part of the grid area has shifted completely from the northeast corner to the southeast corner of the grid. The slope to the northwest has built up completely and appears almost dune-like, with the depression in the lee area. There is a definite small ridge-like feature building along the northwest side of the grid, which is visible in Figure 13 and 14 (the middle map projections), depicting the Mazama contact surface.

SUMMARY

A number of approaches to view the site have been detailed in this chapter. The material provided here is primarily descriptive rather than analytical. This project did not complete any analyses on the sediments to identify them further nor that may have served as comparative material to other sites. The entire site excavation methodology and bulk sediment sampling scheme were established and controlled by the project director for the duration of the Tuscany Project. With no further analyses completed, the only options available for this research has been a descriptive analytical approach. This has combined sediment and soil descriptions with wall profile drawings and details of depth measurements to understand the deposition and evolution or change that has taken place over time within the Tuscany site depression.

As noted in this chapter, a number of soils have developed within the sediments of the site through time. The upper soil, determined to be chernozemic, was not included in the present research. The focus of the flotation analysis has been directed to a series of buried paleosols determined to date between approximately 6700 and 10,000 years BP. These buried paleosols have been amalgamated into four separate soil forming events. Within the four paleosols there are two areas where cultural materials are associated. The uppermost or most recent paleosol has yielded evidence of possible human activity in the depression at some time prior to the overlying Mazama tephra deposition event. This evidence of human activity remains inconclusive at this time.

The fourth pedogenic zone, amalgamated and comprised of Layers H/I and J, contained the cultural materials that are of interest to this research. The lithic, faunal and charred botanical remains identified in this pedogenic zone have yielded patterns of deposition that indicate definite occupation of the Tuscany site depression for a short time period at approximately 7800 years BP.

Chapter 4
THE MACROBOTANICALS

CHARRED WOOD

Thousands of pieces of charred plant remains were recovered from the Tuscany site flotation assemblage. The following discussions concern all of the remains found below the Mazama tephra at the site, but also particularly focus on the cultural occupation and the associated charred plant materials. Each plant that has been identified to the species or genus level of classification in this analysis is discussed in this chapter. Ethnographic information available for the use of these plants is incorporated into the discussion, where possible and applicable, with the intent of exploring ideas related to the landscape of the past and assessing possible reasons why people may have camped at the Tuscany habitation site during the past.

While a considerable amount of time was spent identifying the charred seeds, only a few pieces of the charred woody plant remains have been examined in detail. Most of the pieces of charred plant remains were less than five millimeters in size and they thus appeared to represent small twigs or roots that had been charred. However, it was not completely clear whether or not they were really wood or whether they were simply pieces of plant material that had preserved because of the charring process which had rendered them less susceptible to disintegration compared to non-charred plant material. Clearly, important information about the paleoecology of the Tuscany site may be derived from identification of wood. Unfortunately, it was only possible to devote a limited amount of research into the identification the charred woody plant remains. This was primarily directed to determining if these items were actually small pieces of wood. The most important information such identification might potentially reveal is what species of tree(s) or shrub(s) may have been burned at the Tuscany site during the past.

If the charred plant remains are from wood producing plants and not simply charred or carbonized plant structures that resemble wood, they should exhibit specific structures that may help in their identification. According to Core *et al.* (1979:1), "wood is the cell wall material produced by cells of the cambium in a living tree". The cell walls combine to produce the structure of trees and the morphology of these cell walls can exhibit unique patterns that make wood identifiable. If we take a theoretical slice out of a tree, like a piece of pie, cross-wise, we would have a piece of wood that displays three important identification views. The top of the piece is the "transverse surface" or "cross section", the side view of the piece from the bark to the center, is the "radial surface", and the view that parallels the outer bark in an arc is the "tangential surface" (Core *et al.* 1979). Each view of wood reveals structural characteristics that can be species specific.

Two main types of wood are differentiated in north-western North America. There are "hardwoods" from the deciduous trees, which generally have broad leaves, and "softwoods" from coniferous trees, which generally have needle leaves. As trees grow each year, a pattern known as annual growth rings is produced by the cell structures. Within the annual growth rings the growth pattern of the cells usually changes from large, early growth cells, to small, late growth cells. This pattern is fairly regular in the conifer species. The deciduous species have addition-nal structures known as vessels, which run longitudinally through the tree and are for conduction. In cross section these structures are visible as pores. The combination of cell and pore structures produces variable patterns as annual growth rings cycle from spring to fall and winter.

This brief background provides enough information to examine pieces of wood that have been imaged using the ESEM (Environmental Scanning Electron Microscope) to get a good quality magnified view of some of the cell structures. Small pieces of wood that were similar in appearance were selected from excavation Unit 14 Levels T and U.

The first view (Figure 15) of the piece of charred wood from Level T is at a low magnification and demonstrates

Figure 15: An example of a charred plant fragment from Unit 14 Level T from the Tuscany habitation site

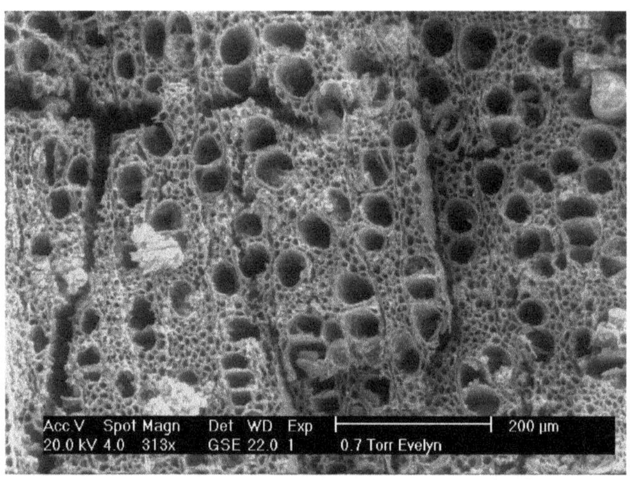

Figure 16: A magnified view of the cross section or transverse view of a piece of wood from Unit 14 Level U, from the Tuscany habitation site

that some structure is apparent but not diagnostic at this magnification. The second view (Figure 16) shows a piece of charred wood from Level U, magnified 313x. It shows clear cell and pore structure indicative of a deciduous, or hardwood tree/shrub. The next view (Figure 17) shows the same piece of charred wood at 672x magnification to detail that cell structure more clearly. The pattern of the larger pores among the cells is "diffuse", one of the classification types described by Core *et al.* (1979). The diffuse pore and cell pattern of the level U charred wood looks quite similar to *Populus* sp. examined by Core *et al.* (1979). It is possible that the charred wood is from a *Populus* sp. tree or from a genus of the Salicaceae (Willow) Family. This brief examination indicates that there may have been some deciduous trees or shrubs present at or near the Tuscany habitation site during the past. The presence of the charred wood could be due to natural growth of vegetation in the locale and subsequent burning of the vegetation cover. Alternatively, wood may have been introduced to the site for use in a fire. The spatial distribution and patterning of the charred plant fragments will address these issues. It is apparent with the charred

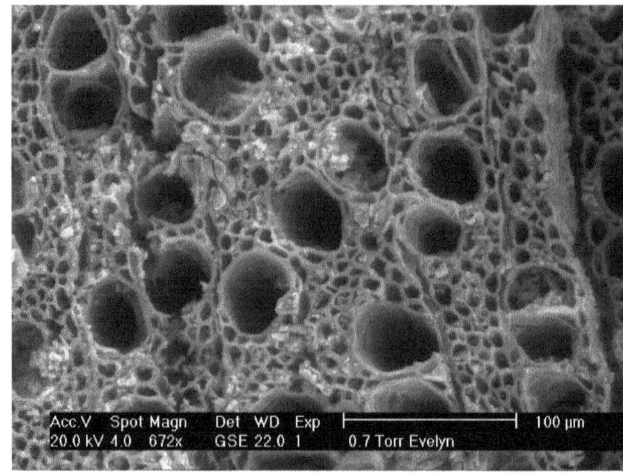

Figure 17: Another view of the piece of wood from Unit 14 Level U, at a higher magnification to vew the pore and cell wall structure

plant remains that there is a distinct area, of 3 x 4 meters diameter around the hearth, where very few charred plant remains appear to have accumulated. This pattern is visible in the figures that will be presented in the spatial analysis in the following chapter.

CHARRED SEEDS AND SEED FRAGMENTS

There are two sets of information that have become apparent from examination of the charred seed assemblage. First, the charred seeds provide an inventory of some of the plant species that grew locally at the Tuscany site during the early Holocene. The second set of information is related to spatial patterning and will be addressed below. This chapter will focus on the inventory of identified charred seeds. The discussion will focus on the growth habits and environment and possible economic or cultural significance that the plant species identified here may have had for people living within the landscape.

Juniper Seeds

There were two plant species, *Juniperus* sp. (juniper) and *Arctostaphylus uva-ursi* (kinnikinnick or bearberry), that were extremely well represented throughout the site. Charred seeds from these plants appear to be ubiquitous, both across the site and down through the depths of the paleosols within the site. This ubiquity implies that these two plants have played an important role in the appearance of the site through time. Their salient characteristics will be discussed using descriptions from personal experience and observation along with information from Farrar (1995), Vance *et al.* (1999) and Wilkinson (1991, 1999).

Two types of juniper are found growing in the vicinity of the Tuscany habitation site today: *Juniperus horizontalis* (creeping juniper) and *Juniperus communis* (common juniper). Both species are considered to be

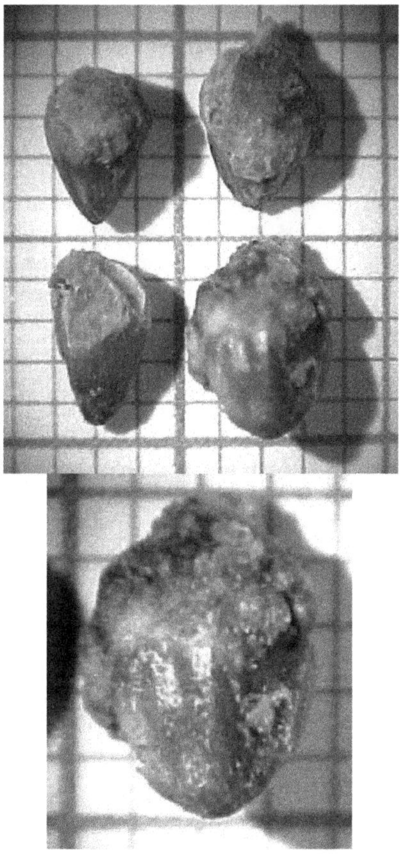

Figure 18: Juniperus horizontalis *seeds from the PMA comparative collection*

transcontinental shrubs that never reach tree size. *Juniperus communis* is noted to have one of the most widespread distributions compared to any other tree or shrub across the Northern Hemisphere. Junipers are members of the Cupressaceae or Cypress Family and are evergreens with two leaf forms, needles and scale leaves. *Juniperus horizontalis* is a flat-lying shrub with long main branches running across the ground and secondary branches that stand up and look bushy. *Juniperus communis* grows as a shrub with no trunk and its long main branches down, close to the ground with upturned ends. Junipers do not do well in shady places but thrive on dry, sandy or rocky slopes and ridges and flood plains in today's environments. Junipers are noted to grow well in limestone soils. Junipers produce female seed cones, <10 mm, which are berry-like, resinous and fragrant, and appear dark blue in color with a kind of powdery coating. While the seeds are edible, can be used as a flavoring when cooking wild game and can be used to produce fermented beverages, they have diuretic properties and should not be eaten in any great quantity (Kuhnlein and Turner 1991). Each seed cone can produce anywhere from one to six seeds. Figure 18 shows examples of modern juniper seeds (*Juniperus horizontalis*) from the comparative collection housed at the Provincial Museum of Alberta (PMA) in Edmonton. Comparison of photos of modern seeds of *J. horizontalis* and *J. communis* in Montgomery (1977) show very little difference between the seeds of these plants. There is almost no discernable difference between the PMA comparative seeds of *J. horizontalis* and *J. scopulorum* (Rocky Mountain Juniper). However, *J. scopulorum* grows only on mountains today, so the assumption can be made that the Tuscany juniper seeds must be either *J. horizontalis* or *J. communis* or perhaps even both. Since this is not certain, identification of the Tuscany juniper seeds can be only to the genus level. Figure 19 shows two examples of charred juniper seeds from the Tuscany habitation site.

Figure 19: Two views of two different Juniperus *sp. seeds from the Tuscany habitation site (EgPn-377)*

Kinnikinnick Seeds

The other type of charred seed that is ubiquitous throughout the Tuscany site is *Arctostaphylus uva-ursi* (kinnikinnick; bearberry), a member of the Ericaceae (Heath) Family. While kinnikinnick is a perennial shrub, it grows so low to the ground that many might not realize that it is actually a woody plant. Like juniper, it is an evergreen, prostrate shrub that sends out long stems as runners trailing across the ground. These stems can produce new plants at their endpoints that set down roots and then send out their own running stems. In this manner, kinnikinnick can cover many square meters of ground completely, forming a fairly dense mat. Kinnikinnick is found widespread through several ecoregions, including the Aspen Parkland, Boreal Forest, Montane, Subalpine, and Alpine. It does have preferred habitats, which include rocky outcrops, dry open woods, eroded slopes and well-drained sandy or gravelly soils. In the plains region, it can be found on the slopes of coulees like Twelve-Mile Coulee, which is close to the Tuscany site. Kinnikinnick produces groups of three to ten flowers, which develop into berries that are 6-10 mm in diameter. These berries are edible but do not have very much taste and are mealy and dry in texture. They contain high amounts of carbohydrates (Wilkinson 1999), which would make them a good potential source of food. Each berry contains approximately five nutlets as seeds, which have a distinctive morphology. The nutlets fuse together as a single stone (Kuhnlein and Turner 1991). Figure 20 shows examples of modern kinnikinnick seeds from the PMA comparative collection. Figure 21 shows two ESEM views of kinnikinnick seeds. The first picture is of a single kinnikinnick seed, which has separated from its original clump. The next picture is of a small clump of fused kinnikinnick seeds, as they would be found within the ripe berry. Importantly, these two figures (recovered charred seed vs modern sample) show the kinnikinnick seeds displaying the same characteristics for the dispersal of their fused nutlet clumps.

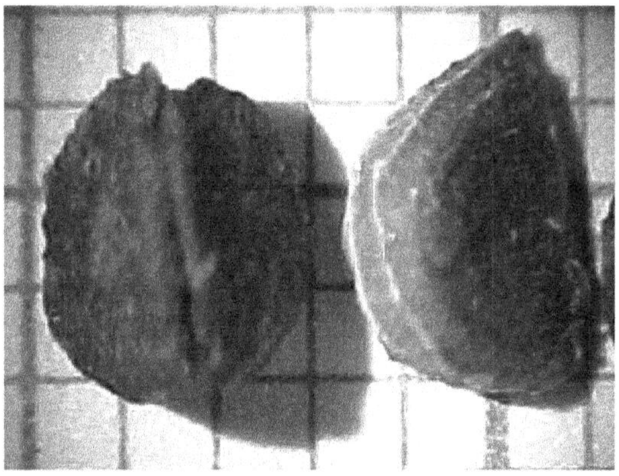

Figure 20: Modern kinnikinnick seeds from the PMA comparative collection. The seeds on the left are still clumped together as they would be found in the berry. The seed on the right has separated from its original clump

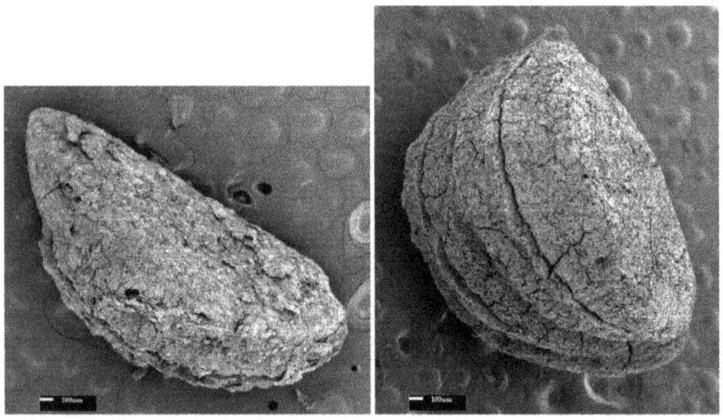

Figure 21: Two different ESEM views are presented of charred kinnikinnick seeds from the Tuscany site. Importantly, we can see that the seeds on the right are found still clumped together, as they would be found in the berry. On the left is a seed that has separated from its original clump. These are exactly the same characteristics apparent with the comparative collection seeds above

The combination of juniper and kinnikinnick in the past environment of the Tuscany habitation site is certainly interesting. They are plants that seem to prefer similar combinations of soils and precipitation. Their affinity in the environment near the Calgary area today has been demonstrated in the previous chapter. This relationship appears to have a prior history, existing through the early Holocene in what is now the greater Calgary region. There appears to be a preference for rocky or sandy soils that are well drained. Other charred seeds have also been identified from the Tuscany flotation samples. Their characteristics and environmental preferences are included in this discussion to expand the portrait of the past environment at the Tuscany site.

Chenopodium sp. Seeds

One of the other widely dispersed seeds found throughout the Tuscany habitation site was identified as a probable *Chenopodium* species. *Chenopodium* species are known commonly as Goosefoot or Pigweed. They are described as weed-like annuals that produce lens-shaped seeds (Moss 1983). Figure 22 shows one example of the seeds identified as *Chenopodium*, from Unit 66 Level Q at the Tuscany site. The characteristic beak on the upper left part of the seed (at 11 o'clock) distinguishes it as a probable *Chenopodium* species. This seed is approximately 1.0 mm in diameter and is quite comparable to a seed identified as *Chenopodium* cf. *C. berlandieri* (false lamb's quarters) in Yansa's paleoenvironmental work at a prairie pothole in southwestern Saskatchewan (Yansa and Basinger 1999:164-165). *C. berlandieri* is described as an annual which can often be ill-smelling. It produces seeds that are 1.0-1.4 mm long and 1.1-1.3 mm wide, with black, shining seed coats (Moss 1983:242). In the Alberta area the variety found is *C. berlandieri* ssp. *zschackei,* which is generally distributed and occurs in grasslands and disturbed areas (Moss 1983:242). This is one, possible identification of the seed in Figure 22. A closer view of the seed coat, looking for characteristic reticulations associated with *C. berlandieri* and visible on the seed examined by Yansa and Basinger (1999), showed that the Tuscany seed was very smooth coated by comparison. This is a feature of *Chenopodium album* (Lamb's quarters) (Moss 1983).

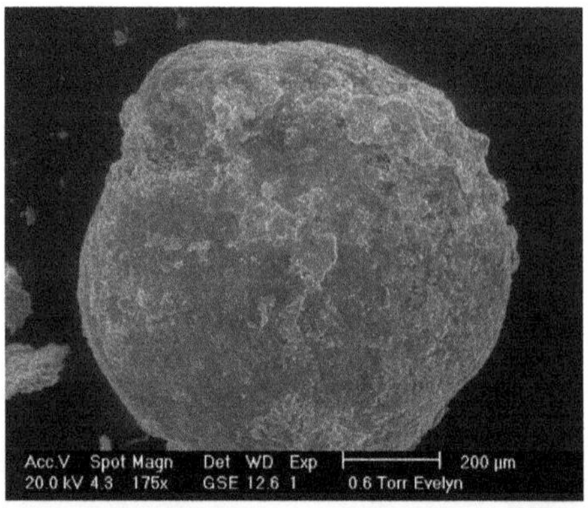

Figure 22: An example of one of the Chenopodium *sp. seeds found in the flotation samples at the Tuscany habitation site. This seed was from Unit 66 Level Q*

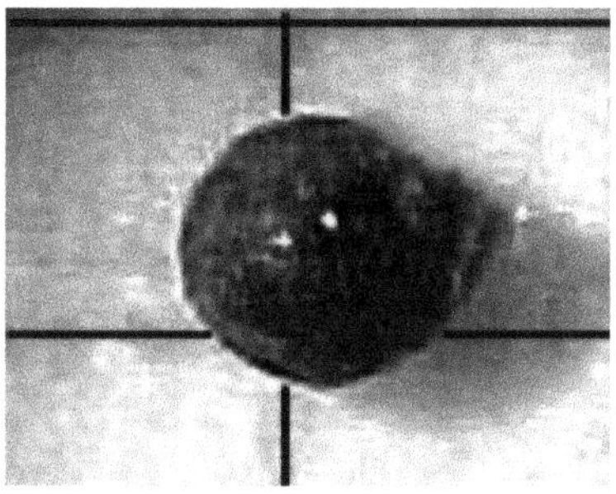

Figure 23: One other example of a Chenopodium *sp. seed from the Tuscany site. This example is from Unit 73 Level BB*

Figure 23 shows another example of a *Chenopodium* sp. seed from Unit 73 Level BB. While the photograph is not particularly clear the size of the seed is discernable and is very similar to the previous seed mentioned, being approximately 1.0 mm wide with a beak present as well. Level BB was near the bottom of the excavation at 1139.62 m.a.s.l. and close to the Glacial Lake Calgary sediments which began at 1139.44 m.a.s.l. in this unit.

The requirements for a more unequivocal identification of *Chenopodium* species are fairly rigorous. Ideally, all the seeds identified as *Chenopodium* sp. should be measured. Their variability in size and differences in characteristics should be quantified statistically, which would yield data by which they could be more positively identified, perhaps even to the species level. This would require a comparative collection from this area, which would have to be assembled and the variability of seeds for each *Chenopodium* species would need to be compiled.

Another problem with these seed specimens and many others in the assemblage was the fact that they were not very clean and physical characteristics were harder to see clearly. Unfortunately, many of the items recovered through the flotation analysis retained clay on their surfaces. This clay was difficult to remove later, especially if the item was fragile as the *Chenopodium* seeds proved to be. Often, all that remained of a seed was a fragment of the shiny seed coat. One option for dirty specimens in the size range of this seed assemblage is to use a sonic cleaner to remove the unwanted materials. However, fragile items can actually be very easily destroyed in a sonic cleaner and cleaning was not attempted in this case.

The presence of a *Chenopodium* species in the Tuscany site does yield some information about the local environment. *Chenopodium* species are part of the Goosefoot family, Chenopodiaceae. Many species are noted to be capable of growing in soils that are more alkaline (Royer and Dickinson 1999). Soil alkalinity is very common in the Calgary area and throughout the Plains to the east. *Chenopodium* species are opportunistic, weedy plants, which move quickly into disturbed areas and establish themselves. This was observed at the Tuscany site, EgPn-375, which was located in the bottom of 12 Mile Coulee, about 1.5 km north of the Tuscany habitation site, EgPn-377. A *Chenopodium* species took quick root in the backdirt piles and grew to large proportions (approximately 70 cm tall). While it is clear that there appears to be one occupation of the Tuscany habitation site, sandwiched between the Glacial Lake Calgary sands and the Mazama tephra, it is clear that the site was not occupied continually during the deposition of sediments and the development of paleosols. The site does appear to have experienced disturbance, however, which may be due to animal activity, ongoing deposition, or the human activity that took place at the site, all of which may have taken place at different times. The presence of one or more *Chenopodium* species through time indicates this possibility of disturbance.

One of the problems with a genus like *Chenopodium* is that many of the seeds produced by the different species actually look very similar and unless very rigorous measurements are taken and characteristics carefully noted, as discussed above, it is difficult to assign a species name to them. Further, there are classification problems with some plants. In the case of *Chenopodium,* for example, Moss (1983) describes *Chenopodium album* (Lamb's quarters) as a stout annual that is introduced and common in Alberta. Royer and Dickinson (1999) describe *C. album* as an annual introduced from Europe and first reported in Canada in 1858. However, they also note that *C. album* may be a native to North America. Obviously, ideas about the origin of plants change through time as more research is completed. It is important to know whether or not *C. album* was introduced or is native to North America from both an archaeobotanical and a paleoethnobotanical perspective and this is not certain at this time.

Kuhnlein and Turner (1991) describe the Goosefoot Family (Chenopodiaceae) from an ethnobotanical perspective based on historical and ethnographic accounts on the use of plants for food by Canadian Indigenous people. Under the common names Lambsquarters, Pigweed or Goosefoot, is the *Chenopodium album* complex in which they include *C. berlandieri*, *C. bushianum* and *C. macrocalycium*. Kuhnlein and Turner (1991:151) note that it "...is a highly variable complex of closely related species and varieties... ...and the seeds are black, rounded, and flattened." Interestingly, Kuhnlein and Turner (1991) point out that in Canada the complex has been described as being introduced whereas *C. berlandieri*, *C. bushianum*, and *C macrocalycium* are classified as indigenous to the United States. This is a very important and even critical point as *Chenopodium* seeds have figured prominently in discussions on the early domestication of plants in the American southeast (e.g. Asch and Asch 1977, 1985; Fritz 1984; Fritz and Smith 1988; Gremillion 1993, 1994; Smith 1984, 1985a, 1985b, 1989, 1995; Smith and Cowan 1987; Wilson 1981).

There is an interesting report of very high quantities of *Chenopodium* seeds (in amounts up to 70%) being recovered from dessicated feces found in Dust Devil

Cave in southern Utah (Reinhard *et al.* 1985). The feces remains are associated with the cultural remains of the Desha Complex which dates from 6800 to 4800 BC (approximately 8800-6800 BP). There is an interesting property of *Chenopodium* discovered from investigation of the feces. *Chenopodium* are anthelmintic and serve well to rid the body of parasitic worms such as roundworms and pinworms (Reinhard *et al.* 1985). The conclusion in this work was that the high rates of consumption of *Chenopodium* in this Archaic population was a result of its availability as a food. Its de-worming effects were just a lucky or serrendipitous bonus.

At the Barton Gulch Paleoindian site in southwestern Montana, a number of features were identified: hearths, roasting pits, and earth ovens (Armstrong 1993). Identified as remains of the Alder complex, dating to 9410 ± 140 years BP (laboratory number not listed), these features contained a variety of plant remains from 28 different species. The most prominent species in the assemblage was identified as *Chenopodium leptophyllum* (slimleaf goosefoot) (Armstrong 1993). The seeds of this plant were found primarily (87%) in the roasting pit contexts. As noted by Armstrong (1993) the occurrence of this *Chenopodium* species at Barton Gulch was the only known example from an early period site in North America. Meadowcroft Rockshelter (southwestern Pennsylvania) is noted to have the earliest archaeological example of goosefoot remains from contexts ranging between 19,650 and 735 years BP (Armstrong 1993).

There is an archaeological record of the recovery of what appears to be a cache of *Chenopodium* seeds from the Ross site (DlPd-3) in southeastern Alberta, a few miles northeast of Lethbridge. Johnston (1962, cited in Vickers 1989) reports the recovery of *Chenopodium* seeds identified as either *C. album* or *C. lanceolatum* from the Ross site in 1960. The seeds appear to have been originally discovered by local people looking for artifacts at the riverbank site. The seeds appear to have been cached in a straight-sided, cylindrical hole, excavated from a floor identified as Layer III (Vickers 1989:8). Layer III has a radiocarbon date from charcoal of 540 ± 134 BP (S-2038), which places it at approximately 1410 AD in calendar years (Vickers 1989:56). This places the possible use of *Chenopodium* well before the historical or Contact period in western Canada.

Ethnographic accounts record the use of Lambsquarters as an edible green and seed by various Indigenous peoples such as the Iroquois and Ojibwa (Kuhnlein and Turner 1991:152). There is an account of the Ojibwa boiling the young leaves of the plant with fat and grinding the seeds into a flour to make a mush bread (Stowe 1940, cited in Kuhnlein and Turner 1991:152). Most interestingly for this part of the country, there is a record of the Blackfoot of Montana gathering the seeds of Lambsquarters for food during historic times (Blankinship 1905, cited in Kuhnlein and Turner 1991:152). This validates, to some measure, the information above on the precontact use of *Chenopodium* seeds revealed in the archaeological record. This gathering and use of *Chenopodium* seeds may extend back through time in this part of the world.

The information presented here is not intended to imply that the people who occupied the Tuscany habitation site were making the same uses of *Chenopodium* as historic period people. The occupation level of the Tuscany site dates to 7800 years BP. Given the obvious knowledge of plants present in older populations, as evidenced at Barton Gulch and Dust Devil Cave, it is not unreasonable to suggest that a *Chenopodium* might also have been a food for the people who occupied the Tuscany site in the past.

Rose Seeds

There were a number of *Rosa* sp. seeds recovered from the flotation samples. They were not that numerous but their representation would fit in well with the kind of cover that would be found in a more open, Parkland type of environment. This kind of environment would be appropriate for the inclusion of all the other plants identified thus far, as well.

Three species of *Rosa* are found in Alberta landscapes today, including *R. acicularis, R. arkansana*, and *R. woodsii*. There are some differences in the environments preferred by each species. *Rosa acicularis* (Prickly rose) grows as a bushy shrub and can be found in woods, on banks near water and near roadsides. *Rosa woodsii* (Common wild rose) also grows as a bushy shrub. It can be found in open woods and plains and is often in association with one of the other two species (Moss 1983:365). *Rosa arkansana* (Prairie rose) grows as a low shrub on the plains and prairies (Moss 1983) and is considered weedy today where it is found in pastures, rangelands and on roadsides (Royer and Dickinson 1999).

Figure 24 shows two views of *Rosa acicularis* seeds from the modern, comparative collection at the PMA. The morphology of these seeds is unique. One end is distinctly thicker than the other, though this is not that visible in the picture. The enlarged view (on the right) of one of the seeds has it oriented with its attachment point at the top. This is the thicker end of the seed, which contains the body of the embryo. *Rosa arkansa* seeds are similar in size (5-5.5 mm long) but are described as being angular in shape (Royer and Dickinson 1999:358). Figure 25 shows two different seeds, identified as *Rosa* sp. from the Tuscany site. Again, the seeds were not completely clean and this remains problematic for pictures. A species designation could not be determined for the Tuscany specimens.

Seeds newly removed from the ripened fruit (rose hip) of a rose plant, are covered with fine, light colored hairs that cause rectal irritation when they are eaten and excreted by humans. The seeds or achenes also have robust seed coats. The skins of the fruits (hips) are high in vitamin C and A, calcium, phosphorus and iron (Wilkinson 1991:118), but the seeds are not recommended. The leaves can be used for tea (Kuhnlein and Turner 1991).

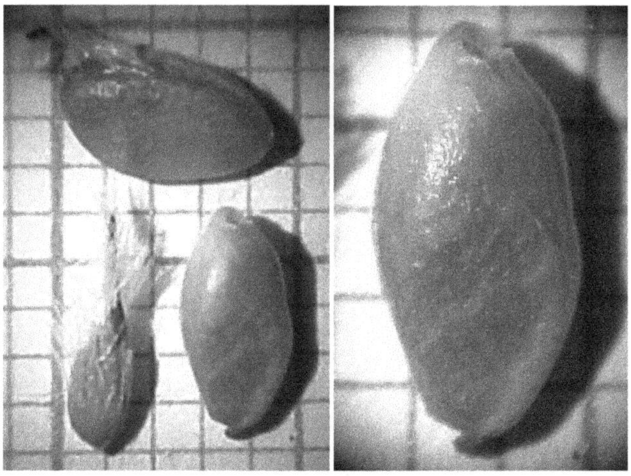

Figure 24: Two views of Rosa acicularis *(prickly rose) seeds from the comparative seed collection at the PMA*

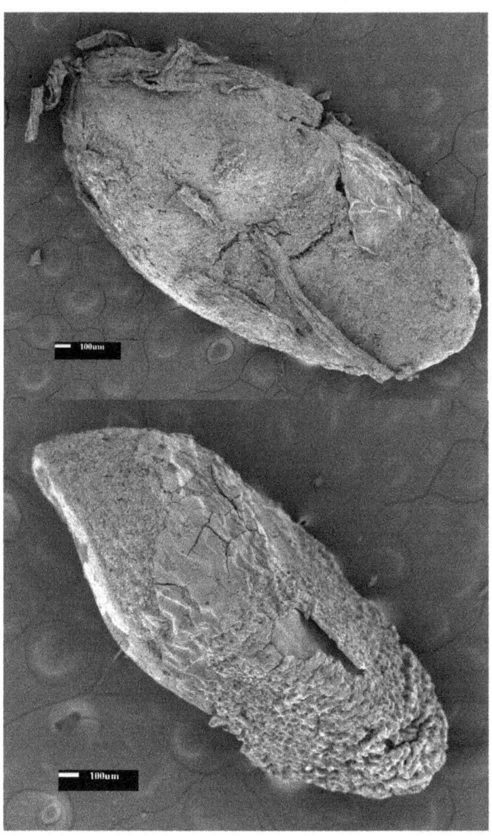

Figure 25: Two charred Rosa *sp. seeds from the Tuscany site*

Rose seeds were also found in the features (just described above in the *Chenopodium* discussion) from the Alder complex at the Barton Gulch site in Montana. However, the seeds were considered discard materials as the seeds are known to not be useful for consumption (Armstrong 1993).

Picea sp. Seeds and Cone Scales

There are a few charred seeds identified as *Picea* sp. (spruce) from the Tuscany paleosols. Identification of the *Picea* sp. seeds is far more tentative than any of the seeds examined thus far. As well, there are a number of interesting charred structures, which are interpreted as charred cone scales from *Picea* sp. cones. Initially, there was concern that the *Picea* sp. seeds could be misidentified juniper seeds, which were so ubiquitous in the assemblage. However, examination of the comparative seeds at the PMA revealed that *Picea glauca* seeds are consistently smaller than juniper seeds. *Picea glauca* seeds, also, never have the ridges that are found on most juniper seeds. There is variability in this phenomenon as well. Some juniper seeds appear very smooth. Many factors can cause variability in seed production from one plant to the next, even when the plants are in close proximity to each other. There was a range of variability in size in the comparative set of juniper seeds, probably due to natural effects at different locations related to amount of moisture received, soil richness, amount of sunlight received, and so on. Figure 26 compares a modern juniper seed to a modern *Picea glauca* seed to clarify the differences in their morphology. When juxtaposed in this manner, it becomes quite clear that these two species have very different seed morphology. There is a noticeable difference in seed size, as well. The juniper seed is approximately 4.5 mm in length (the background graph paper is in a millimeter scale). The *Picea glauca* seed is only 2 mm in length (the background scale is a close-up of the millimeter scale). Figure 27 shows two different views of charred *Picea* sp. seeds from the Tuscany habitation site. The seed on the left is from Unit 66 Level M, next to the Mazama tephra boundary. There is the possibility that it is intrusive from rodent activity but it was the best example of a *Picea* sp. seed from the flotation analysis. It is clear that the charring process is very degrading to botanical materials and can make identification difficult and uncertain. In some instances, the seeds are so badly damaged that the identification is quite tentative.

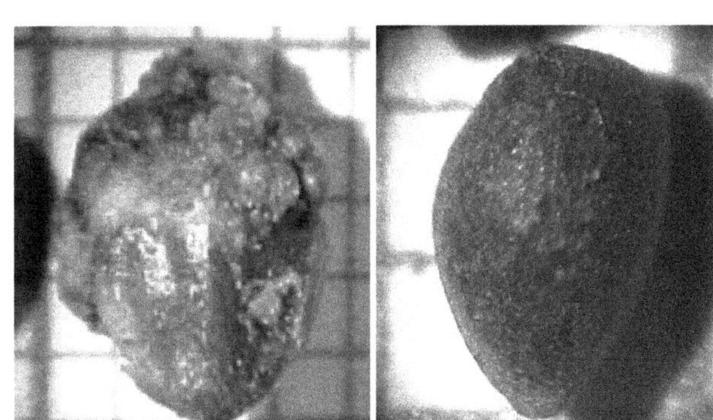

Figure 26: The seed on the left is from Juniperus horizontalis. *It has a distinct ridge and very shiny seed coat. The seed on the right is from* Picea glauca. *It has a characteristic, offset oval shape, no ridges and a smooth seed coat*

Initially, the *Picea* cone scales were identified as small knots or branching twig nodes that maintained part of their structure after charring. It became clear that they

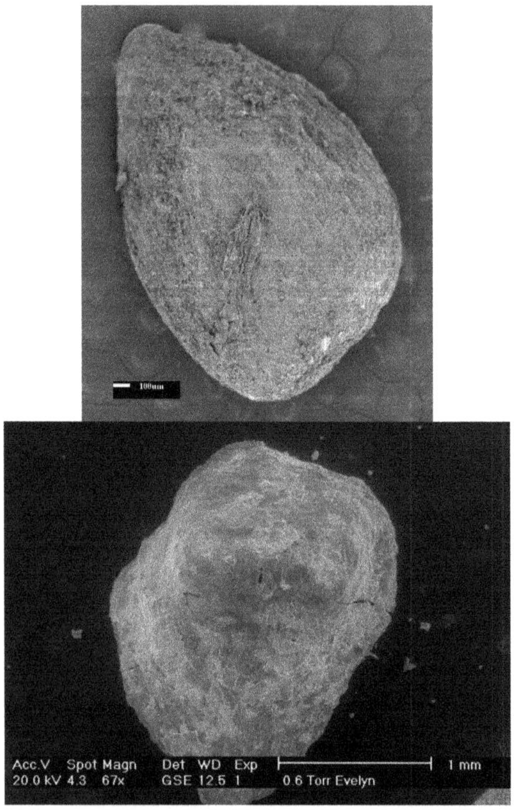

Figure 27: Two examples of Picea *sp. seeds (spruce) from the Tuscany habitation site. The seed on the left does not appear to have charring damage. The seed on the right is damaged fairly severely from charring effects*

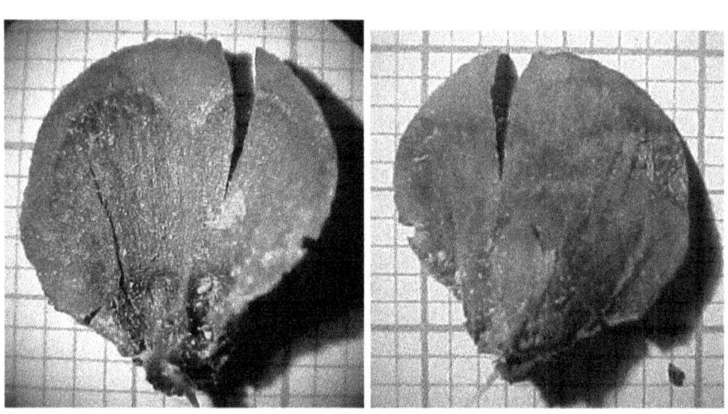

Figure 28: Two views (front and back) of a Picea glauca *(white spruce) cone scale from an identified specimen from the PMA comparative seed collection. Note the background graph paper is in millimeter scale. This complete cone scale is ~12 mm long and ~11 mm wide*

were more numerous and all looked alike—not something one would expect from random plant growth. They are compared to cone scales from specimens from the PMA comparative collection. Figure 28 shows two views of a *Picea glauca* cone scale. On the left is the inside view of the scale, where the seed would rest with its wing extending out to the edge of the scale. On the right is the outside view of the cone scale against which a seed would press from a scale below it. Cone scales are characteristically disk shaped with an inner bowl in which two seeds are nestled. There is a kind of partial 'septum' which divides the scale in half and forms a ridge in the center. This ridge extends to the point where the cone scale attaches to a central core. The cone scales are actually similar to bird feathers in the manner in which they attach and spiral on the cone. The scales are fan-shaped and woody and their surfaces actually have furrows, which radiate out from the center like the ribs of a hand-held fan. These are the characteristic structures that preserve even when these items are charred. The modern *Picea glauca* cone scale is approximately 12 mm long and 11 mm wide. The outer perimeters of the scales are very fragile and it is easy to imagine how most of the cone scale would not survive the charring process. In Figures 29, 30 and 31 are views of what are being interpreted as charred cone scales from *Picea glauca* cones. Figure 29 shows a closeup of two different charred cone scales from two different units and levels within the Tuscany habitation site. Figure 30 shows 11 cone scales from Unit 10 Level AA in the excavation grid at the Tuscany site. These are the charred remnants of *Picea glauca* cone scales. The parts that survived the charring process are the innermost sections of the cone scales, which attached to the central structure of the cone. They are only 4 mm long, on average. They still retain their fan structure and the furrows can be seen, radiating outward from the center, like they do in the modern, comparative cone scale. The fragile, outer parts of the cone scales did not survive the charring process. The final figure of a cone scale structure, Figure 31, shows an ESEM view of a charred remnant. The resolution of this picture reveals the remnant fan shape and the radiating furrows which are found in each one of the specimens identified from the Tuscany flotation samples.

There are no known uses ethnographically for *Picea* sp. seeds. They are actually quite small, unlike some seeds produced by some low-growing pine trees (e.g. *Pinus edulis* and *Pinus flexilis*) in western North America, which are edible and provide a significant food resource for Aboriginal populations through time. The cones of Picea do have uses recorded for them in ethnographic accounts—for tasks like smoking hides and dying cotton fishnets (e.g. Leighton 1985; Siegfried 1994). Different Aboriginal populations (Leighton 1985; Siegfried 1994) have demonstrated many uses for various parts of the entire spruce tree.

The question of importance is related to the issues of how and why the spruce seeds and cone scales ended up in the site sediments. Thus far, there is no evidence of any wood from a conifer tree in the samples. This is because there has been only preliminary random examination of the charred plant fragments from the flotation samples. The entire cone structure of *Picea glauca* is actually not very large, being in the range of 4 cm (from observation, with great variability between trees). Cones usually do not travel very far from the tree of origin. If there is any kind of grassy or herbaceous vegetation in the vicinity of a tree, cones would be quickly trapped in place. Though *Picea glauca* cones are smaller structures, they are heavy enough that wind transport is not an important factor in their movement. However, cones can be carried away by

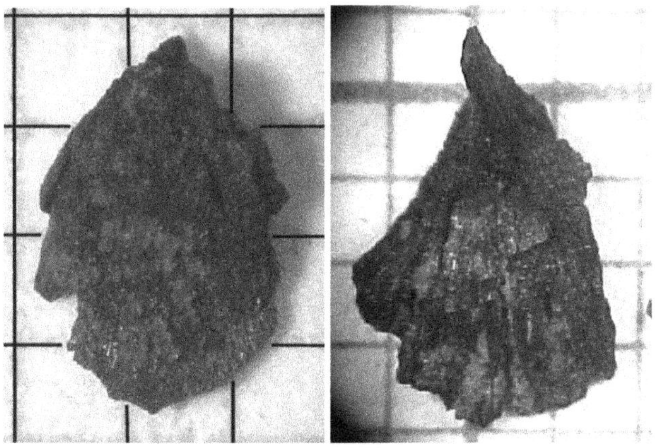

Figure 29: Two charred cone scales from the Tuscany habitation site proposed to be from cones from Picea glauca *(white spruce) conifer trees. The one on the left is from Unit 45 Level O and the one on the right is from Unit 10 Level Y. The background scales are in millimeters*

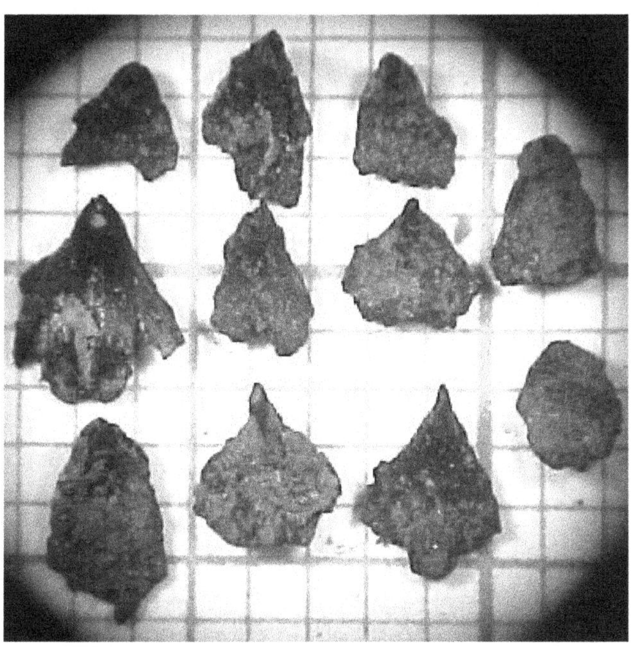

Figure 30: A view of 11 charred cone scale structures from Unit 10 Level AA from the Tuscany habitation site. The background grid scale is in millimeters

Figure 31: An ESEM view of a charred cone scale from the Tuscany habitation site (Unit 10 Level AA)

animals like squirrels and may have come into the habitation site from somewhere close by where spruce trees grew. People who brought in wood for their fires may have also brought the cones to the site. Thus, even though we may assume that cones would probably not have blown into the site from somewhere else, there is still the possibility that they were introduced by some other means.

The conclusion that may be drawn from the evidence available from the present analysis is that there was at least one white spruce tree present in close proximity to the habitation site, and it was a tree that had a life span of several decades. The spatial analysis of the distribution of the spruce seeds and cone scales may indicate the possibility of more trees through time; this possibility will be discussed further in the next chapter.

Type-D Charred Seeds

There is a type of charred seed from the Tuscany site that has proven problematic to identify. Many seeds have been examined in the PMA comparative collection and Montgomery (1977) to try to identify the species of plant this seed type represents. One of the problems encountered during the identification of seeds for this dissertation has been the lack of comprehensive reference material for Western Canada. The seed collection available at the Provincial Museum of Alberta contains many seeds related to agricultural crops and weeds. Montgomery's work (1977) is useful but focuses on the plants of eastern Canada, which are often not important to western landscapes.

The Type-D charred seeds were actually quite widespread in their distribution, both across the site and through the unit levels. They were generally quite small, usually <1 mm long, but ranged up to 1.5 mm in length. They were extremely fragile, often being hollow, charred seed coats with no internal structure left. Unfortunately, little else can be said of the Type-D charred seeds. They did show an interesting pattern in the spatial analysis that fit in well with the other charred seeds and will be discussed further, below. Figure 32 shows two examples of typical Type-D charred seeds. At one point, seeds from *Potentilla fruiticosa* (shrubby cinquefoil) were considered possible candidates for the identification, however, close comparison (e.g. see Figure 33) demonstrates that they are not really the same shape at all, though they have a similar size range. The attachment point is located in approximately the same spot on the charred seeds and the shrubby cinquefoil, but the overall morphology does not compare well. Unfortunately, the Type-D charred seed remains unidentified for this analysis and may be an area where future research is directed.

Other Charred Seeds

There are many charred seed remains that have not been identified from the samples. There are also a number of charred seeds found in very small amounts in the suite of samples, for which tentative identifications have been

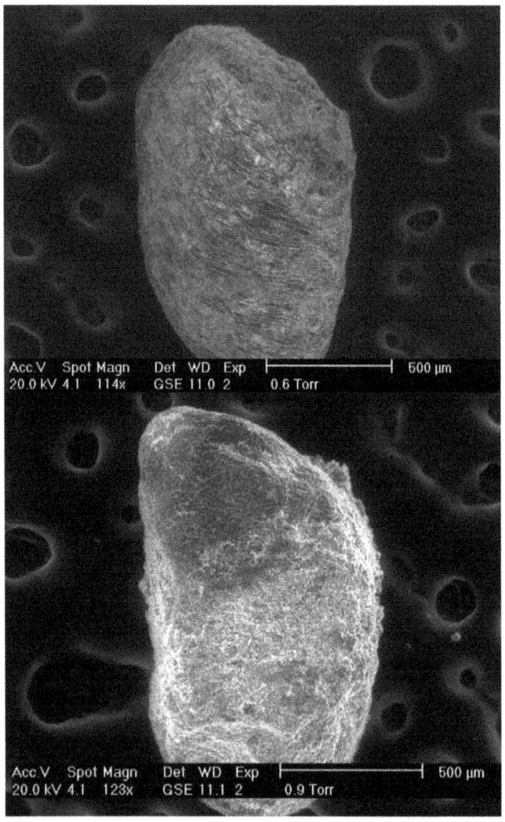

Figure 32: Two examples of the Type-D charred seed. The one on the left is from Unit 11 Level Q and the one on the right if from Unit 14 Level T

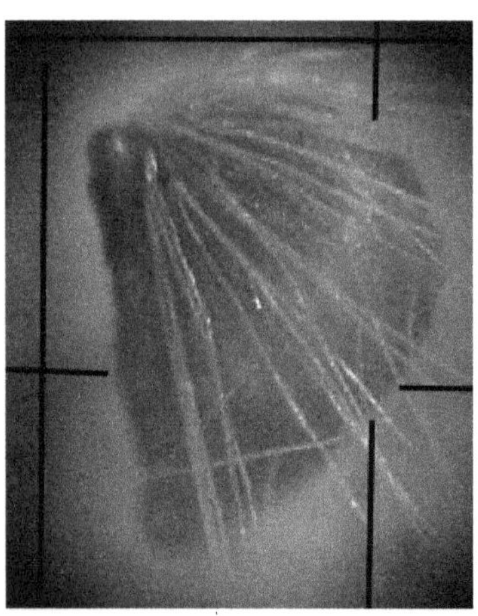

Figure 33: An example of a Potentilla fruiticosa *seed from a plant within 50 meters of the Tuscany habitation site grid*

made. The word tentative is used here because the sample sizes are too small to be representative of the variability that can exist in seed sizes from one flower to the next on a single plant or within a patch of the same species.

The first of these poorly represented charred seed types is *Symphoricarpus occidentalis* (buckbrush; western snowberry) which is a member of the Caprifoliaceae, or Honeysuckle Family. This plant is found widely scattered throughout Alberta though it does seem to prefer the Peace River valley in the northern part of the province. The preferred habitats of *S. occidentalis* include dry open woodlands, prairie slopes like coulees and ravines and open hillsides (Moss 1983; Wilkinson 1991). Distribution maps show that this plant is found widely spread through the Prairies and Aspen Parklands (Moss 1983; Wilkinson 1991).

Buckbrush grows at the Tuscany habitation site today, covering a 10 m x 10 m patch of ground on the west slope (facing east) of the grid. In fact, Buckbrush is very prolific and widespread throughout the entire area surrounding the Tuscany site. It is easy to picture this plant occupying the landscape around the site through time. Though it forms quite dense ground cover, there are surprisingly few charred seeds from the buckbrush plant in the flotation samples. This may be a function of the seeds themselves and the relatively non-robust seed coats, making them less likely to survive the charring and deposition processes through time.

A similar species, *Symphoricarpus albus* (snowberry) is recognized by the Blackfoot of southern Alberta, but the berries of this plant and buckbrush are noted to not be edible (Peacock 1992). Peacock's (1992) only known use of the snowberry plant was to tie it up into a twig broom to use when people were camped. Johnson (1987:55) lists a number of uses for this plant: as a broom, for arrow shafts, as a tea, as medicine and also notes that the berries were eaten during times of scarcity. Figure 34 shows an ESEM picture of a charred seed from the Tuscany site. This can be compared to modern *S. occidentalis* seeds from the PMA comparative seed collection shown in Figure 35. While there is great variability in the size of the seeds pictured, there is consistency to their overall morphology.

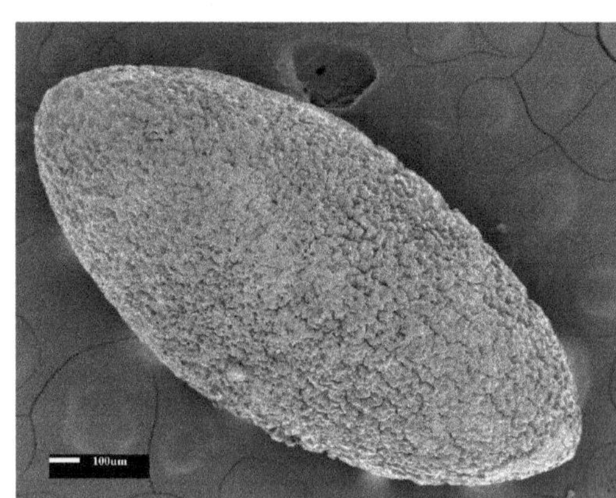

Figure 34: A charred Symphoricarpus *sp. (snowberry) seed from the Tuscany habitation site paleosols. While the seed does look similar to both* S. albus *and* S. occidentalis*, it is proposed to be the latter, which grows abundantly around the site today*

Another plant which is well represented in today's landscape around the Tuscany habitation site is *Shepherdia canadensis* (soapberry), a member of the

Figure 35: Two views of modern Symphoricarpus occidentalis *seeds from the PMA comparative collection*

Elaeagnaceae, or Oleaster Family. Though this plant produces fairly robust-looking seeds, very few were recovered through flotation analysis of the paleosols. This may indicate that this plant did not grow in the same relative abundance during the past, or that the seeds do not survive charring and burial processes very well, despite their appearance.

The soapberry plant is a spreading shrub, ranging between 1 and 3 meters in height. Each bush can produce numerous small red berries, 5-6 mm long. The preferred habitats of soapberry plants include open woods, shorelines and riverbanks. Wilkinson (1991:148) notes that it is often found on limestone soils and is abundant and characteristic in semi-open and dry Montane spruce and pine forests throughout Alberta. Though widespread, it is not found in the southeastern parts of the province except in the Cypress Hills, an upland feature of the southeastern corner of Alberta.

The berries of this plant are edible and have become a feature food of Indigenous people in what is now British Columbia. The juice of the berries can be whipped or frothed to make a confection called "Indian ice-cream", which many groups considered a special treat for parties and gatherings (Kuhnlein and Turner 1991:162). It is the saponin present in the berry juice that gives it the foaming property (Kuhnlein and Turner 1991).

The seeds found in the flotation samples at the Tuscany site do not appear to be charred in this case. This is anomalous. Given this circumstance, there is some question as to their origin. They may be intrusive. It is possible that the original bulk soil sample contained a rodent (e.g. Richardson ground squirrel (*Spermophilis richardsonii*)) disturbance that was not noticed at the time of its removal. Given the rarity of these seeds in the flotation samples, this explanation does not seem likely. The soapberry seed in Figure 36 is from excavation Unit 34 Level P. There were very few charred remains at this level and the uncharred seeds may be a reflection of the lack of intense burning of the landscape just prior to the Mazama tephra deposition event. They may constitute unusual preservation. The seeds could also have been carried in on the shoes of people working on the site excavation since they often explored the area around the site during lunchbreaks. The soapberry seed from excavation Unit 34 (Figure 36) is compared to an example from the PMA comparative collection shown in Figure 37.

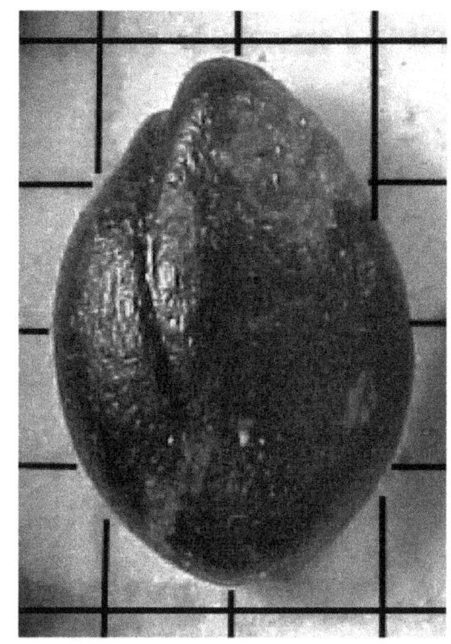

Figure 36: A seed from a Shepherdia *sp. from Unit 34 Level P from the Tuscany site grid. It compares closely to* Sherpherdia canadensis *(soapberry) seeds*

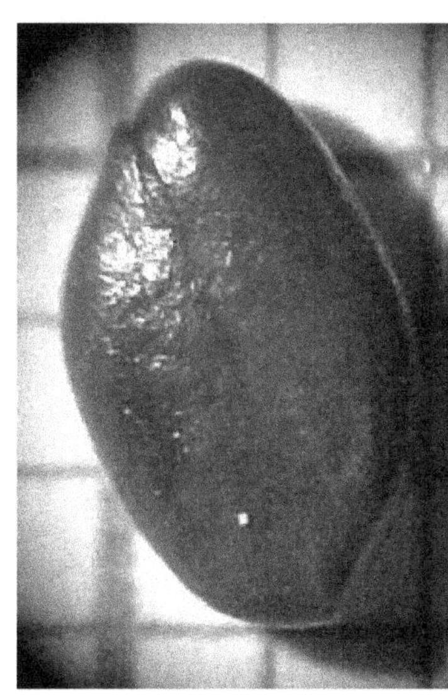

Figure 37: An example of a Shepherdia canadensis *seed from the comparative seed collection at the PMA*

Another species of plant that is present but poorly represented in terms of seed recovery is *Monarda fistulosa* (western wild bergamot), a member of the Lamiaceae/Labiatae, or Mint Family. Bergamot is an erect perennial that prefers habitats with some moisture, including open woods, semi-shaded woods like aspen poplar groves, moist waste places, open fields, moist meadows, thickets, coulee slopes and floodplains (Moss 1983; Vance *et al.* 1999; Wilkinson 1999). This plant grows fairly abundantly in the greater area surrounding the Tuscany habitation site today. It grew noticeably better away from the site depression, however, growing closer to the edges of 12 Mile Coulee and near the small stands of scrubby *Populus tremuloides* (trembling aspen) and *Elaeagnus commutata* (silverberry; wolf-willow) that dotted the landscape to the north of the site. This growth pattern demonstrates the preference for areas with a bit more moisture than offered by the sediments in the Tuscany site depression today.

A number of different Aboriginal populations used wild bergamot leaves for tea (Iroquois and Kootenay) and others used it for a seasoning, scent and a food preservative (Flathead and Apache) (Kuhnlein and Turner 1991:211). In this area the Blackfoot people of southern Alberta had a number of uses for wild bergamot leaves and blossoms including an eyewash, a herbal water for stomach pain, a tea for childbirth, and a poultice for acne pustules (Johnson 1987:51). Figure 38 shows an example of what has been determined to be a seed from *M. fistulosa* (wild bergamot) recovered from flotation. Figure 39 shows examples of modern seeds of *M. fistulosa* from the comparative collection at the Provincial Museum of Alberta.

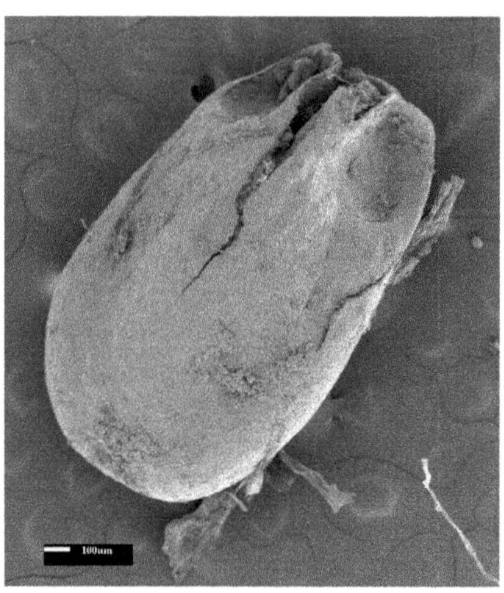

Figure 38: An example of one of the seeds of Monarda fistulosa *(western wild bergamot) recovered from flotation analysis of the bulk soil sample from Unit 28 Level S at the Tuscany habitation site*

Figure 39: Examples of seeds of Monarda fistulosa *var.* menthaefolia *from the comparative seed collection of the PMA*

The next species to be discussed represents a tentative identification as well. The charred seeds differentiated from Type D and identified as *Potentilla fruiticosa* (shrubby cinquefoil) were scattered sparsely throughout the Tuscany site. They are very small seeds and are extremely fragile, much like the *Chenopodium* seeds. There is some variability in size, with a length range of .5 mm to 1.5 mm for these seeds.

Potentilla fruiticosa is a freely branched shrub that grows to heights of approximately 1 meter. The preferred habitats of this plant today are dry slopes and open woods to Subalpine elevations (Wilkinson 1991). It can be found widely throughout the greater Calgary area and is noted to be common in the Foothills and Prairies. In the city of Calgary it is used as an ornamental and slope stabilizer along major thoroughfares and intersection exchanges. It thrives well because it is indigenous to this region.

Interestingly, there are ethnographic accounts on the use of the leaves and stems of the shrubby cinquefoil to make a beverage tea. Both the Fisherman Lake Slave and Inland Tanaina of Alaska used the plant in this manner (Kuhnlein and Turner 1991:242). This demonstrates the vast, modern, regional distribution of this plant. In southern Alberta, the Blackfoot are noted to have mixed the leaves with dried meat as a spice and also used them as a deodorant (Kuhnlein and Turner 1991:242). Figure 40 shows two examples of *Potentilla* sp. seeds recovered from flotation samples from the Tuscany habitation site. The smaller seed on the left is from excavation Unit 51 Level P, while the seed on the right is from excavation Unit 67 Level W. There is great variability in seed sizes, both within a plant and between plants, as mentioned before. This is well demonstrated in the two examples of modern seeds of *Potentilla fruiticosa*, taken from a shrub within fifty meters of the Tuscany excavation grid. In Figure 41, the seed on the left is almost as wide as it is long, though it has a very asymmetrical, nearly crescent shape. The attachment point is on the upper left side of the seed and numerous long hairs flare out from around the attachment point scar. The seed on the right is much longer than it is wide. Though it is asymmetrical as well, it is almost tubular in shape. The attachment point is on the upper left side with numerous hairs flaring out around it. The two comparative seeds are very different in morphology from one another. This is a problem for identification. Until the second seed was viewed, which looks much like the recovered charred seed from excavation Unit 51, *Potentilla fruiticosa* did not seem to be a very likely identification. However, given this seed's morphological variability, this shrub is a good candidate for occupation of the Tuscany site landscape during the past, as well as the present.

Three plants tentatively proposed to have a history of occupation in the Tuscany habitation site landscape are *Galium boreale* (Northern bedstraw), *Castilleja* sp. (Indian paintbrush) and *Actaea rubra* (red and white baneberry). There are only a few seeds of these plants present in the whole suite of flotation samples and they really cannot be included in the tallies for spatial analysis

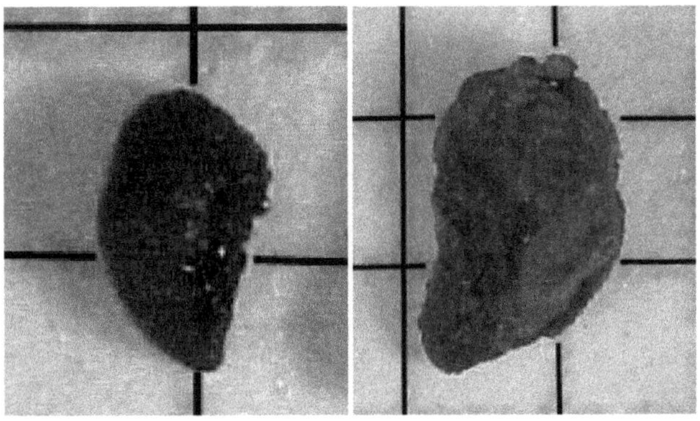

Figure 40: Two examples of charred seeds tentatively identified as Potentilla fruiticosa *(shrubby cinquefoil) from the Tuscany flotation samples. The left seed is from Unit 51 Level P. The right seed is from Unit 67 Level W*

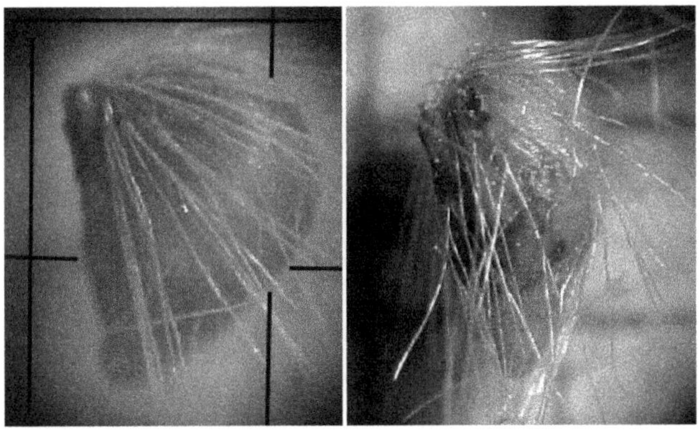

Figure 41: Two examples of modern Potentilla fruiticosa *seeds from a plant growing close to the Tuscany excavation grid. Note the great variability in seed shape. These seeds were from two separate fruits from the same plant*

– they must be considered in terms of presence-absence. The presence of seeds from these plants may be through sheer serendipity. Most of the plants in the discussion thus far include plant species that produce more robust seeds that can survive the charring and other deposition processes they have been subjected to through time. Northern bedstraw and Indian paintbrush are herbaceous forbs with very small seeds that do not have harder seed coats like the seeds of many of the shrubs in this region. Red and white baneberry, however, is a herbaceous forb that produces seeds with a hard, nut-like seed coat that could easily be mistaken for an *Arctostaphylos uva-ursi* (kinnikinnick) seed.

In the Prairie regions *Galium boreale* can be found in aspen poplar groves and shrubby areas of the transition zone between Parkland and forest (Vance *et al.* 1999). *Galium boreale* is a member of the Rubiaceae or Madder Family which are noted to have 1-seeded nutlets or a dehiscent capsule. This may be why a few seeds have survived within the Tuscany site sediments. *Castilleja* sp., members of the Scrophulariaceae or Figwort Family, are perennial herbs found growing in essentially the same habitats as *G. boreale* in the Prairie regions. *Actaea rubra* (baneberry) is a member of the Ranunculaceae or Crowfoot Family. It is an erect perennial that prefers habitats like the moist woodlands of mixed forests and aspen poplar bluffs (Vance *et al.* 1999:75). It can also be found in wooded ravine areas in the Prairie region (Vance *et al.* 1999:75). It produces either red or white berries that are poisonous. The habitats of all three plants indicate the preference of some soil moisture, as in mesic to subxeric moisture regimes.

Figure 42 shows examples of seeds identified as *Galium boreale* from the Tuscany flotation samples. Figure 43 shows seeds from the comparative seed collection at the Provincial Museum of Alberta. Figure 44 shows examples of seeds identified as *Castilleja* sp. from the Tuscany flotation samples. Figure 45 shows an example of a modern *Castilleja* sp. seed from the comparative seed collection of the PMA. Figure 46 shows an example of a seed identified as *Actaea rubra* from the Tuscany flotation samples. Figure 47 shows an example of a modern *Actaea rubra* seed from the comparative seed collection of the PMA.

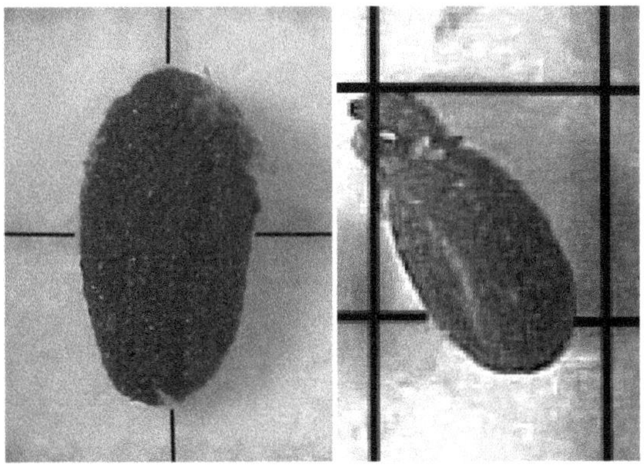

Figure 42: Two examples of charred seeds tentatively identified as Galium boreale *(northern bedstraw). Each of these seeds has a distinct groove down the center of one side, just as the comparative seeds do. The seed on the left is from Unit 90 Level X and the seed on the right is from Unit 41 Level U*

Figure 43: Examples of modern seeds of Galium boreale *(northern bedstraw) from the comparative seed collection at the PMA*

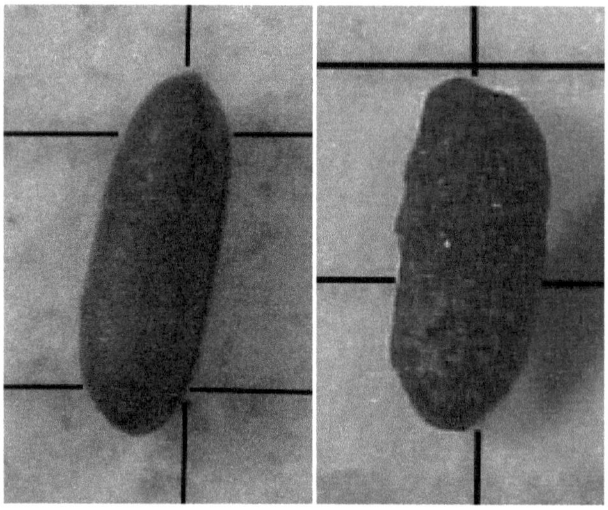

Figure 44: Two examples of charred seeds tentatively identified as Castilleja *sp. from the Tuscany flotation samples. The seed on the left is from Unit19 Level Y and the seed on the right is from Unit 50 Level S*

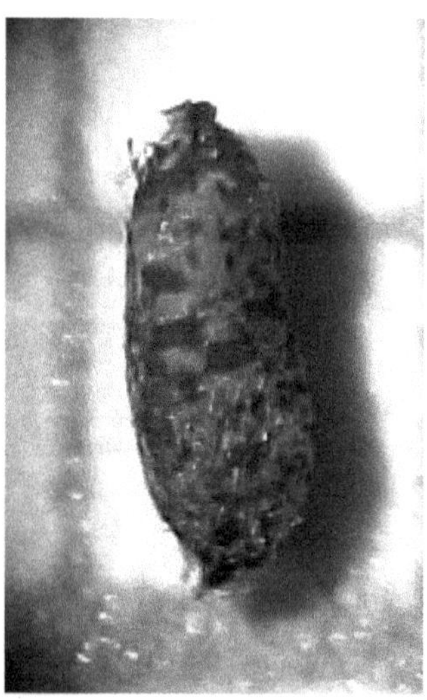

Figure 45: An example of a modern seed of Castilleja *sp. from the comparative seed collection at the PMA*

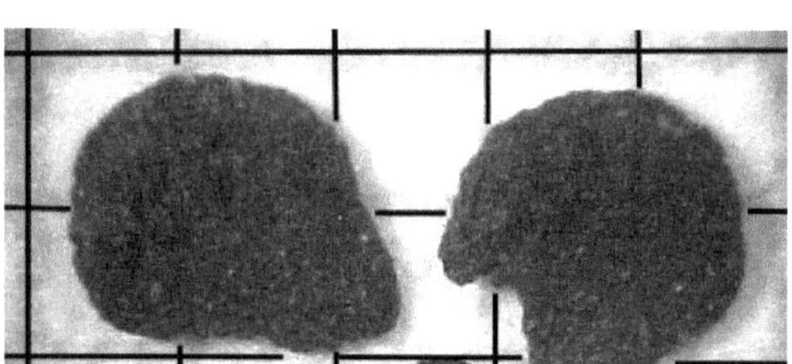

Figure 46: Two halves of a charred baneberry seed from the Tuscany habitation site from Unit 50 Level S

There are a couple of important points to make note of with the last three plant species. First, there is the fact that they represent forb plant species rather than shrubs, as are most of the charred seeds recovered from the Tuscany

Figure 47: Example of a modern baneberry seed from the comparative seed collection at the PMA

flotation samples. As forbs, they seem to require more moisture for growth than the shrubs and that may have been a problem in the paleo-Tuscany landscape. This may, indeed, be the reason why there were few seeds from these plants in the assemblage—we may simply be 'lucky' to have recovered the ones that we have. Examination and identification of the charred seeds from the flotation samples indicates the presence of plants that thrive in Parkland and Plains environments.

INTERPRETING THE CHARRED SEED INVENTORY

Archibald *et al.* (1996) have provided information on the plant species that are most common to the ecosites they describe for southwestern Alberta. Included with a picture to help identify each of these plant species is an 'edatopic grid', which graphically represents the range of moisture and nutrients, along with a range of elevations which each plant preferred. The edatopic grids show that Juniper would be the plant that would probably first find limits in the landscape restricting its ability to occupy a locality.

For this study we must assume that the moisture, nutrient and elevation requirements for the plants inventoried through flotation recovery have not changed through time. The perturbations in climatic regimes in this region reflect adaptive migration patterns rather than adaptive evolution. Thus, when the climate has changed from cooler and moister, to hotter and drier, plants have come and gone within the landscape—they have not evolved to adapt and remain. This is demonstrated with changing boundaries for forest margins during the past. When conditions were not optimal for the different tree species, they died out locally and there is an impression of them moving within landscapes. Thus, it appears that the Boreal forest margins moved, receding northward as the climate became warmer and drier and returning southward with cooler and wetter conditions. These gross patterns have been discerned with pollen studies.

The question that arises with our attempts to understand how the vegetation landscape has changed through time is, whether we can find similar sets of vegetation schemes today, compared to those that show up in pollen and macrobotanical studies. The problem with macrobotanicals is that some seeds will survive better through time because they have tougher seed coats. The charring process may not help in this case either. Though charred seeds will survive better through time in sediments, there may still be a bias inherent in representation because the more delicate seeds did not even survive the charring process in the first place.

This may be a problem inherent with the charred seeds and seed fragments that have been recovered from the Tuscany habitation site. Both juniper and kinnikinnick have fairly robust seeds, with nut-like seed coats. In the Tuscany samples, the kinnikinnick seeds have survived the charring process quite well and they are well represented in all the units. Other seeds do not have such robust seed coats. This was very visible when a comparative seed collection was used to identify the seeds from the Tuscany data set. Seeds from a plant like *Galium boreale* (Northern bedstraw) are very small and far more fragile than kinnikinnick. Only one or two seeds of this plant species have been tentatively identified from the Tuscany samples compared to the dozens of kinnikinnick seeds identified. A natural bias is thus likely inherent in the data set before we have even counted one seed.

Another natural bias may be inherent in the data set due to the natural propagation of one species compared to another. Archibald *et al.* (1996) outline a set of characteristic species present in each of the identified ecophases for southwestern Alberta. The lists include plants from four classes of vegetation found in the landscapes; these are trees, shrubs, forbs, and grasses. The average percent cover for each plant species is listed. In some cases, kinnikinnick represents an average of 16% of the shrub cover in an ecophase, whereas juniper represents only 1% average cover. The number of seeds that will be produced by each species will be very disproportionate, greatly reducing the chances of seeds from the poorly represented species even surviving through time. In the case of robust seeds from plants like the junipers, preservation of a few seeds through time can be reasonably expected. In the case of plants having seeds with less robust seed coats, the likelihood of preservation is remote though they may also have been present in the landscape in low numbers. Thus, though a small representative sample may indeed appear in the initial distribution of seeds in the landscape, there may still be a natural bias because of preservation factors, as to which set of remains are finally recovered.

When we begin to list all the inherent problems and bias that can be present in a data set, it is tempting to conclude that we will get such distorted results that the whole exercise might be considered a total waste of time. The positive view of this kind of study, though, is to recognize the problems and count every identification of a seed as a positive result, in the sense of presence/ absence results. This is the assumption made here.

The ethnographic information provided here confirms that each plant represented in the site could be used in at least one way by people. There is a problem with extrapolating this information any farther than acknowledging known uses for these plants. The problem with the data lies in the distribution of the seeds through the paleosols. The seeds are well-distributed through the pedogenic zones three, four and five. They are not exclusive to zone three which is associated with the Lusk cultural component. If use was made of these plants during the occupation of the site it would be reasonable to expect that there should be a visible difference between that zone and the zones of non-occupation. Human activities of collection and processing should affect the distributions of the seeds within the site. That does not seem to have happened here. Thus, though we are well aware that most plants in Indigenous landscapes have at least one use and often multiple uses (e.g. see Siegfried 1994), there is no certainty that the plants represented in this site were actually used by the people who camped in it during the past. The spatial analysis discussed next in Chapter 5 should help clarify this dilemma. What is clear from the information derived from the charred macrobotanical remains is that there were a number of plants that were potentially useful to people in the Tuscany locale during the past and this may represent one of the reasons why they chose to camp there.

Chapter 5
THE SPATIAL ANALYSIS

THE EXCAVATION GRID

The contents of the flotation samples from the Tuscany habitation site were summarized on simple tally sheets for the creation of a database. It is important to remember here that correction calculations were applied to all the sums to make each sample equivalent to 1000 ml because of the great discrepancies in the original bulk sediment sample volumes. One of the first views from the database was a simple summation of all the charred plant fragments recovered from each level of each unit. The purpose of this was simply to explore the possibility of patterning in terms of where the highest numbers of charred plant fragments might be located relative to the former glacial sand surface revealed in the map generated from the elevation data from the excavation units of the grid.

The first spatial view of the charred plant fragments plots the sums of the pieces found in each unit (all the levels are added together) of the Tuscany grid. The intention of this map is to show the planview distribution of the charred plant remains throughout the site and answer the question of where most of the pieces were found. In Figure 48, two views of the data are presented. The upper map is a wireframe plot presentation of the totals for the charred plant pieces found in each unit in the grid. Interestingly, the greatest accumulation of charred plant remains is in the northeast quadrant of the map. This is the area that the geoarchaeology discussion showed was downslope from the rest of the excavation area. It is the area that is proposed to have been the lowest part of the channel scar attributed to drainage patterns in the landscape. It would represent an area where material would accumulate from both aeolian sources and downslope movement of colluvium (slopewash).

Three scenarios are presented that explore the differences observed between the tallies of charred plant remains throughout the grid. In the first scenario, the burned plants were located upslope from the low spot and their remains washed down into the hollow by traction and other movement. In this case, we should expect higher accumulations of charred plant remains to occur in the lowest part of the grid and the lower accumulations should occur upslope. In the second scenario plants were located randomly throughout the site and their burned remains were washed downslope. Again, we should expect higher accumulations of charred plant remains in the lowest spots. In the third scenario plants were located all over the site, however, they were more prolific in the low spot because more moisture accumulated there. Though burned remains were scattered randomly throughout the site there were more remains left behind where more prolific plants were burned, in the low part of the site. Thus, again, we should expect more charred remains in the low spot without any necessary movement of materials downslope. The post map plot of the charred plant sums (Figure 48) reveals that there is some inconsistency to the locations of the highest sums of charred plant fragments.

It is apparent that there are a number of high points in the charred plant sums. In the case of Unit 34, this is where the ground slopes downward toward the lowest point in excavation (see Figure 6 for a map of the grid and the thickness of the paleosols). In the case of excavation Units 11, 12, 13, 8, 3, and 41, the highest concentration of charred plant material is on the actual slope itself. The slope is not dramatic in terms of size or steepness. The scale of this site should be considered as micro-topography as the distance covered ranges between 5 and 8 meters. The drop in elevation referred to here is in the range of 40 to 50 cm over 8 meters distance. Oddly, excavation Unit 34 actually has a lower count for charred plant remains compared to excavation Units 33 and 35 on either side, which both have relatively high counts. This pattern suggests that the distribution is not related to the present ground surface contours per se. Rather, there appears to be a relationship between the higher charred plant tallies and the location of the deep portion of the channel scar surface visible in the wireframe surface trend maps from the Chapter 3 discussion on the paleosols. In Figures 16 and 17, the original surface of

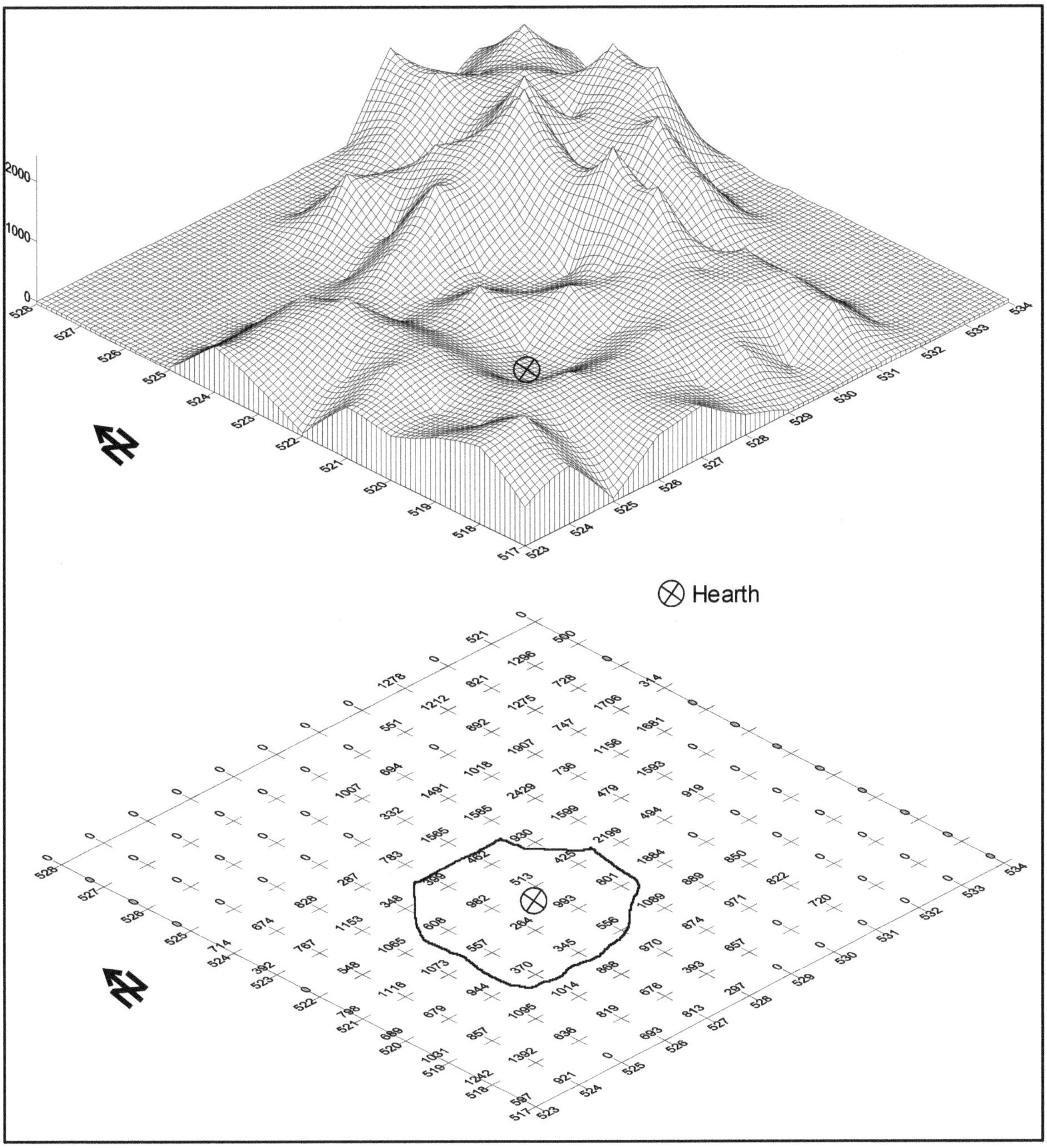

Figure 48. Density peak and post maps of the charred plant fragment sums for the Tuscany site (EgPn-377)

the site was interpolated through time, from the original post-Glacial Lake Calgary topography to the Mazama tephra contact and then to that of the present day. The higher tallies of charred plant remains occur on the slope of the channel scar as it dips down toward excavation Units 33, 34, and 35 in the far northeast section of the grid. Through time, the channel scar became less evident in the surface topography of the site. By the time of the Mazama tephra deposition event the surface of the site had become much more even in appearance (see Figure 16 of the trend surface changes) as the lower part of the channel scar filled with sediment. It can be suggested that the existence of the slope associated with the original channel scar was already becoming a more negligible feature in the evolving landscape of the site by the time of formation of the third pedogenic zone.

There is no doubt that traction and saltation movement of the fragments of charred plant remains must have played a small role in the location of materials. Relative amounts of soil moisture must also have played a role in the location of plants in the landscape. As suggested in the third scenario above, plants may have grown larger in the channel scar where more moisture probably accumulated. This may have led to more abundant and/or larger plants. These would have left behind more charred material once they burned off the landscape. One other contributing factor may be attributed to the pieces of bone recovered from the paleosols in the northeast quarter of the site grid, which might have contributed to enrichment of the soil. Enrichment combined with increased moisture would have lead to more productive plant growth at that locality and there would have been more material left behind when the vegetation burned. The lower elevation to the northeast would also have aided some downslope accumulation of charred plant remains as part of a natural drainage pattern. There are no definite conclusions that may be drawn to explain the observed patterning of accumulating charred botanical remains. The most fitting conclusion at this level of analysis is that several of the factors discussed probably worked together to produce the overall pattern observed. Thus, the low area in the northeast quarter of the grid provided a location where plants thrived due to increased moisture accumulation. These plants produced higher amounts of charred plant material when they burned. At the same time, some downslope movement of charred plant fragments contributed to increased proportions on the low-grade slope of the channel scar as well. These processes, working together, produced the higher frequencies of charred plant fragments observed in the area of lower elevation in the northeast quarter of the site grid.

A point for consideration in these proposed scenarios is the scale on which these observations are based, notably that the bulk sediment samples represented only 1% of the volume of each level excavated. The data can, realistically, indicate only broad trends.

Density variability for the charred plant fragments within the excavation levels has been explored further. There is the question of whether or not there was variability between the levels in terms of charred plant fragment totals. Further, did these level totals match up in any way with the changes observed within the sediments. The subsequent development of horizons via pedogenic processes within the paleosols indicates periods of landscape stability and the establishment of vegetation in the area. The reddish 'B' horizons in the two middle paleosols indicate the possibility of some forest development. It was decided, that though the site was excavated in arbitrary levels, discernable patterning might be visible if the densities of the charred plant fragment totals were plotted into profiles drawn along the northing and easting grid lines of the excavation. Peaks in the charred plant tallies were visible in the database but it was not clear if these matched across the sloping topography of the site from one unit to the next. Though each level was excavated on the horizontal plane and crosscut the sloping stratigraphy of the site, it was hypothesized that a 'stepwise' pattern should be visible in the profiles.

The charred plant summations of N519, N520, N521, N522, N523, N524, N525, N526, and N527 are presented first with Figures 49 to 57. Each figure presents the data in a consistent manner. First, the surface trends are plotted across the top of the profile. None of the excavated levels within the upper chernozem have been plotted as they were not relevant to this study. Thus, the next line down from the surface is the top of the Mazama tephra contact and the third line from the top is the bottom of the tephra boundary and top of the first paleosol. Thereafter, each level is drawn in as measured until the excavator reached the Glacial Lake Calgary sediments, which were not excavated.

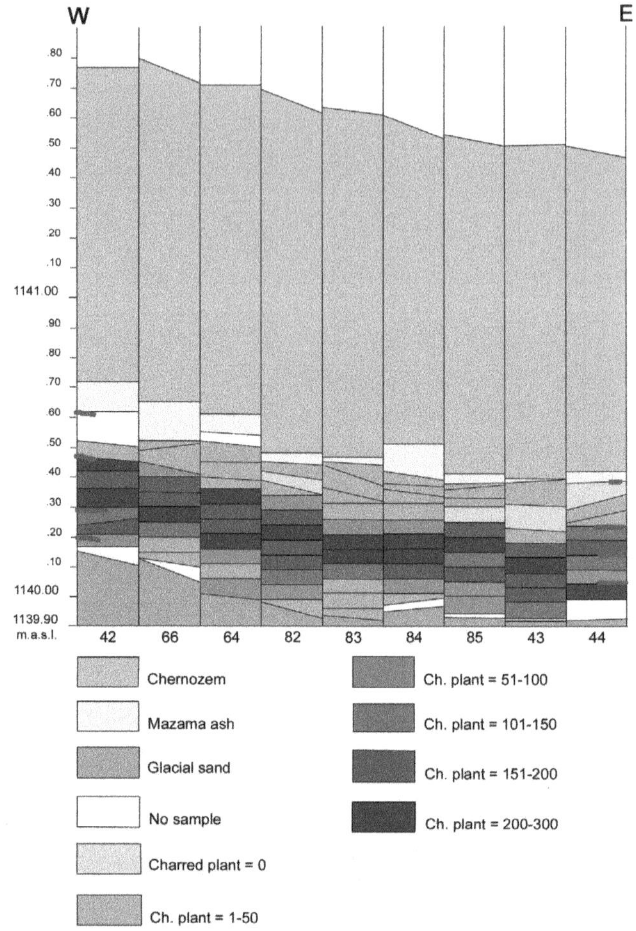

Figure 49: Greyscale density plot of charred plant fragment count for units 42 to 44 W-E in Northing 519. The red lines indicate the projected division of the 4 pedogenic units within the paleosols

A pattern is indeed visible. In each unit, in general, there were very few charred plant remains in the first three or four excavation levels immediately below the Mazama tephra boundary. There was then a dramatic increase in the density of charred plant pieces in the next level, beginning at 1140.47 masl in unit 42 in Northing 519 (Figure 49), for example, and sloping down to the east in the rest of the profile. In Northing 519, the concentrated charred plant layer was 15 to 20 centimeters below the Mazama tephra contact boundary on average in each profile. As we move north through the site from one northing to the next (Figures 49 to 57), this concentrated charred plant layer appears at lower points in each profile. There are high concentrations of charred plant near the bottom of the profile in northing 522 (Figure 52), at 1139.80 masl in excavation Unit 3. This appears to represent a second charred plant concentration zone.

Northing 519 is on the south end of the site grid. The paleosols were beginning to thin out significantly as the excavation moved south in the last season of fieldwork. Some of the differences observed in the charred plant densities from the upper levels to the lower levels within the paleosols may be partially due to translocation of

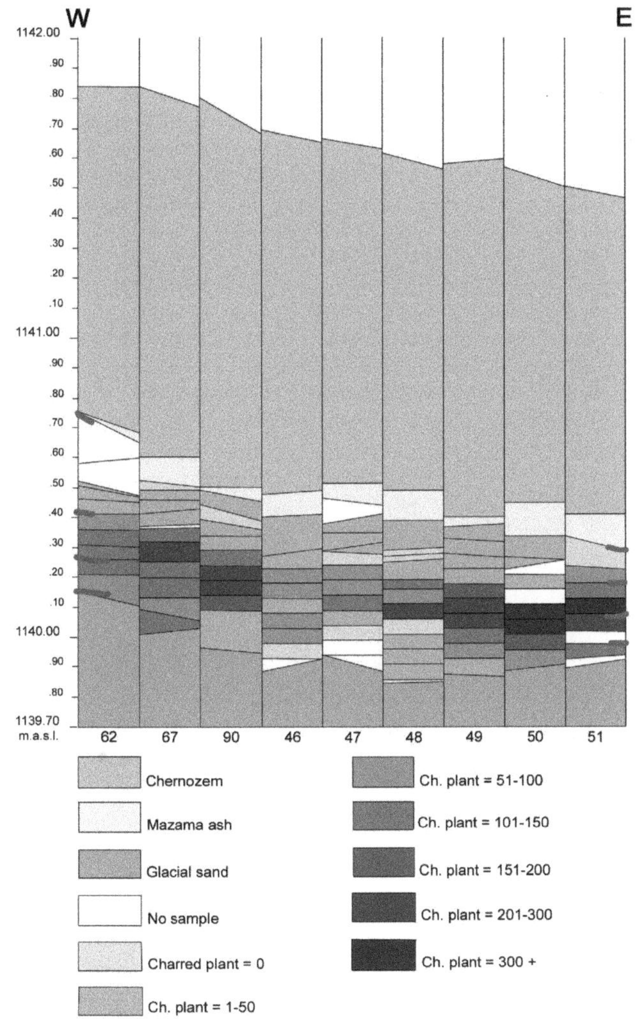

Figure 50: Greyscale density plot of charred plant fragment count for units 62 to 51 W-E in Northing 520. The red lines indicate the projected division of the 4 pedogenic units within the paleosols

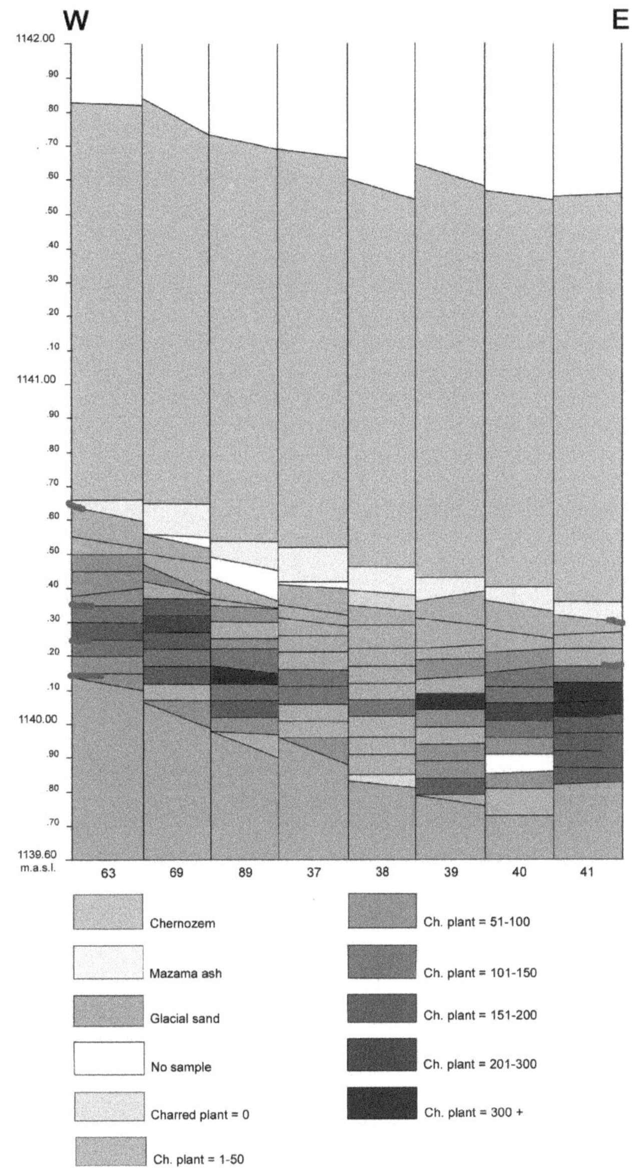

Figure 51: Greyscale density plot of charred plant fragment count for units 63 to 41 W-E in Northing 521. The red lines indicate the projected division of the 4 pedogenic units within the paleosols

materials through processes like freezing and thawing. The differences in densities through the different levels may also be due to repeated burning and regrowth episodes of vegetation at the site location. There may also be blurring effects from animal and insect burrowing through time. The length of time of the exposure of the landscape, which led to pedogenic processes and successional vegetation, also gave time for materials like the charred botanical fragments to accumulate at varying rates.

Another phenomenon that must be accounted for in this sloping landscape is the thinning out of layers up the slopes of the depression. At a certain point, the upper and lower accumulations of charred plant fragments are so close together that they are hard to differentiate. All the problems differentiating layers "H/I" and "J/K/L" in the soil profiles in Chapter 3 (Figures 10, 11, and 12) are probably related to compression of the stratigraphy due to thin deposition of sediments on the slopes of the depression. An extremely fine-grained sampling scheme might help differentiate these layers.

Figures 58 and 59 have been drafted to differentiate the concentrations of charred plant fragments within the upper and lower parts of the paleosols visible in Figures 49 to 57. Figure 58 shows a diagram of the Tuscany grid. This diagram represents the concentration of charred plant fragments that occurs within the paleosols just above the Glacial Lake Calgary sands. Units with the highest concentrations of charred plant pieces in the lower layer are colored black, while units with low/no concentrations are gray, and unexcavated units are white. The pattern of charred plant fragment deposition is somewhat odd. There is a higher concentration of material in the northeast part of the site, which has the lowest elevation. However, there is a small break of two to three meters distance in the grid, oriented SE-NW, where there appears to be a real dropoff in the number of charred plant fragments in the lower burn layer. There is

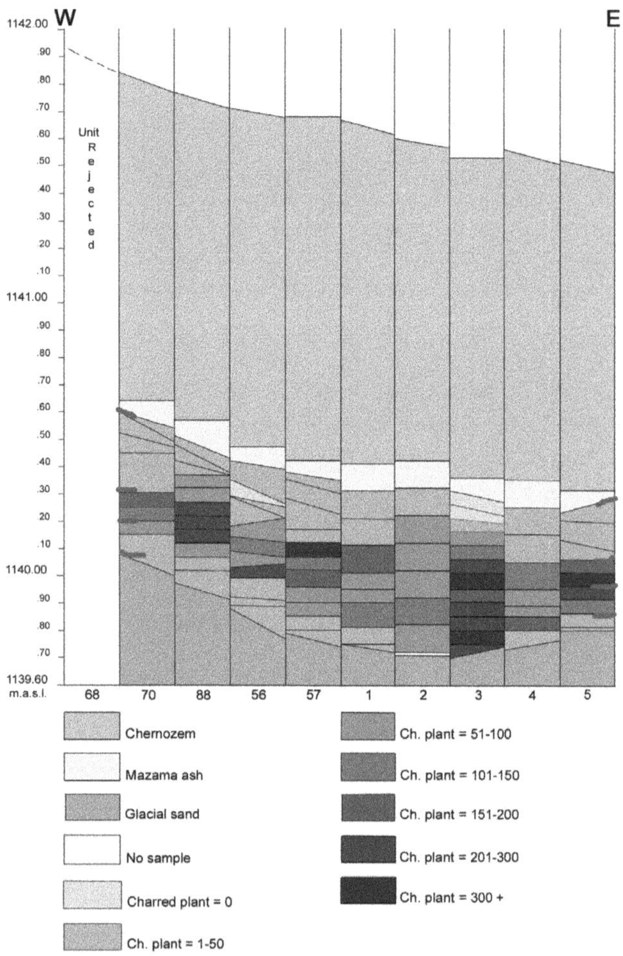

Figure 52: Greyscale density plot of charred plant fragment count for units 68 to 5 W-E in Northing 522. The red lines indicate the projected division of the 4 pedogenic units within the paleosols

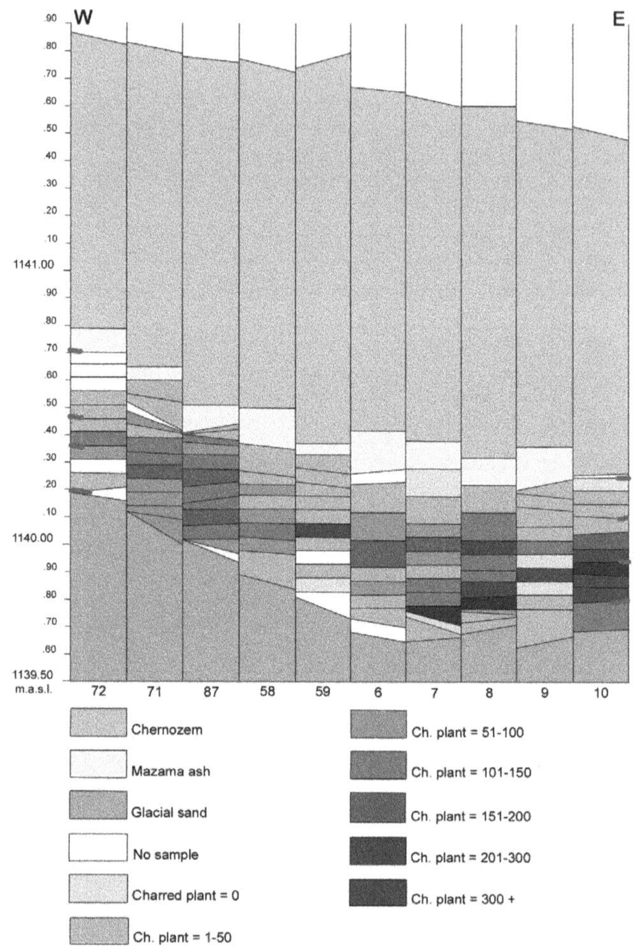

Figure 53: Greyscale density plot of charred plant fragment count for units 72 to 10 W-E in Northing 523. The red lines indicate the projected division of the 4 pedogenic units within the paleosols

then another patch of charred plant concentration in the southwest corner.

Attempts to understand any patterning visible with the charred botanical remains are restricted to low-level hypotheses that are essentially not testable. The data are not robust enough to push them to another level of analysis. At this level, suggestions can be offered and those yielding the most reasonable explanation(s) can be considered. With these problems in mind, one possible explanation for the pattern plotted for the charred plant remains from the lower excavation levels is that the vegetation during this time period was not as dense as it may have been later, in the upper part of the paleosols. In general, the first 10-15 centimeters of sediment that deposited overtop of the Glacial Lake Calgary sands, contained few charred plant remains and few charred seeds as well. This can be interpreted to be the result of a lesser occurrence of shrubby types of vegetation like kinnikinnick, juniper, buckbrush, soapberry and shrubby cinquefoil. A lower occurrence of these plants combined with a sparse distribution over the landscape would lead to a sporadic or patchy distribution pattern of remains if the vegetation burned off in a fire or series of fires over time. This may be the pattern that is viewed in Figure 58. Alternatively, an argument for differential, poor preservation of plant remains may also be made to explain the lower amounts and patchy distribution of the remains that are represented. Given the ubiquity of the charred plant fragments throughout the bulk sediment samples from the site, preservation does not seem to be problematic and this suggestion is considered less tenable.

One of the problems with this effort to understand the dynamics of the vegetation cover that appears to have existed at the site is the fact that nearly all the recovered remains from the paleosols are charred. There are <10 seeds that are not charred recovered from the paleosol samples. The amount of botanical remains recovered from the sediments above the paleosols is also represented by an extremely small number of charred plant remains from the samples. Non-charred seeds appear near the surface levels, which is expected. Thus, the major difference visible between the upper chernozem and the lower paleosols is in the amounts of charred plant remains recovered. This difference is dramatic. There are thousands of charred botanical remains within the

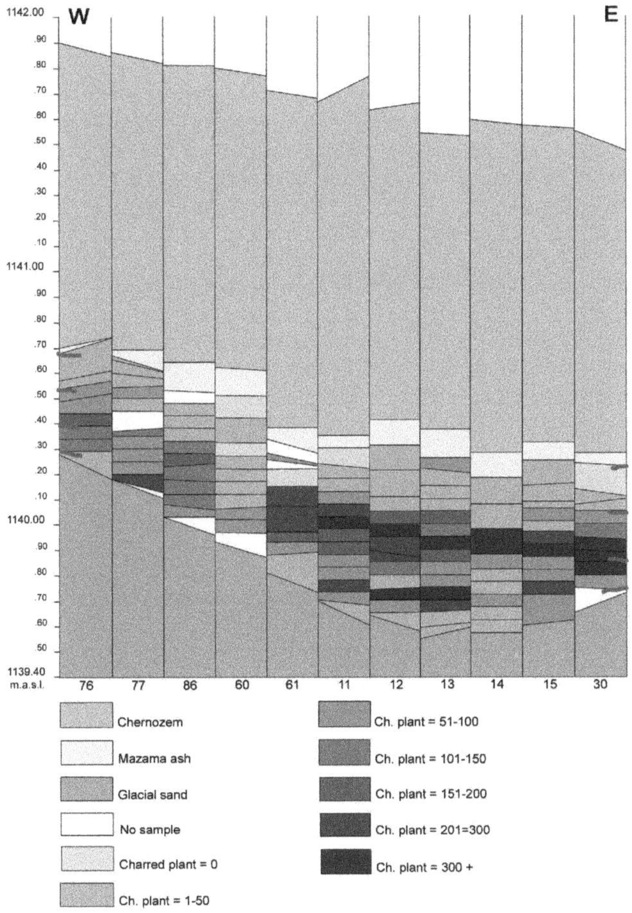

Figure 54: Greyscale density plot of charred plant fragment count for Units 76 to 30 W-E in Northing 524. The red lines indicate the projected division of the 4 pedogenic units within the paleosols

paleosols and very few within the chernozem above. Chernozems are soils that form in prairie/plains settings. They are associated with grassland and forb vegetation cover. This is the vegetation cover at the Tuscany site today. The well-developed chernozem above the Mazama tephra has no observable differentiation within either its 'B' or 'C' horizons, indicating that there has been no obvious sedimentation deposition of any kind. Thus, there is no visual/visible evidence of shrub or forest vegetation growing in the depression, dying off, becoming buried by continuing deposition of sediment, which subsequently developed into a deep chernozem soil. There is no evidence of organic banding within the chernozem that might indicate that type of scenario taking place. One more point of interest in this discussion is the composition of the charred botanical remains that have been recovered from the bulk sediment samples located above the Mazama tephra contact. The remains do not appear to consist of the same woody structures that occur within the paleosols. These remains will eventually need to be examined and identified further and compared carefully to those from the paleosols.

The primary evidence for past vegetation cover within the confines of the depression is the charred botanical

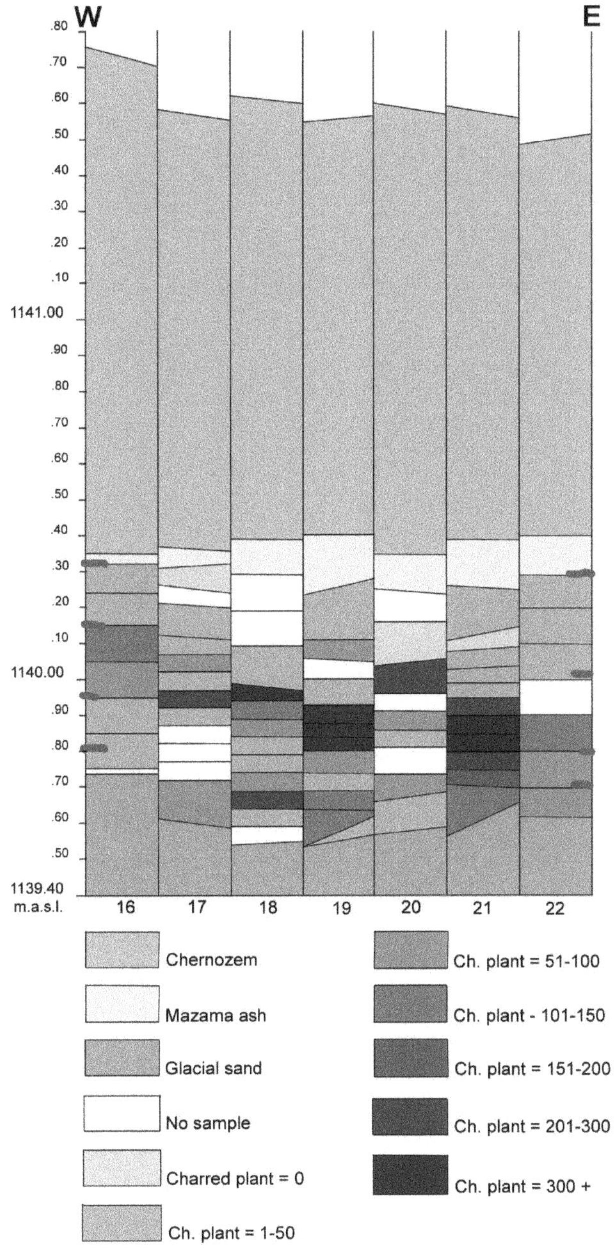

Figure 55: Greyscale density plot of charred plant fragment count for units 16 to 22 W-E in Northing 525. The red lines indicate the projected division of the 4 pedogenic units within the paleosols

fragments and seeds recovered via flotation analysis. The lack of charred botanical remains within the chernozem above the tephra boundary can be hypothesized as the result of a continuous grass and forbs landscape from the Mazama tephra event to the present. In this scenario, the presence of more charred plant remains below the Mazama tephra would be indicative of more shrubby vegetation in those earlier times, these woody charred remains being more able to survive various post-depositional taphonomic processes. Alternatively, the evidence of other vegetation like trees and shrubs may have simply been obliterated with time and subsequent pedogenic processes. If vegetation does not become charred or preserved by some other natural process, most remains do not survive through time.

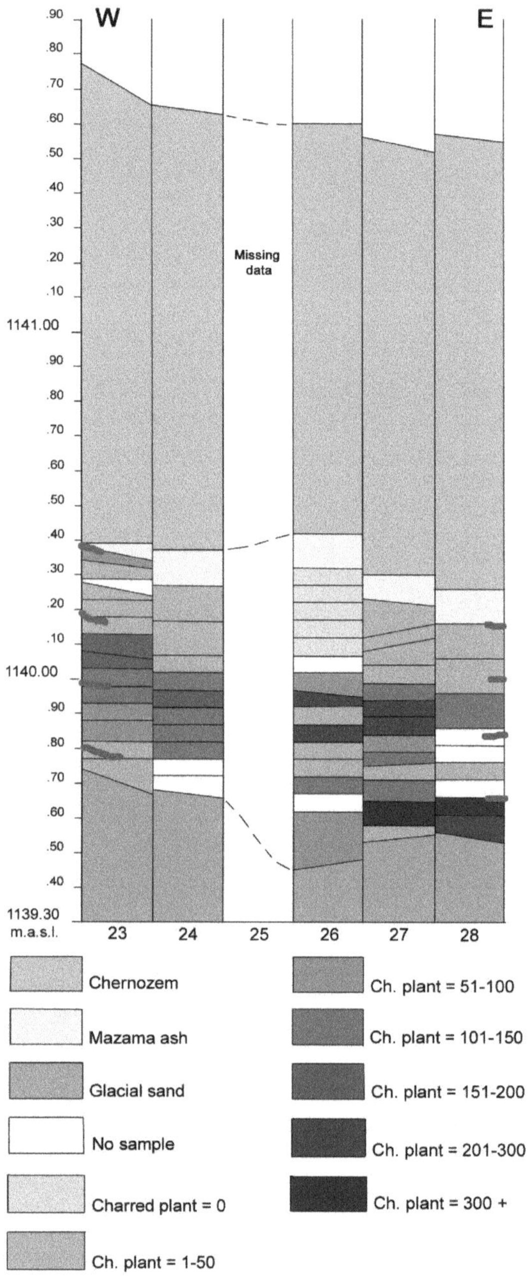

Figure 56: Greyscale density plot of charred plant fragment count for units 23 to 28 W-E in Northing 526. The red lines indicate the projected division of the 4 pedogenic units within the paleosols

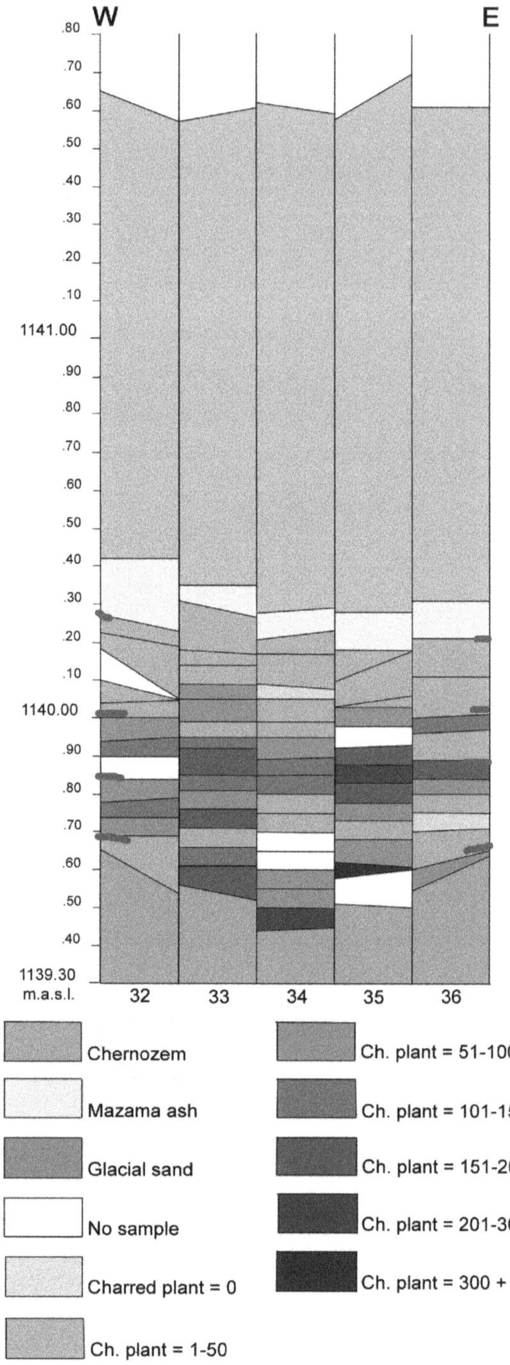

Figure 57: Greyscale density plot of charred plant fragment count for units 32 to 36 W-E in Northing 527. The red lines indicate the projected division of the 4 pedogenic units within the paleosols

The problems noted for the lack of charred plant remains in the first 15-20 cm of the paleosols can fit into the same set of arguments just presented for the chernozem. Absence of charred remains may indicate the possibility that a grass and forb dominated vegetation cover extends back even further in time. Alternatively, the absence of charred remains may simply indicate the lack of preservation of any botanical remains through time. When the charred plant remains do begin to appear in the profiles, the change is startling. The charred plant remains appear and continue to occur right down to the top of the Glacial Lake Calgary sediments. However, kinnikinnick and juniper seeds do not occur very often in the last 10-15 cm of the paleosols.

We must refer back to the chapter discussing the paleosols and the profile descriptions provided there. The first, earliest soil (fifth pedogenic zone) that developed in the depression was described as a "black layer". It appeared to be the first A, organic soil horizon which was labeled layer "M" in the profiles. Layers "K" and "L" were interpreted as an A and B pair that made up another soil (the fourth pedogenic zone) that had time to form above "M". Layers "H/I" and "J" were also interpreted as an A and B pair that had time to form a soil (the third pedogenic zone).

In comparison to the upper paleosols there appears to be fewer charred remains found in the very dark organic layer "M", overlying the Glacial Lake Calgary deposits. While there are occasional seeds of kinnikinnick and juniper, they are more rare than in the sediments above. The occurrence of charred botanical material that appears to be wood, is far more patchy or sporadic in Layer 'M'. This landscape is interpreted to have had a vegetation cover dominated by grasses and forbs with occasional shrubs.

The contact between Layer 'M' (in the profiles) and the glacial sands immediately below it remains problematic for this site, although it does not effect the interpretation of the material higher up in the profile. As outlined above, the lowermost radiocarbon date of 9950 ± 120 years BP was derived from charred botanical material from excavation Unit 34 Level DD. The charred botanical material came from the bulk sediment sample, a 5 cm x 5 cm x 5 cm cube of sediment overtop the glacial sands. The problem with this particular date, the sample's provenience and the glacial sands immediately underlying, is in the timing of events between the draining of Glacial Lake at 13,900 calendar years ago and the initial establishment of vegetation at 9950 ± 120 years BP (the oldest Tuscany date), which is a non-calibrated radiocarbon date. These dates are not comparable and this makes the chronology confusing. Using the program OxCal v. 3.5 (Ramsey 2000), the date from Tuscany can be calibrated and ranges between 12,200 and 11,200 years ago at 95.4% probability, which can now be compared to the 13,900 calendar years of the chlorine-36 date derived by Fisher (1999). This comparison reveals an unaccounted-for gap of time with a range of 1700-2700 years between the Glacial Lake Calgary sands and the first evidence of vegetation on the Tuscany landscape. It does not seem likely that there would be no vegetation for such a long period following the draining of the glacial lake, especially with dates of ~11,300 ^{14}C years BP derived for extinct faunal remains in the Calgary area, animals whose existence implies the availability of grazing and browsing vegetation.

Two hypotheses can be offered at this time to address this problem. The first hypothesis is the possibility of an erosional event of some kind that destroyed/removed the previously existing stratigraphy and created a new surface upon which the vegetation eventually established. In fact, the drainage channel scar discussed in Chapter 3 may represent such an erosional event within the site. There was a layer of small, pea-sized gravel just beneath the surface of the Glacial Lake Calgary sands that appeared to be associated with the channel scar, according to Oetelaar and Zaychuk (1997). This may represent an erosional unconformity upon which more sand material was deposited by slopewash, which was then followed by the establishment of vegetation and the eventual formation of the first organic soil of Layer M in the wall profiles.

The second hypothesis is that there may be a problem with the radiocarbon date itself, such as contamination of the charred botanicals that were dated. This could be tested by deriving several dates from charred plant remains from the lowest excavated level at different locations in the excavation grid. While the dates derived thus far for the site are in good stratigraphic context, a number of dates would make a stronger case and may clarify some of these problems.

Overlying Layer "M" was the fourth pedogenic zone. This is the part of the profile labeled Layers "K" and "L" in the wall profiles described in chapter 3. This is where kinnikinnick and juniper really begin to become established in the landscape and also appear to be burned off by fire(s) in the locale. The burning of vegetation may have taken place with one effective local fire or a series of local fires. The continuous occurrence of charred plant remains from one level to the next favours the suggestion that fire played a major continuous role in the appearance of this landscape. The other possibility for the continuous occurrence of charred fragments would be an extraordinary amount of translocation of materials by some means through time. One of the ways translocation can take place is through seasonal freezing and thawing. Since the water table at the site was encountered just below the paleosols, some frost heaving may be occurring. However, there were no odd soil structures present in any of the profiles that could be related to water saturation and freezing-thawing cycles. Landals (1998:61) encountered interesting ice-related structures at the Minnewanka site, called 'drop features,' which were attributed to saturation of the site because of its location on the shore of the Lake Minnewanka Reservoir that has a yearly rising and falling shoreline. Serious translocation occurred at the Minnewanka site as a result of this phenomenon. There is nothing remotely similar to this phenomenon visible in the Tuscany profiles.

The third pedogenic zone in the site was labeled "H/I" and "J" in the wall profiles. It was in "J" that a very visible increase in charred plant remains could be discerned as the units were being excavated. The sediment in layer "J" was visibly reddish in color even when it was moist and white deposits of calcium carbonate were visible in it as well. The small flecks of charred plant were visible in the sediment but they degraded quickly on exposure and were essentially not recoverable (by hand) from the sediment matrix. Some samples were taken but they crumbled to pieces in the foil packets. It was apparent that three point proveniencing samples was not worth the effort; there were too many pieces being exposed and they were problematic to recover by that method.

The increase in the number of charred plant fragments in the sediments showed up dramatically in the bulk sediment samples. The charred fragments appear to have fared better when they were recovered through flotation of the sediments. Most of them retained their original size and structure and did not crumble or powder. They looked like small pieces of burned twigs or roots. There is a significant difference between the concentration of charred plant remains in this section of the profile (Layer

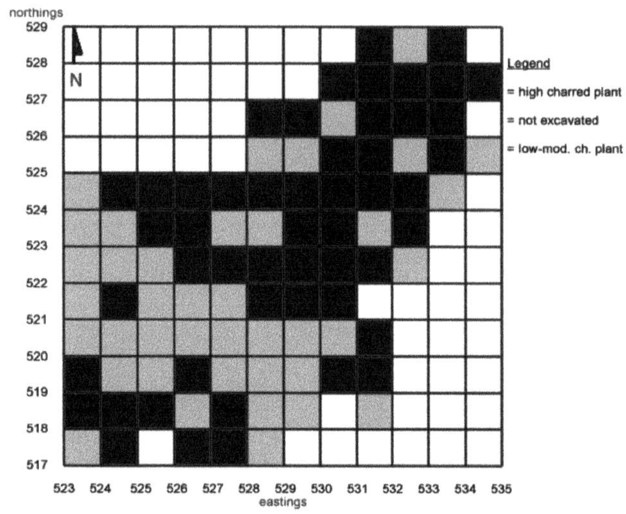

Figure 58: This diagram highlights the lower burn layers (K, L, and M) within the Tuscany paleosols. The units containing higher numbers of charred plant pieces are colored black

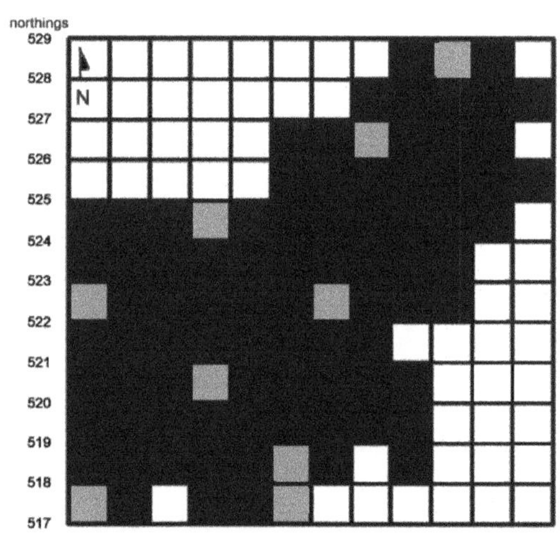

Figure 59: This diagram highlights the upper burn layers (H, I and J) found within the Tuscany paleosols. The units containing higher numbers of charred plant pieces are colored black

"J") compared to the lower concentration layer (Layer "K" and "L"). This is visible in the comparison between Figures 58 and 59. In Figure 59, nearly every excavated unit in the whole site had a distinct concentration of charred plant remains—the concentration showed up in every bulk soil sample from layer "J". There are only nine excavation units that do not have indications of a layer "J" in the profile. These are the squares that are colored grey in Figure 59. Two of these excavation units (68 and 74) were rejected from the database (unrecoverable errors) and the field notebook was lost for excavation Unit 25, which was not included in any of the site analysis. There were only six units with problems such as missing bulk sediment samples for the critical part of the profile. The white squares represent unexcavated units in the site grid in Figures 58 and 59.

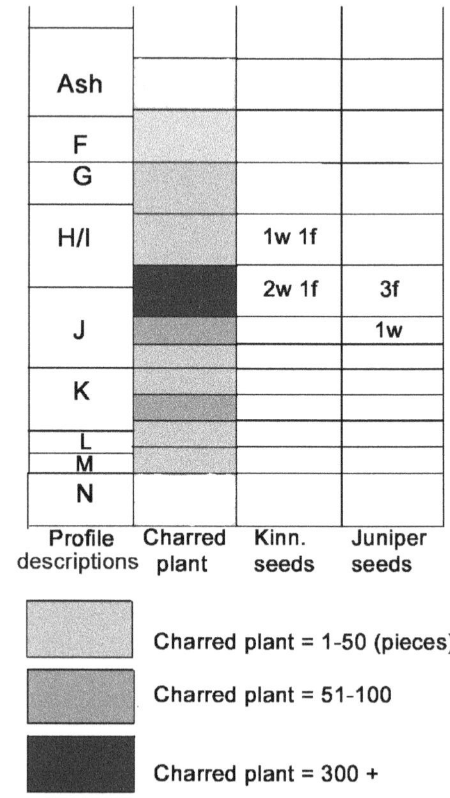

Kinn. = Kinnikinnick; w = whole seed; f = seed fragment

Figure 60: This figure summarizes information for Unit 14. The first column, on the left, shows the profile layers within the paleosols that were defined in the field, based on color, texture, and acid reaction. The second column shows a color gradient differentiating the number of charred plant pieces found within each excavated level of the paleosols. The final two columns show the totals for the important charred seed species found in each level of Unit 14. All the totals are based on the items recovered from the 1% bulk soil samples taken for each level of the unit as it was excavated. These totals have also been adjusted/recalculated to standardize the sample volumes for comparative purposes. Thus, each total is based upon projections for a 1000 ml bulk soil sample

The relationship between the analyzed and described sediment profile and the excavated and sampled profile is illustrated in Figure 60, which summarizes information for excavation Unit 14 (N524, E531) in the Tuscany site grid. The apparent trend with the charred plant remains is visible in this diagram. Layers "H/I" and "J" represent the third pedogenic zone at the site and layers "K" and "L" represent the fourth. What is interesting in this view of the excavation is that both levels that contained the higher concentrations of charred plant materials are at the bottom of what appears to be the A horizon of pedogenic zones 3 and 4. Thus, in both cases, the burned vegetation remains ended up in the transition zone between the A and B horizons. During a fire, the above-ground vegetation would have burned and depending on the moisture content of the site, the leaf litter layer may have been consumed as well. The charred plant remains, including the charred seeds, would have ended up lying on the bottom of the former litter and A horizon.

Following the fire, continuous/rapid deposition of sediments must have taken place, along with continued formation of a leaf-litter horizon burying the charred plant remains. They remained incorporated in the continuing development of the paleosol through time. In the first cycle of soil development, a B horizon eventually developed and was later excavated by us as Layer "L". A similar cycle of deposition, vegetation growth, fire regimes and eventual soil formation must have taken place with Layers "H/I" and "J".

After the last concentration of charred plant remains in the paleosols our visible evidence of vegetation at the site drops off dramatically with the drop-off of charred plant remains. Moving further up the profile, there are fewer charred plant remains compared to quantities recovered at the top of the last concentration layer. Charred seeds of both kinnikinnick and juniper continue to occur above Layers 'H/I/J', which would appear to indicate that these two plants may have continued a very reduced occupation of the landscape of the depression. Alternatively, translocation may account for the presence of these charred seeds above the concentrated charred layers. A reduced occupation of the landscape by these plants is considered the more tenable proposal.

Another pedogenic zone developed above the one containing the major burning event. This zone was made up with Layers "F" and "G", interpreted to be an A and a B horizon, respectively. As noted, there were few charred remains in this pedogenic zone. The B horizon (Layer G) was very sandy in texture and notably sterile of animal bones, which were abundant in the A horizon (Layer F).

The sudden dropoff in the number of charred plant remains primarily takes place during the formation of the second pedogenic zone at the Tuscany habitation site. This sediment was found immediately underneath the Mazama tephra contact boundary. The charred plant dropoff is visible in every profile created from the grid. The dropoff is poorly defined in some units but is fairly consistent overall. In excavation Unit 14 (see Figure 60), layers "F" and "G" are both 10 centimeters deep, making the second pedogenic zone (which is the uppermost buried paleosol) 20 centimeters thick in total.

After the last charred plant concentration very few kinnikinnick seeds, juniper seeds and charred plant fragments are found in the second pedogenic zone despite the fact that the sediment of the B horizon was sandier and the type of soil that appears to be preferred by both of these plants. Interpretation becomes problematic. It is apparent that at least kinnikinnick and juniper shrubs, and possibly other shrubs, continued to occupy the landscape. They appear to have continued to be burned off by fire in the locale as there are a few charred seeds and charred woody fragments in the sediments. Thus, it does not appear to be the case that there were no fires for an extended period of time. It appears that there was at least one fire, possibly more, but far less material was left behind as evidence of these events. There is a question of preservation of material. If there were no fires in the locale, vegetation would simply proliferate through time and no evidence of it would be left behind, which is the case in most soil profiles. Thus, for the uppermost pedogenic zone of the buried paleosols, it is still not clear whether the relative absence of charred plant material is due to a relative absence of shrubby vegetation or a low fire frequency resulting in little preserved plant material. Given the certain continued evidence of at least some shrubs, a change in fire frequency is also plausible.

The question of a proposed change in vegetation cover prior to the Mazama tephra boundary may be answerable through the application of pedological techniques such as organic carbon content and humic and fulvic acid content to evaluate differences between A and B horizons. These methods were applied to buried paleosols at the Fish Creek site, south Calgary, and appear to have successfully differentiated evolved horizons (Kingston 1982). Application of a phytolith analysis to soil samples from the Tuscany site could also answer the question of changing vegetation from shrubs and trees to a predominant grass cover. A phytolith analysis of the Fish Creek buried paleosols by Waters (1979) indicated the presence of a landscape dominated by Gramineae, the Grass Family.

The example from south India (Morrison 1994) is worth mentioning here. Morrison's (1994) pollen analysis summed charcoal particles and summed total area of charcoal per cubic centimeter per sample, which was charted against the stratigraphy of the core. The evidence of burning correlated with the pollen analysis that showed that a more open vegetation existed in the landscape after burning episodes (Morrison 1994). While there is no corroborative pollen analysis with the Tuscany site data, a similar pattern of local vegetation recovery can be proposed. One means of testing this hypothesis is through similar analyses of other sites with comparable stratigraphy. Phytolith analysis may also yield a potential means of determining the general vegetation cover changes from trees/shrubs to grasses/forbs (or vice versa) in these sites through time.

THE HEARTH AND STRUCTURE

There is another spatially related pattern from the sums of the charred plant fragments and seeds that is not immediately apparent on the map projections. It is related to the hearth feature located in excavation Unit 37 (N521, E526), on the east side (see Figure 48). There is a drop in the raw totals of charred plant fragments and charred seeds in a 3 x 3 meter vicinity around the hearth locality (see the post maps in Figures 48 and 61). The drop in raw totals is actually far more dramatic with the identified charred seeds from the site, which will be discussed below. The sums for the charred plant fragments do not drop off completely—there are remains in the area around the hearth and the upper charred plant concentration layer is still evident in the profiles. The dimensions of this space within the site are very interesting. It is proposed that the space may be a trample

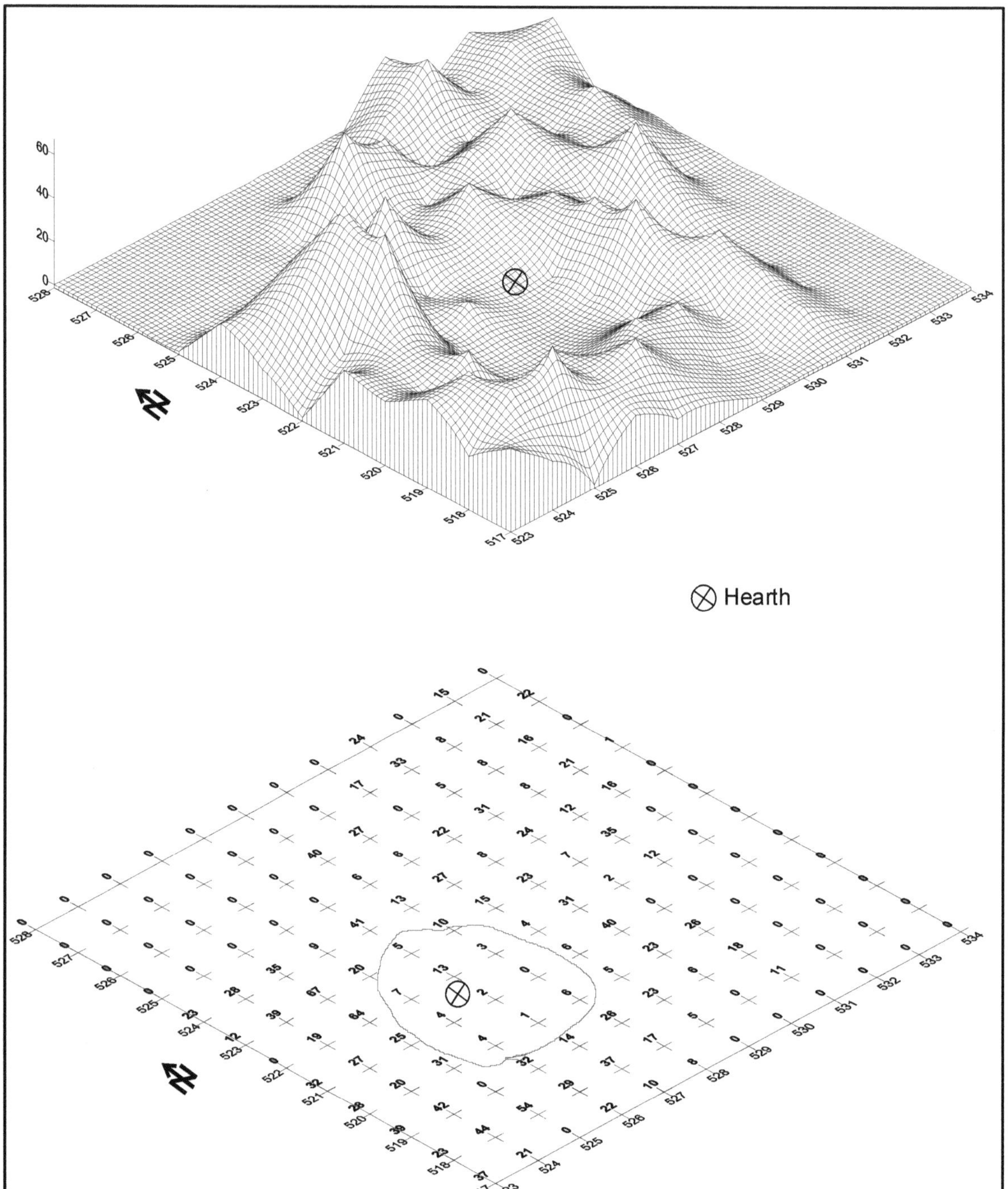

Figure 61: Density peak and post maps of the charred seed totals for the Tuscany site (EgPn-377)

zone around the hearth where charred plant fragments were crushed and powdered to some degree. The circular area may even represent a structure location though no posthole features were ever found.

Two sites, Vermilion Lakes (Alberta) and Hell Gap (Wyoming) have been previously discussed. Each presents evidence of small structures of comparable size to the dimensions of the hypothesized trample zone identified in the Tuscany site. These two examples do not appear to be directly comparable to the Tuscany habitation site. The distribution patterns of materials are different and they both contain evidence of postholes, which does not occur at the Tuscany site.

The timeframe for all the sites appears to represent the PaleoIndian period between 10,800 to 7800 years BP. The projectile point that was found at the Tuscany habitation site was located 23 centimeters away from the hearth and was also in excavation Unit 37 Level S (N521, E526) (see Figures 48 and 61). Since the hearth and the point appear to be contemporaneous, the assumption is made that they probably represent contemporaneous evidence of human activity/occupation of the locality. The point is a smaller lanceolate, leaf-shaped point determined to be Lusk. Normally within the Plain's chronology, this would place it in the timeframe around 8,000 years ago, according to the cultural chronology outlined for this region by Vickers (1986). Two samples (one bone collagen, one charred plant fragments) closely associated with the hearth and projectile point from the Tuscany site have yielded radiocarbon dates of 7840 ± 100 and 7610 ± 70 years BP.

SPATIAL DISTRIBUTION OF THE CHARRED SEEDS

Planview Distributions

The charred seeds recovered through the flotation analysis were quantified by unit and by level for the site. These tallies of the seeds were used to create data files with X, Y, and Z coordinates that could be used in the Surfer 7 mapping program. It must be remembered that the map projections are based on standardized probability counts of each seed discussed. The planview maps have a number scale on the right side.

The first view of the spatial distribution of the charred seeds is derived from summing all the charred seeds found in each level of each excavation unit. In Figure 61, the sums of charred seeds in each unit are presented in a wireframe map, which shows each excavation unit's total as a peak point on the grid. Interestingly, a 3 x 3 meter area around the hearth feature in the paleosols shows up quite clearly as a flat zone on the map because there are very few charred seeds. The corrected or standardized tallies of the charred seeds themselves are presented in the post map, below the wireframe. This view of the site grid shows the corrected distribution numbers (standardized to 1000 ml samples) for the charred seeds found within each excavation unit. In this view the unexcavated units outside the grid are counted as zeros.

There is a gradual gradient across the site in terms of the distribution of charred seeds. There is trend toward higher charred seed counts in the southwest half of the grid, compared to the northeast. Two excavation units, 87 and 88 (N523, E525; N522, E525), have very high totals and are on the northwest edge of the south half of the grid. This is quite the opposite distribution pattern than was observed with the charred plant totals where there was an increase in totals in the northeast, in the area where there was a depression or channel during the early Holocene. There may have been more shrubbery or small trees in the northeast section of the grid, possibly related to bone deposition and moisture as previously suggested. An increase in shrubs that did not produce large seeds like willows or poplar tree cover would lead to more charred plant fragments being produced with fewer charred seeds once the vegetation burned.

The next spatial distribution pattern examined is for the juniper seeds and seed fragments. Juniper and kinnikinnick seeds are common, ubiquitous across the site, and clearly identifiable. As a result, they were chosen to examine whether or not there were difference in distributions of whole and broken seeds. Broken seeds could be related to disturbance of the site or cultural activity like food processing. Both kinnikinnick and juniper seeds are edible though kinnikinnick is more palatable as a food. These two species were examined in this manner, however, none of the other identified species were subdivided further. *Chenopodium* seeds are so fragile that breakage is expected and should not be considered culturally meaningful. The other seeds were simply not as well represented as kinnikinnick and juniper and hence did not warrant such an examination.

In Figure 62, two views of the juniper seed data are presented. The top view is a wireframe map of the site grid with the sums of whole juniper seeds for each unit in the site. The bottom view is a density map in planview, which shows the most-dense concentrations of whole seeds as white 'hotspots'. In general the whole juniper seeds appear to be distributed throughout the site. Excavation Unit 12 (N524, E529) shows a high quantity of whole juniper seeds compared to surrounding units.

These maps also reveal an area of 3 x 3 meters around the hearth feature where the number of seeds drops off dramatically. This is particularly clear in the density planview map. There is a dark circle in the center of the grid showing the area where very few whole seeds are present.

Figure 63 presents a peak and density map of the juniper seed fragments distribution within the site grid. Interestingly, there is a distinct high point of concentration of juniper seed fragments in the northwest quarter of the grid, especially in excavation Unit 71 (N523, E524). This corner of the grid was considered disturbed during profiling of the walls. The stratigraphy was extremely hard to discern; there was no Mazama tephra layer to differentiate between the paleosols below and the chernozems above. The 'hotspot' of broken juniper seeds is odd, perhaps corroborating the possibility of disturbance. There does appear to be a higher density of seeds in the southwest corner of the grid, which would have been upslope at the time of growth and deposition. The empty 3 x 3 meter spot in the grid shows up well with the juniper seed fragments, though it is not as distinctly round as it is in some of the other distribution maps that are presented below.

The next spatial distribution figures are for the charred kinnikinnick seeds and seed fragments. Figure 64 presents peak and density maps of the whole/unbroken charred kinnikinnick seeds distribution. The top view is a wireframe map of the peak points where the whole seeds were found. The bottom planview map shows the densities as 'hotspots'. There is a definite trend toward a higher occurrence of whole charred kinnikinnick seeds in the southwest corner of the grid north-south along eastings 525, 524 and 523. The circular area containing low numbers of seeds is more clearly defined in the density map for whole charred kinnikinnick seeds, than in either of the sets (whole or broken) of juniper data.

In the next Figure 65, two views of the charred kinnikinnick seed fragments are summarized. Again, the top view is a wireframe map of the peak numbers of

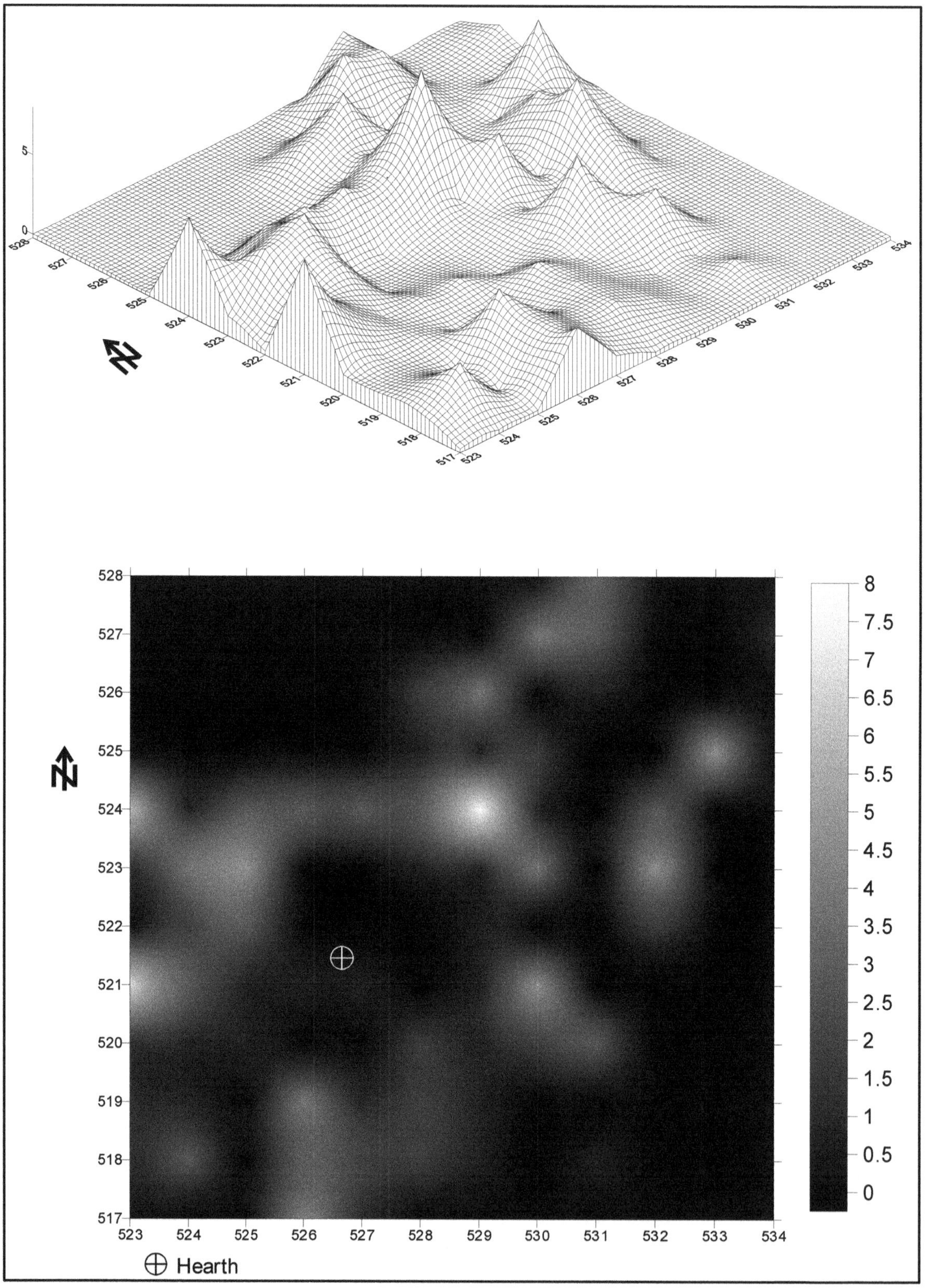

Figure 62: Peak and density maps of the distribution of whole/complete juniper seeds found within the paleosols at the Tuscany site

seeds in each unit in the site grid. There is a distinct concentration of charred seed fragments along the same three eastings as seen for the whole seeds. This plant seems to have preferred the upslope areas in the landscape of the site. Interestingly, the bottom density map of the charred seed fragment distribution shows an extremely clear 3 x 3 meter space where there were very few seeds found in the flotation samples. Another interesting pattern to note is the concentration of charred seed fragments in the northwest corner, similar to the charred juniper seed fragment distribution in excavation Units 87 and 88 (N523E525 and N522E525).

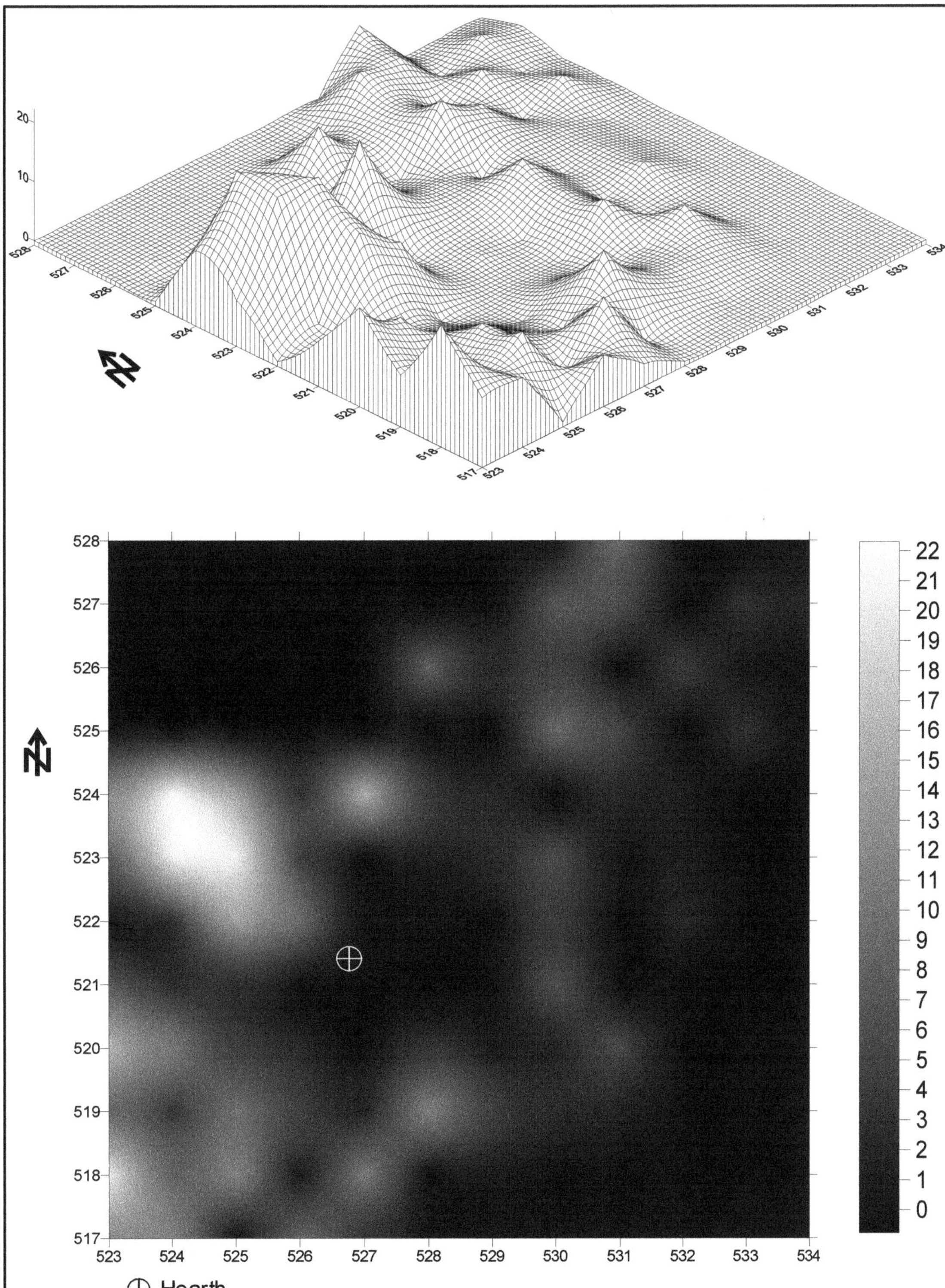

Figure 63: Peak and density maps of the distribution of juniper seed fragments found within the paleosols at the Tuscany site

The circular space on the map is fairly well defined with the seeds that are well represented in the assemblage, like kinnikinnick and juniper. However, it also shows up in the maps that have been generated for some of the other seeds represented in the site.

In Figure 66, two views of the distribution of *Chenopodium* sp. seeds in the paleosols are presented. The top view is a wireframe map of the peak numbers of seeds found in the flotation samples for each unit. The seeds of this plant appear to be distributed throughout the site. It must be remembered that the northwest and southeast quarters of the grid are black in color because they were not excavated. The distribution of the seeds runs from the northeast to southeast corners. Aside from the circular area with no remains, there are seeds

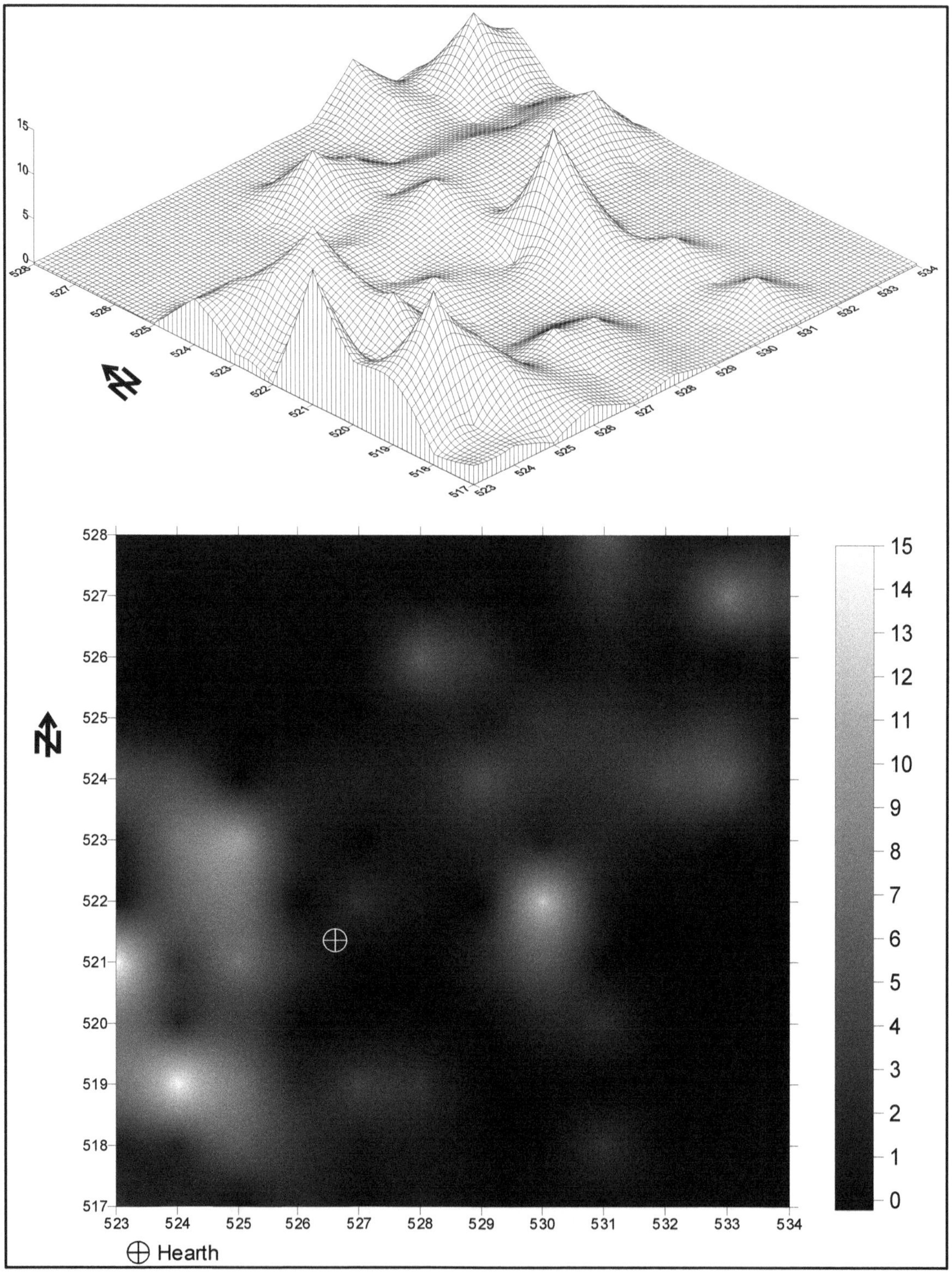

Figure 64: Peak and density maps of the distribution of whole/complete Kinnikinnick seeds found within the paleosols at the Tuscany site

represented in nearly all the excavated areas of the grid. Again, there is a preference for the upslope area of the site in the southwest corner of the grid. The distinct circular space where low numbers of seeds were deposited is extremely clear in the planview density map of the grid.

In Figure 67, the same two views of the distribution of charred shrubby cinquefoil seeds are presented. The seeds of this plant appear to be distributed primarily in the south half of the site grid. Again, the circular space where low numbers of seeds were deposited is very clearly outlined in the planview density map.

In Figure 68, two views of the distribution of charred rose seeds are presented. At this point in the spatial analysis we begin to see a breakdown in the pattern of deposition of the seeds. There were not very many charred rose

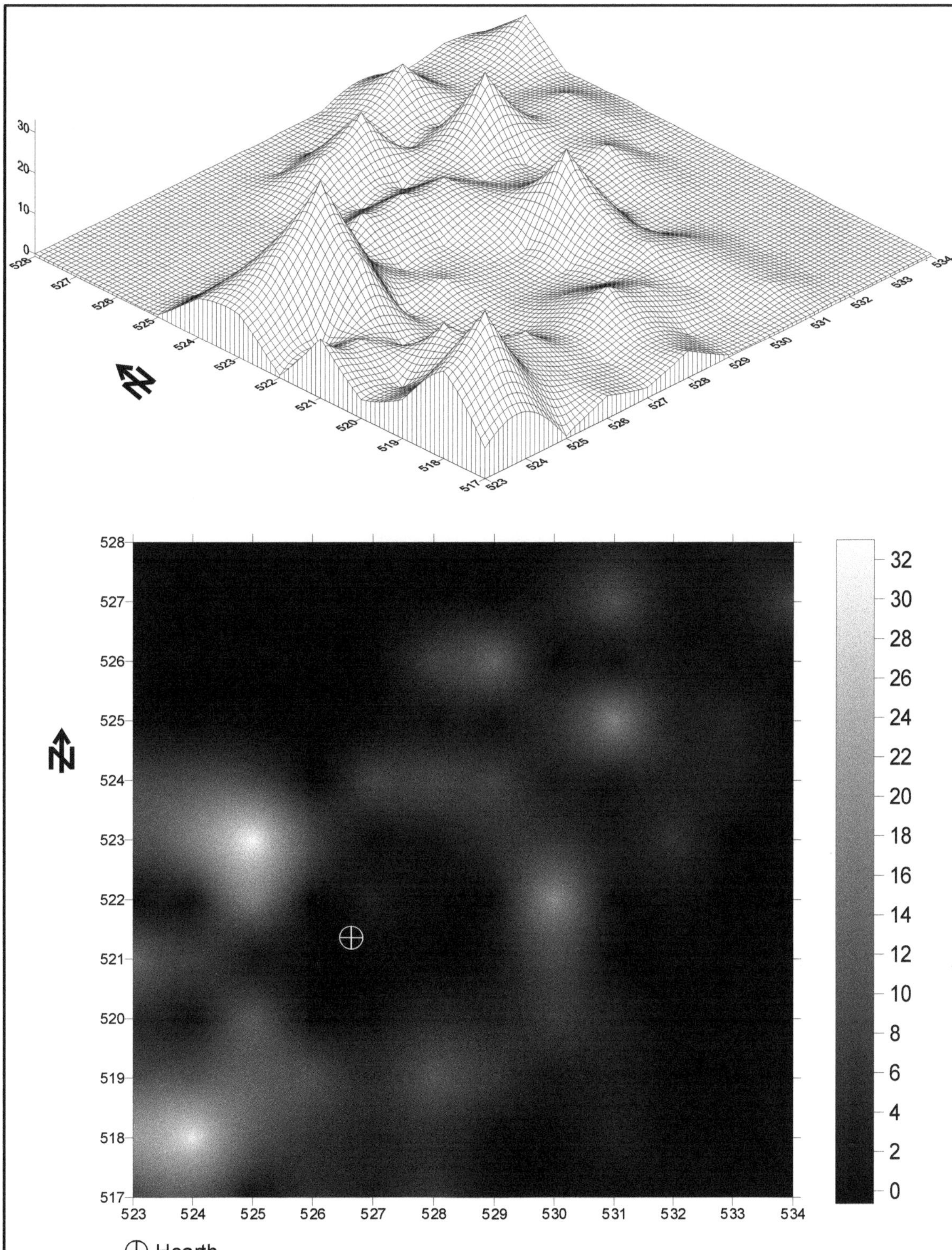

Figure 65: Peak and density maps of distribution of kinnikinnick seed fragments found within the paleosols at the Tuscany site

seeds recovered in the flotation analysis. The charred rose seeds that were recovered appear to be distributed across the site, from the northeast to the southwest end of the site grid. There is a distinct peak in the number of charred rose seeds found in excavation Unit 12 (N524, E529), where there was a peak for the number of charred juniper seeds found as well. However, the circular clear space is not visibly outlined by the quantification of these seeds.

Figure 69, shows the distribution of charred white spruce seeds. Admittedly, some of the identifications of the white spruce seeds are tentative. There were very few of these seeds recovered through the flotation analysis. White spruce seeds are not very large and appear to deteriorate easily not even surviving very well in comparative collections. After viewing modern samples of these seeds from the comparative collection at the

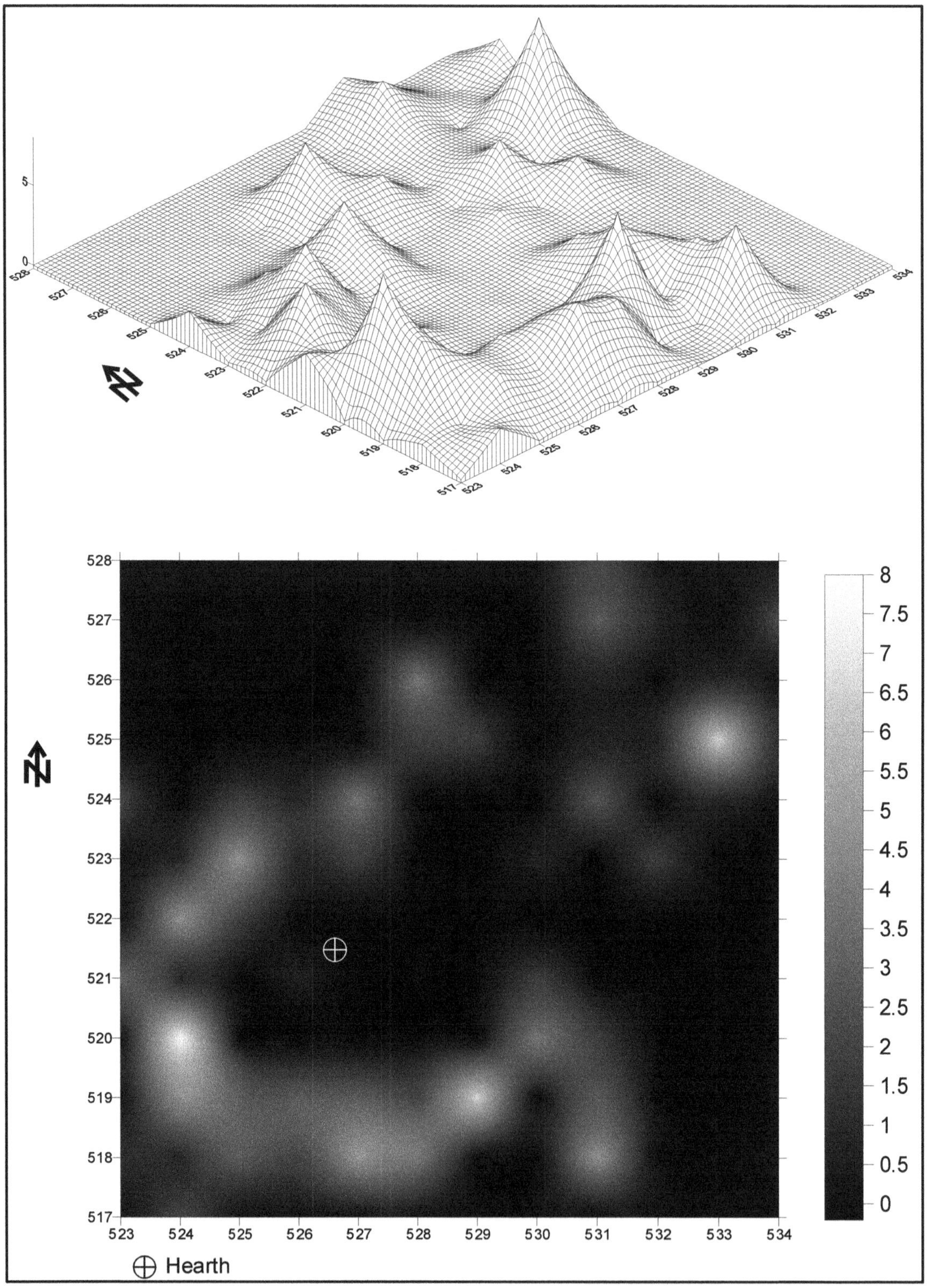

Figure 66: Peak and density maps of the distribution of Chenopodium seeds found within the paleosols at the Tuscany site

Provincial Museum of Alberta, it is easy to understand why they might be so degraded in the charred context that the Tuscany site specimens have been recovered from. Despite their rarity and degradation, the circular absence area in the hearth vicinity is still relatively visible.

In conjunction with the identification of degraded charred white spruce seeds, there were quite a few charred cone scales recovered. The distribution of the charred cone scales is presented in Figure 70. In this case, there is a departure from the patterns observed for the seeds discussed thus far. The charred white spruce cone scales are located in the northeast quarter of the site. Oddly, there is also a concentration of white spruce seeds in excavation Unit 12 (N524, E529), as has been observed for the charred rose and juniper seeds. The charred

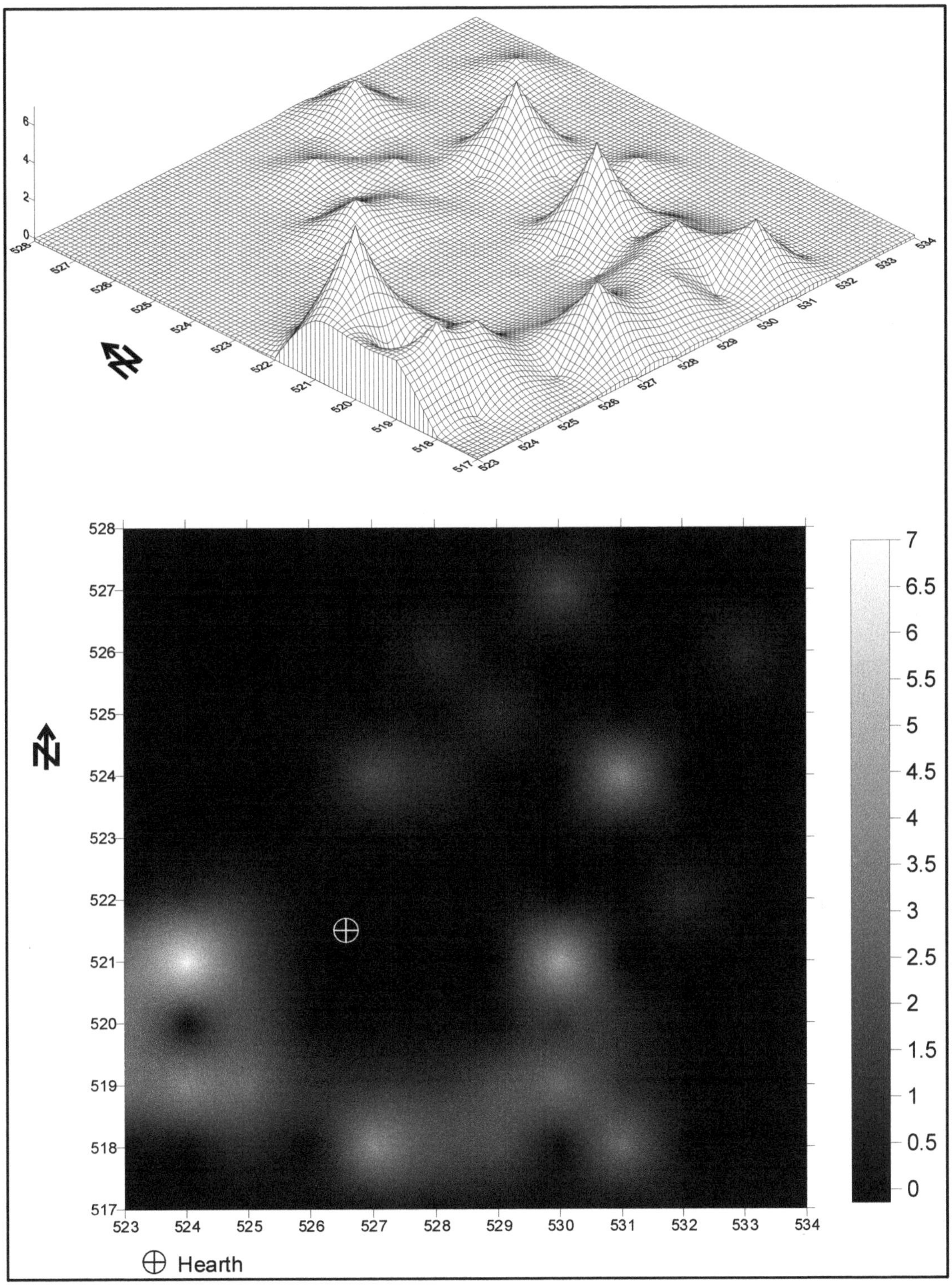

Figure 67: Peak and density maps of the distribution of shrubby cinquefoil seeds found within the paleosols at the Tuscany site

materials associated with white spruce are in the part of the site that was downslope from where the habitation area was located and where the paleosols were the deepest. As a downslope area the bottom of the depression/channel would have been an area of greater moisture as well, a prime location for larger shrubs or trees.

The small, concentrated distributions of seeds are interesting. They may be related to animal foraging activity and excretion. Conversely, they may be related to natural propagation of these plants in the landscape. White spruce and rose bushes do grow together in Alberta, especially in the Boreal forest region. Juniper has been observed growing near spruce and poplar at the edge of Twelve Mile Coulee, just north of the site.

At this stage of the analysis these maps appear to be showing natural distributions of seed deposition in the landscape due to natural, non-cultural patterns of growth,

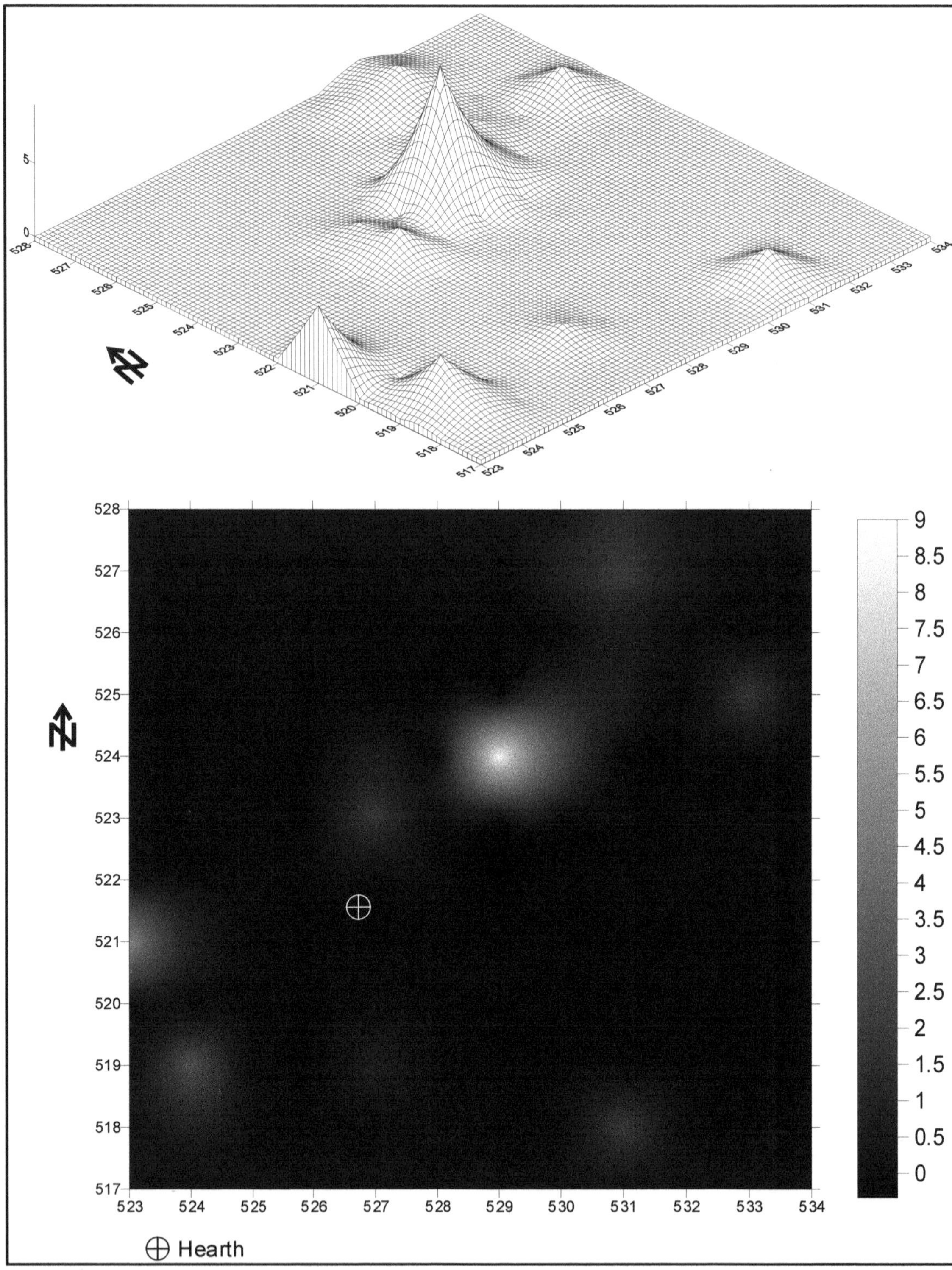

Figure 68: Peak and density maps of the distribution of rose seeds found within the paleosols at the Tuscany site

with the exception of the low frequency area around the hearth. Fragmentary and complete seeds seem to have comparable distributions. The pattern of deposition of the charred kinnikinnick and juniper seeds and fragments and charred plant material may simply be due to preferential growth in the upslope position of the site. When the site experienced a fire in the locality, there were more materials deposited in that location because there was simply more plant material to burn, including seeds and twigs. The data need to be viewed from another perspective. The next part of the spatial analysis will examine the depth range distribution.

The circular area of 3 x 3 meters in the near-center of the grid has been suggested to be a circular trample zone where there may even have been a small structure that could have housed a small family group of 6-8 people. There is a distinct, general pattern of deposition of

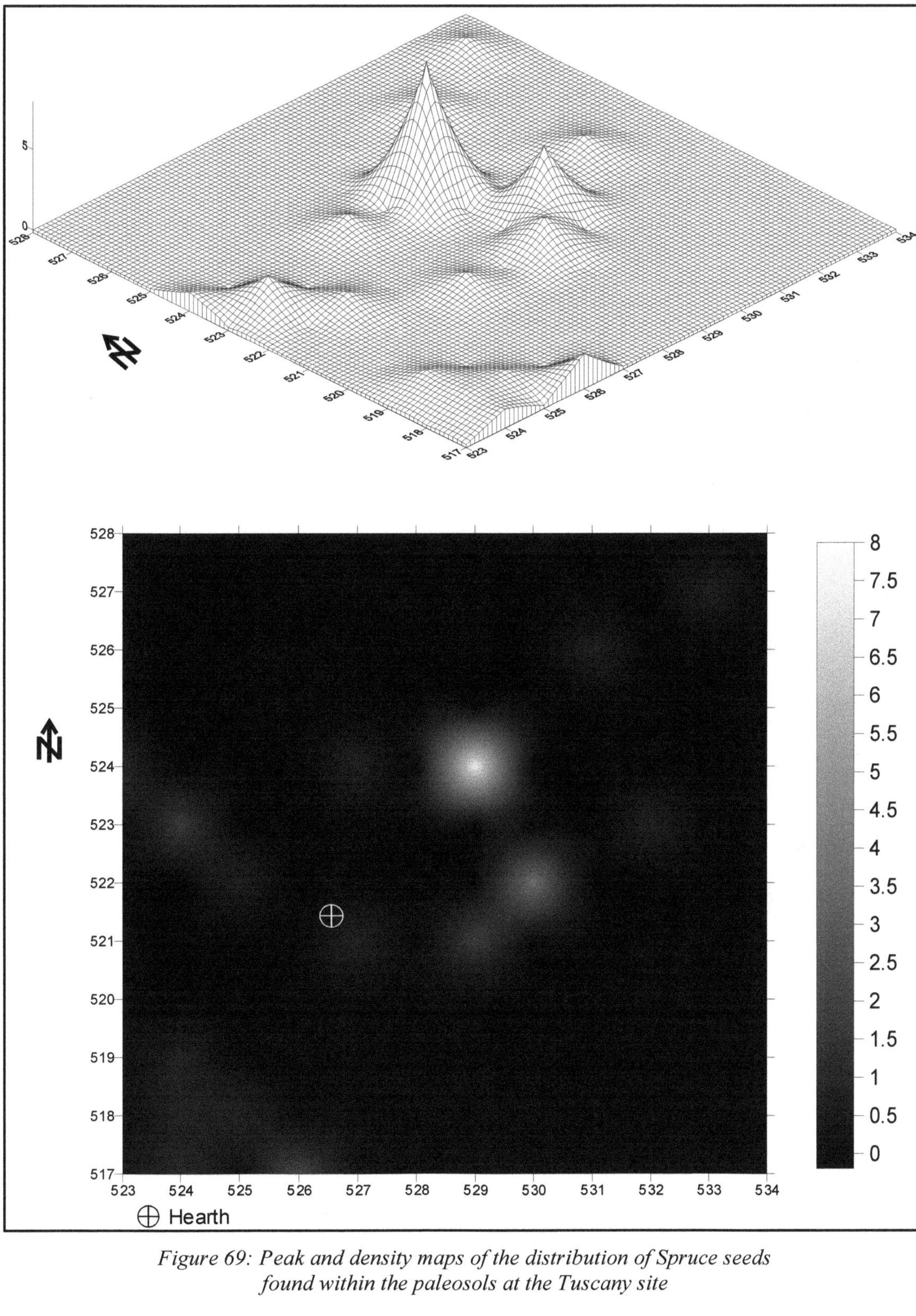

Figure 69: Peak and density maps of the distribution of Spruce seeds found within the paleosols at the Tuscany site

charred kinnikinnick seeds and seed fragments, and charred juniper seed fragments in the southwest corner of the site. This location is upslope from the empty circular space and the hearth feature. People clearing out the space where the hypothesized, small structure was erected may explain this pattern. Alternatively, it can also be suggested that higher frequencies of both plants growing in the southwest part of the site contributed to higher frequencies of their seeds becoming incorporated into the sediments over time. A pattern of clearing or sweeping out the small structure may help explain the lower frequencies of charred plant material within the circular area of 3 x 3 meters around the hearth feature. However, the continued presence of fragmentary and whole seeds in this circular area suggests a pattern that may not be entirely due to trampling. On the one hand there is a general absence of other charred macrobotanical remains in this circular zone along with the presence

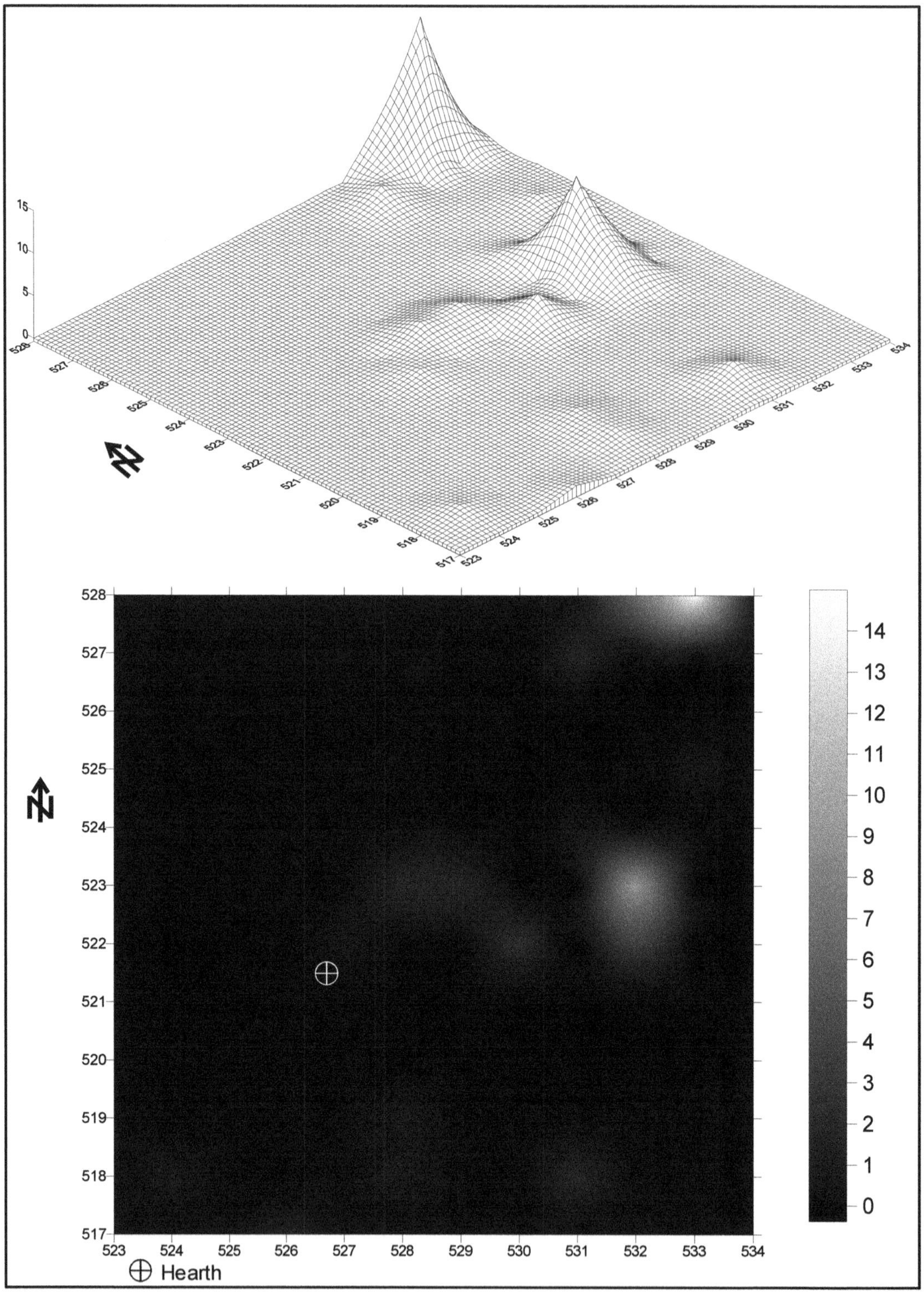

Figure 70: Peak and density maps of the distribution of spruce cone scales found within the paleosols at the Tuscany site

(low frequency) of whole seeds and fragments. This suggests a pattern that might not be entirely attributable to the trampling of seeds to the point where they are no longer recognizable. One explanation for the continued appearance of seeds and fragments may be their inherent robusticity compared to the small woody plant fragments. They are there because some of them are robust enough to survive trampling effects and thus we find them in lower frequencies. As with many of the suggestions for some of the patterns being observed with the data, it is difficult to be conclusive about them. One of the tests that can examine this pattern more closely requires a more rigorous examination of the leftover sample matrixes, looking for evidence of crushed charred plant remains within them for the excavation units in the circular area. These could then be compared to the matrixes of the units surrounding the circular area to see if there are fewer crushed charred plant remains contained in them. If there are differences, the proposal of a trampling area may become more robust. This circular patch clearly indicates a more complex human behaviour pattern.

DISTRIBUTIONS OF SEEDS AND LITHICS THROUGH PROFILES

The distributions of seeds from the site can be viewed in the same manner devised for the examination of the charred plant remains. A series of profiles have been created for several northings in the site. These profiles are all based on the corrected or standarized tallies calculated for all the charred plant material from the Tuscany site. Charred kinnikinnick and juniper seeds at the site provide the most intriguing data because they are both ubiquitous and well represented. The profiles for charred kinnikinnick seeds and seed fragment distributions are presented first with Figures 71 through 79. The profiles for charred juniper seeds and seed fragment distributions are presented next with Figures 80 through 88.

The pattern observed for the distribution of charred kinnikinnick and juniper seeds is not as well defined as it is for the charred plant material distributed throughout the site. The boundary between the charred plant deposition and the point where it ceased is fairly visible in most of the profiles. The charred seeds, however, continue to appear in levels above that fairly defined boundary. In fact, there are small amounts of charred material in the levels above the well-defined concentration layer, right up to the Mazama tephra boundary. For various reasons, perhaps because of moisture content or properties of the soil itself, kinnikinnick and juniper are not found in the depression or even close by today.

The circle of 3 x 3 meters around the hearth feature shows up well in profile lines N520, N521, and N522 for the charred kinnikinnick and juniper seeds. There were no charred kinnikinnick seeds at all found in excavation Units 37, 38, and 39, in profile N521 (Figure 73). However, there are a few charred juniper seeds and other seeds in the area around the hearth in the upper burn layer of the third pedogenic zone (H/I and J). It must be remembered that the 3 x 3 meter hearth area was not completely devoid of charred plant remains. There were charred plant remains but these were fewer in number. This is visible in the charred plant density profiles, N520 and N521 and is less clear in N522 (Figures 50, 51 and 52). The densities of the charred plant fragments around the hearth area are reduced, but they are not obliterated. Thus, the stratigraphy of the upper charred plant concentration layer was maintained in the profiles. Though there is a distinct drop in the numbers of charred plant fragments and seeds, the remains that do occur are still more numerous in the concentrated charred plant layer (Layer J) that is associated with the hearth and the projectile point in excavation Unit 37.

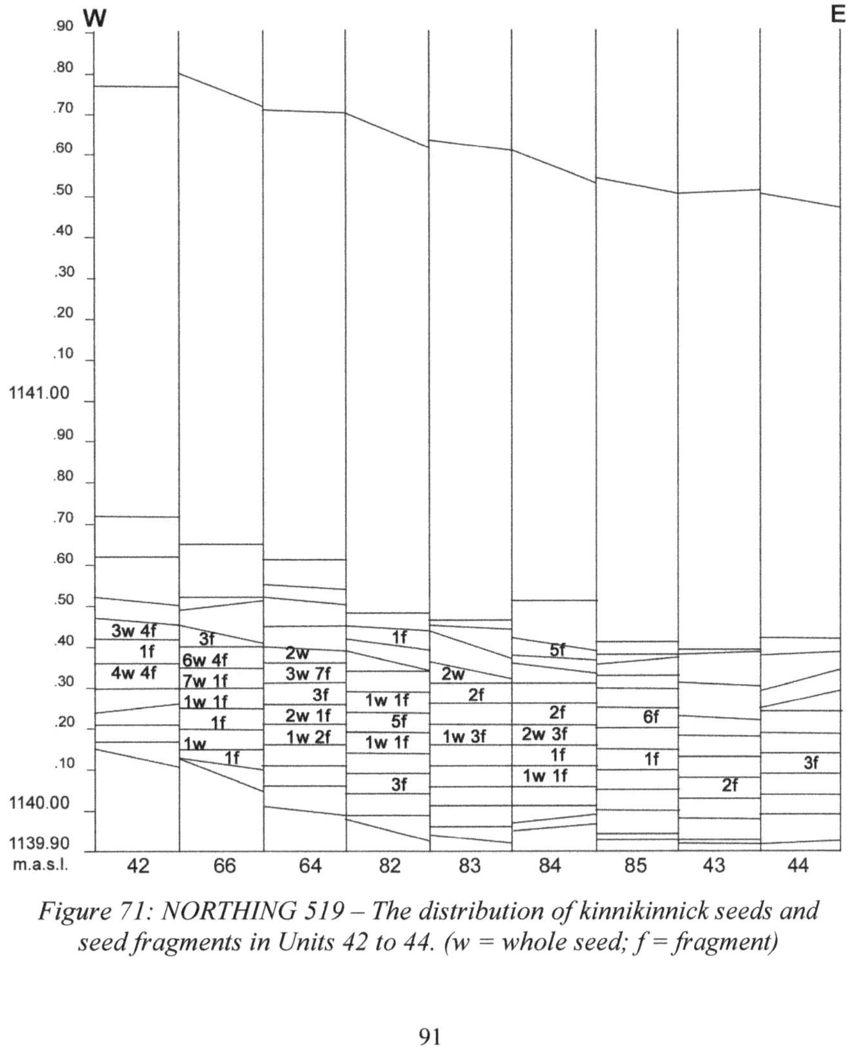

Figure 71: NORTHING 519 – The distribution of kinnikinnick seeds and seed fragments in Units 42 to 44. (w = whole seed; f = fragment)

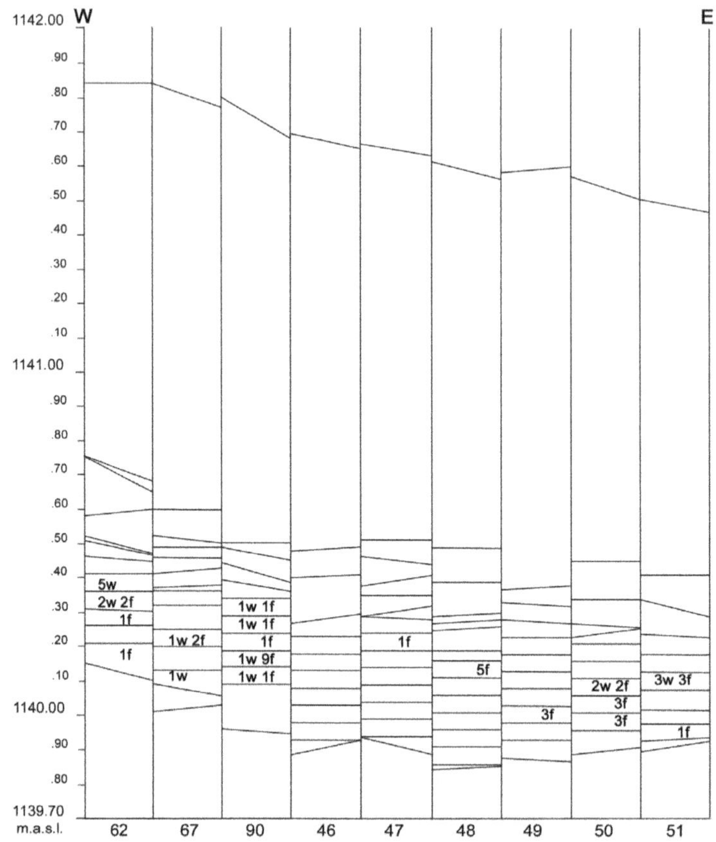

Figure 72: NORTHING 520 – The distribution of kinnikinnick seeds and seed fragments in Units 62 to 51. (w = whole seed; f = fragment)

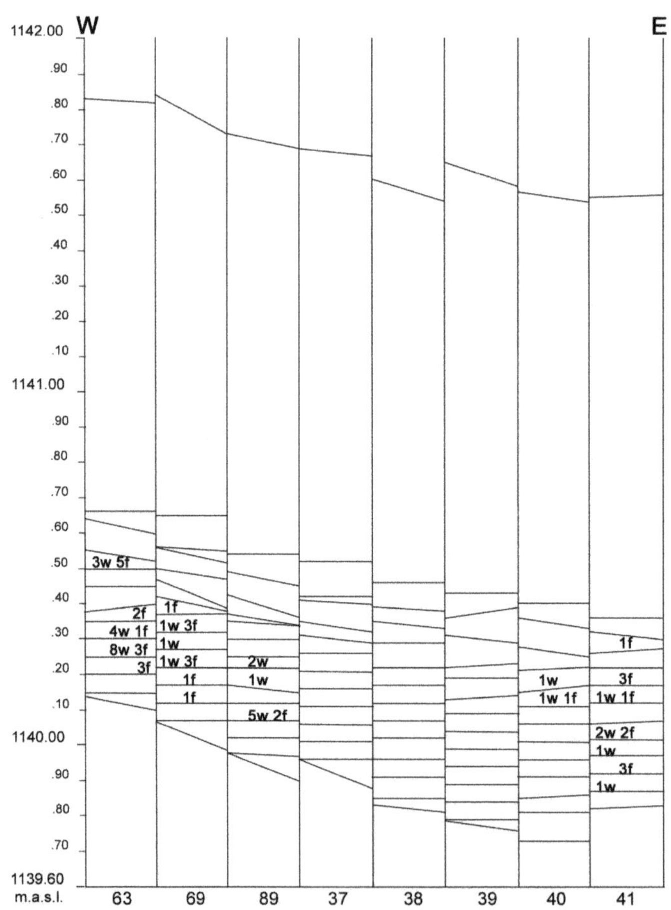

Figure 73: NORTHING 521 – The distribution of kinnikinnick seeds and seed fragments in Units 63 to 41. (w = whole seed; f = fragment)

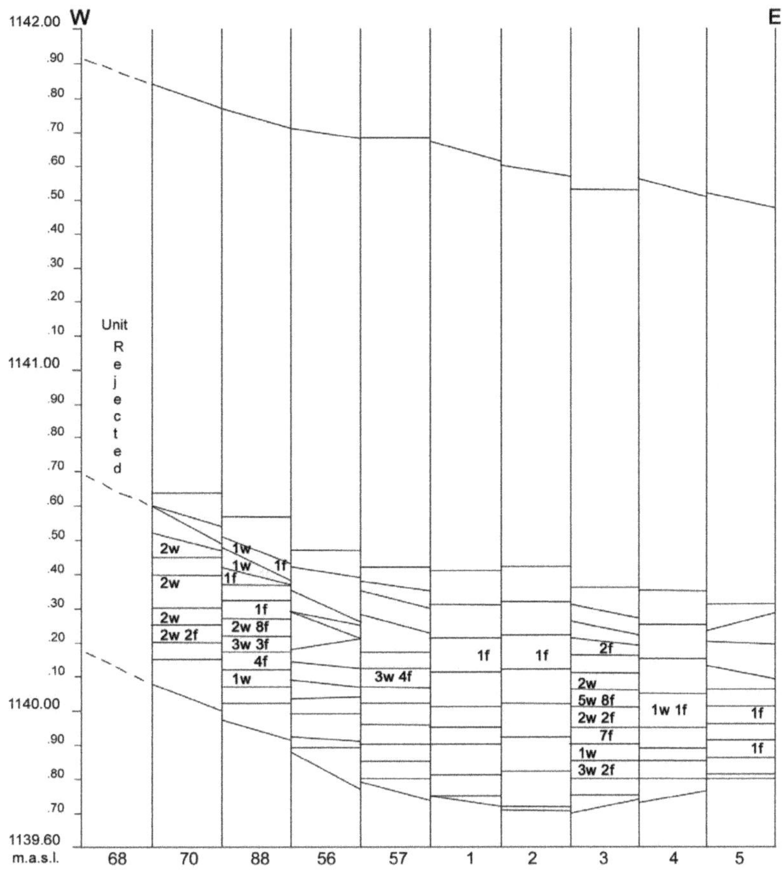

Figure 74: NORTHING 522 – The distribution of kinnikinnick seeds and seed fragments in Units 68 to 5. (w = whole seed; f = fragment)

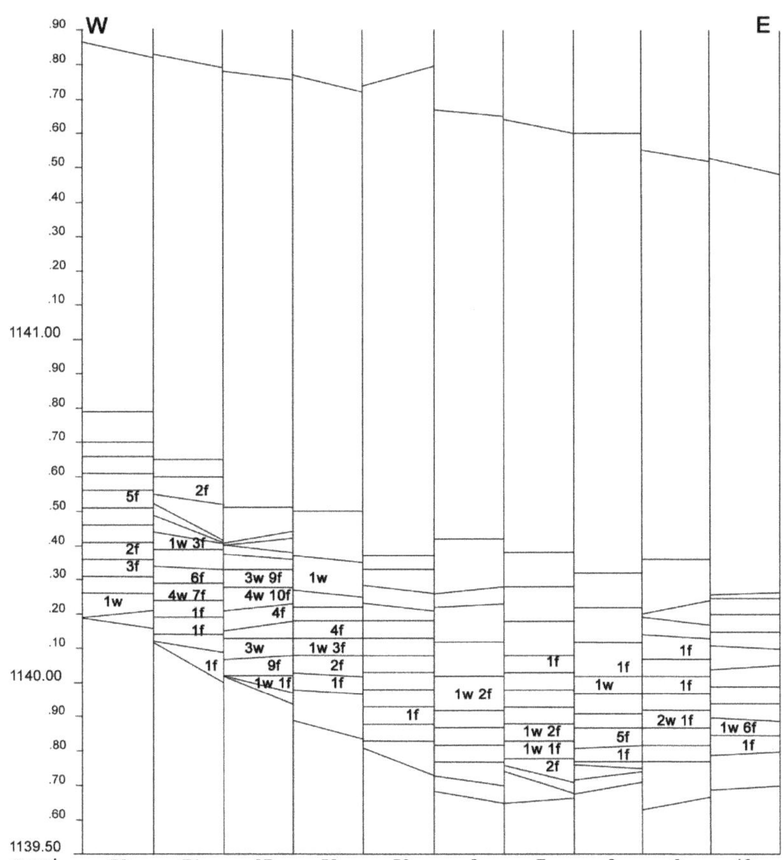

Figure 75: NORTHING 523 – The distribution of kinnikinnick seeds and seed fragments in Units 72 to 10. (w = whole seed; f = fragment)

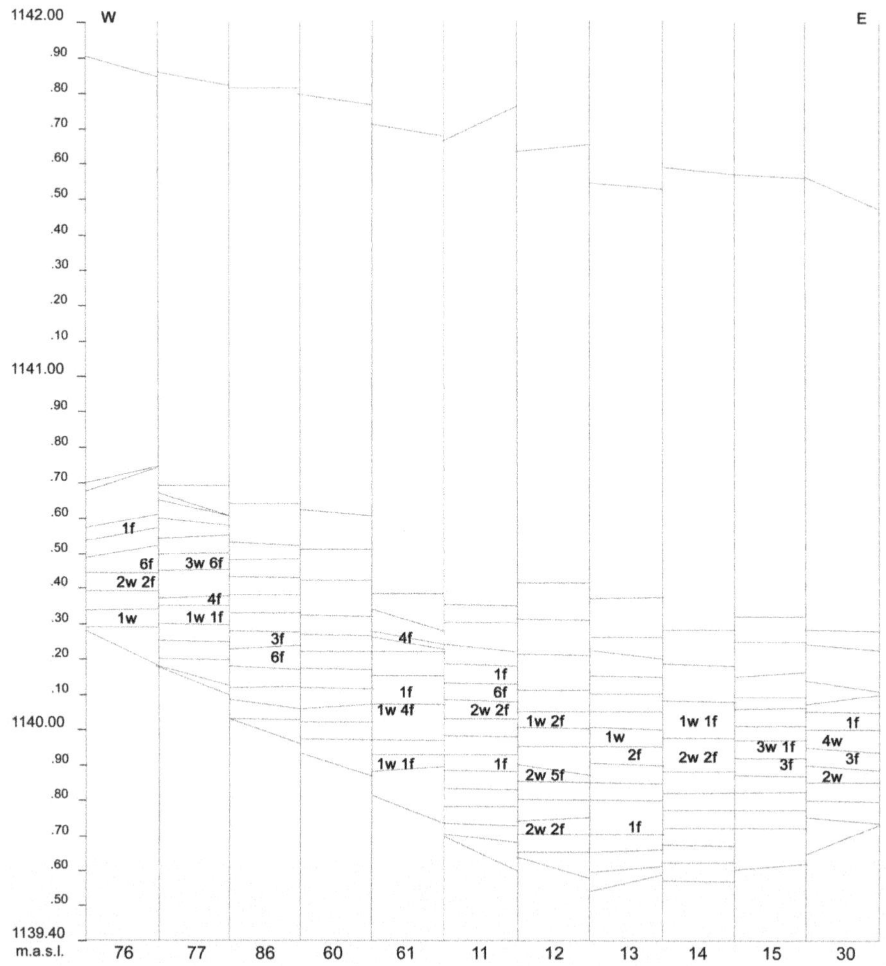

Figure 76: NORTHING 524 – The distribution of kinnikinnick seeds and seed fragments in Units 76 to 30. (w = whole seed; f = fragment)

These profiles confirm the patterns observed in the planview maps generated for this spatial analysis. In the northing profiles kinnikinnick and juniper appear to have left more seed remains on the outer, sloping edges of the depression, which would probably have been drier compared to the depression itself. This pattern is understandable given the habitat preferences described for these plants in Chapter 4.

Other hypotheses can be considered to try and explain the patterns visible in the charred plant maps and profiles. One idea considered to explain the lack of botanical material around the hearth was the possibility that it represented a tree throw. A large tree with shallow, spreading roots would disturb an area approximately equal to the 3 x 3 meter area observed in the planview maps. The obvious problem with this scenario is that the stratigraphy is intact within the charred plant concentration of Layer J. The hearth feature is also apparently intact within the charred plant concentration, along with the projectile point in close proximity. As well, once the lithic debitage data (from the excavation analysis, not from the flotation samples) was plotted into comparable depth profiles (see below), things became somewhat more complex. The conclusion from these lines of evidence is that a tree throw does not provide a reasonable alternative hypothesis to explain the area with reduced charred plant tallies.

The lithic debitage recovered from the excavation grid was included in this spatial analysis to examine whether or not any relationship existed between the pattern of its deposition within the site and that of the charred plant remains. Most of the debitage is concentrated on the east side of the site. The activity area begins in Northing 520 with excavation Units 47, 48 49 and 50. In Northing 521, excavation Units 38 to 41 contain the lithic concentration, which thins out in Northing 522 (excavation Units 1, 2, and 3), and in Northing 523 (excavation Units 6, 7, and 8) (see Figures 89 to 97).

There appears to be some translocation of flakes up and down through the profiles, which is not unexpected. There are a few flakes that are within the boundaries of the proposed living structure. The hearth and flintknapping activity areas do appear to be related. Figure 7 shows the dimensions of the excavation grid with the proposed activity areas outlined. If there was a small habitation structure present, then perhaps there were skins placed on the ground inside which prevented the flintknapping debitage from spreading into it. Or, perhaps the flintknapping activity was simply restricted to

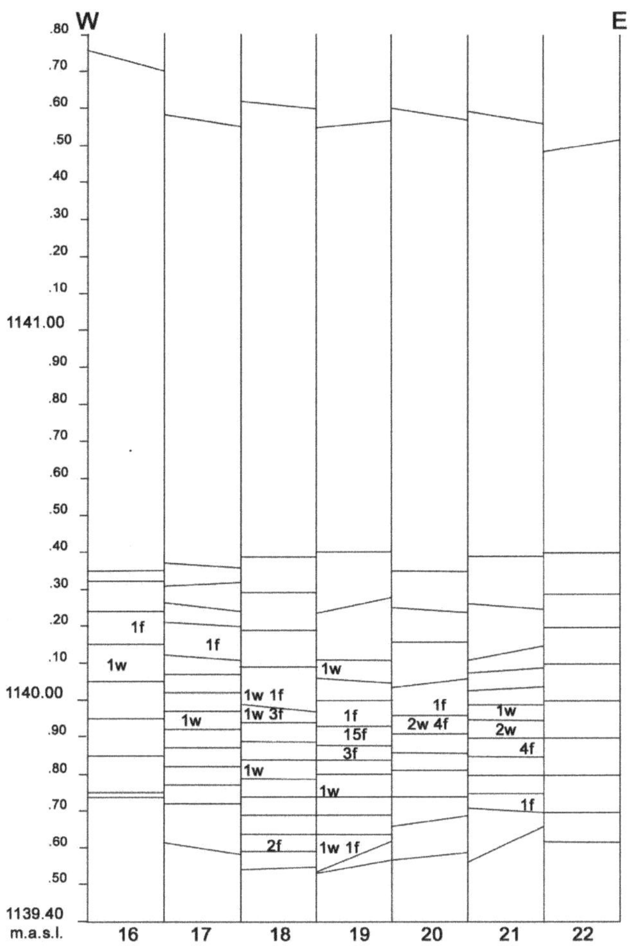

Figure 77: NORTHING 525 – The distribution of kinnikinnick seeds and seed fragments in Units 16 to 22. (w = whole seed; f = fragment)

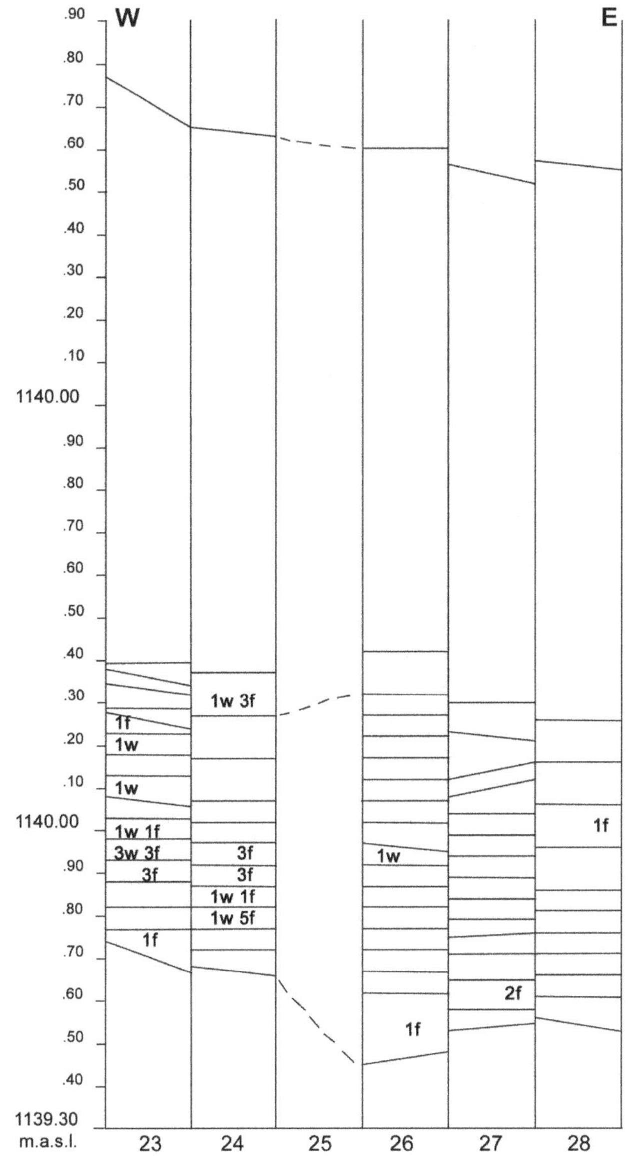

Figure 78: NORTHING 526 – The distribution of kinnikinnick seeds and seed fragments in Units 23 to 28. (w = whole seed; f = fragment)

the small dimensions that it appears to be with a few stray flakes being distributed a bit farther away.

The pattern of flake distribution at the Tuscany habitation site is different from that recorded for the Vermilion Lakes site. At Vermilion Lakes, there were significant concentrations of debitage around the posthole features, which actually delineated the perimeters of the structure at that site. Some of the debitage appeared to have fallen into the postholes (Fedje *et al.* 1995). The quantities of debitage recovered in the Vermilion Lakes site (thousands of pieces) does not compare at all with the Tuscany site (a few hundred pieces), hence this may explain the less distinct pattern at Tuscany.

The Hell Gap site in Wyoming contained evidence of three structures as previously outlined. Irwin-Williams *et al.* (1973:47) note that there were no internal features found within the delineated structures, however, there was some correlation found with the distribution of occupational debris. Unfortunately, Irwin-Williams *et al.* (1973) do not demonstrate this correlation any farther. This pattern appears to correspond with observations at the Tuscany habitation site. However, the distribution pattern of debris inside or outside the structures at the Hell Gap site is not shown. The most obvious difference between the two sites is the hearth feature found in the center of the proposed Tuscany structure. These differences may indicate season of occupation. Perhaps a fire was not used inside the Hell Gap structure because it was summer and heating was not necessary. Perhaps the Tuscany structure was occupied in late summer or fall when a small fire would provide warmth at night. These are speculative hypotheses, although such patterns have been hypothesized at other sites. For example, Van Dyke, *et al.* (1982), in their analysis of the Bow Bottom site, found the remains of fire pits inside tipi ring structures which they propose may be related to the seasonal occupation of the site during the late fall or winter when a fire would have been required to heat the inside of the structure. Frison (1983) also suggests stone circles with stone-filled firepits in the center may have been occupied during cooler weather, implying seasonality. The suggestion for this phenomenon at the Tuscany site may be difficult to test now. What is important here is the

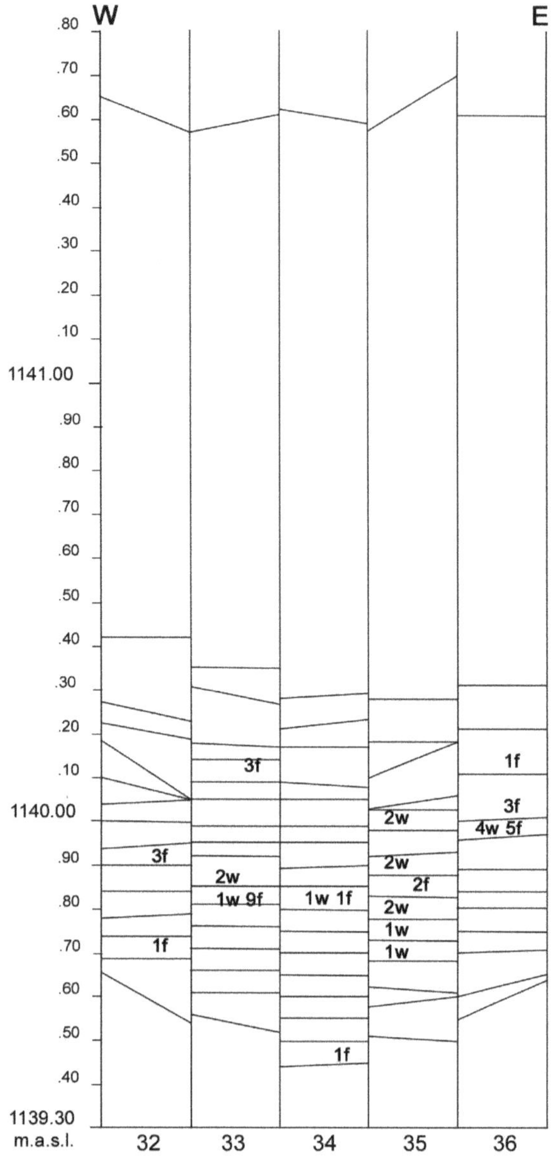

Figure 79: NORTHING 527 – The distribution of kinnikinnick seeds and seed fragments in Units 32 to 36. (w = whole seed; f = fragment)

evidence of similar structures being used in these sites with their associated hearths and proposed seasonality of occupation.

The final analysis of the Tuscany habitation site has not been completed and there is no summary or site report available yet. Thus, the faunal remains from the Tuscany site have not been included in the profile analysis. However, previous work on the faunal remains has shown a distinct pattern of deposition that is different from both the lithic and macrobotanical remains found in the paleosols. Nearly all the large bone remains were found in the northeast section of the grid, the area that would have been slightly downslope from the projected enclosure (see Figure 7). The smaller fragments were in the same area as the debitage and closer to the hearth and proposed trampling area around it (Siegfried and Oetelaar 2000). This disposal pattern is reminiscent of the "drop and toss" zones described by Binford (1978). One other important feature of the faunal remains recovered in association with the Lusk cultural component is the fact that none of them appear to be burned or charred. Two scenarios can be derived for this observation. In the first hypothesis, an intense fire swept through the site, the vegetation recovered, and the occupation during which the faunal remains were discarded took place during the recovery phase. In this case the fire could have been started by a lightning strike or it could have been started deliberately, to burn the berry bushes for better productivity. Such strategies are well known and described by Turner (1991) for Aboriginal people in interior British Columbia. It can take 2-5 years for the site to recover and produce better berry crops, at which time the people who burned the site would return to reap the harvest. The faunal remains discarded during the later occupation were slowly buried and shielded from further charring effects from subsequent fires. In a second scenario, the site was occupied, the bones were scattered, and then leaf litter covered the evidence of occupation. Late spring and early autumn are noted to be the times when natural fires usually occurred in the past (Lewis 1982). A fire burning at the site during one of these seasons may have left the wet leaf litter and the bones underneath them relatively unscathed as these are times when more moisture is present on the ground. Such a fire could leave behind charred plant remains that would deposit along with the bone and eventually all materials would become incorporated into the soils that formed subsequently.

These are relatively simple hypotheses that would require experimental archaeology to begin to resolve some of the questions raised: how deep would the leaf litter cover have to be to protect the bones, how wet/damp does the litter have to be in order not to burn, etc.? In fact, there are studies that have been carried out related to fire ecology and experimental archaeology that reveal this to be a very complex problem that could be a topic of further research.

Complexity begins with understanding the composition of forest floors. The remains of vegetation material that covers ground under forest canopies is known as "duff" and is made up of the plant litter from leaves and twigs (Miyanishi 2001). These remains are composed of cellulose and hemicellulose that decompose fairly readily, lignin which is a fairly robust organic compound that decomposes at a slower rate, a series of compounds like cutin and waxes, and finally, a series of secondary compounds (Miyanishi 2001:439-40). These compounds are present in variable amounts depending upon plant species and duff is thus a complex mixture of materials that degrade at different rates (Miyanishi 2001). Miyanishi (2001) notes that when duff burns, it burns at different rates in different locations and can leave a patchy mosaic of burned out holes and unburned areas that form ridges on the forest floor. This may actually help to explain the patchy distribution pattern that was observed in the lowest concentration of charred botanical remains in the Tuscany depression. There may have been areas where very few remains survived the burning of the

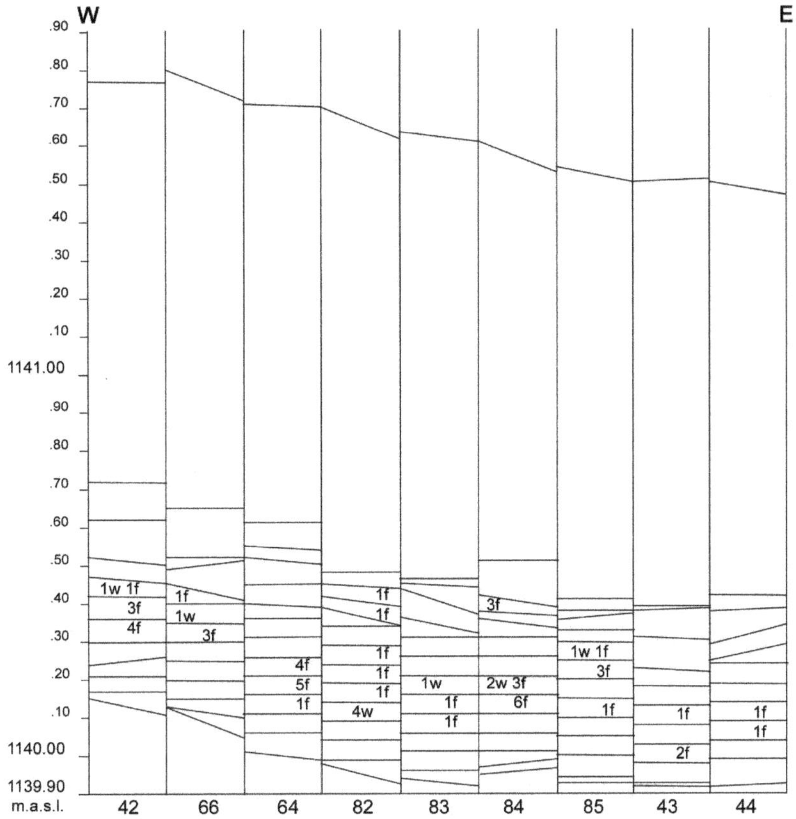

Figure 80: NORTHING 519 – The distribution of juniper seeds and seed fragments in Units 42 to 44. (w = whole seed; f = fragment)

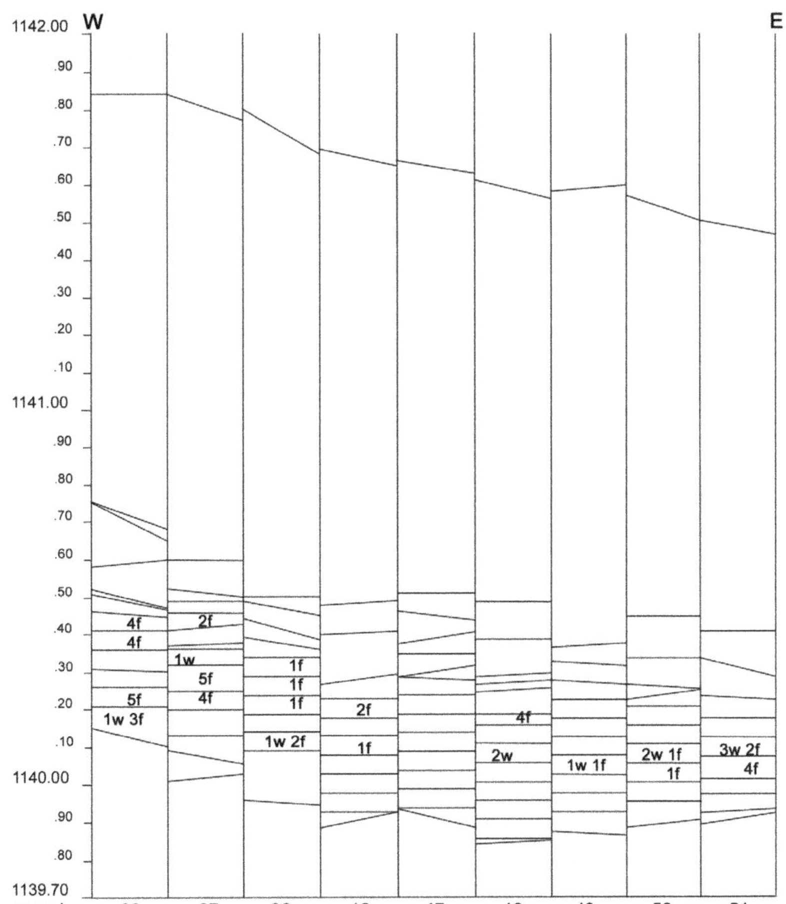

Figure 81: NORTHING 520 – The distribution of juniper seeds and seed fragments in Units 62 to 51. (w = whole seed; f = fragment)

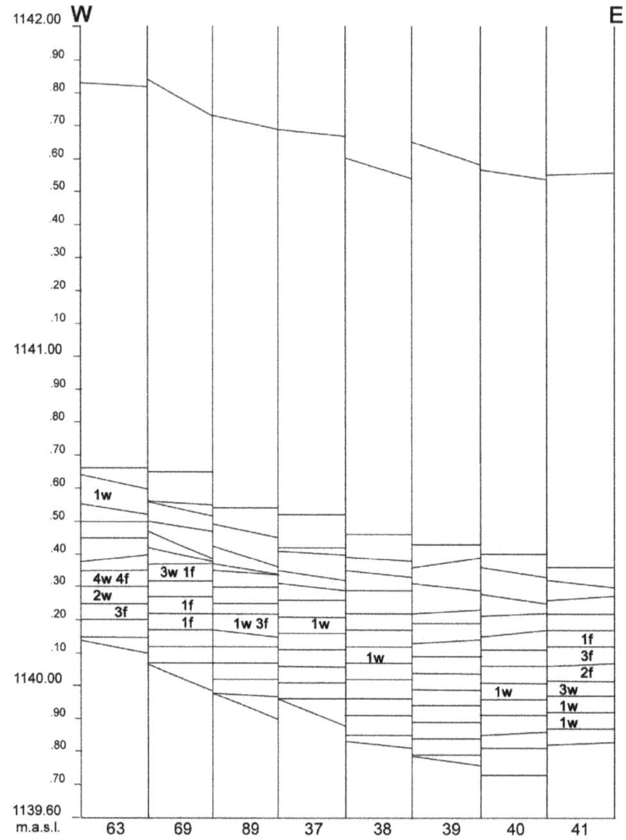

Figure 82: NORTHING 521 – The distribution of juniper seeds and seed fragments in Units 63 to 41. (w = whole seed; f = fragment)

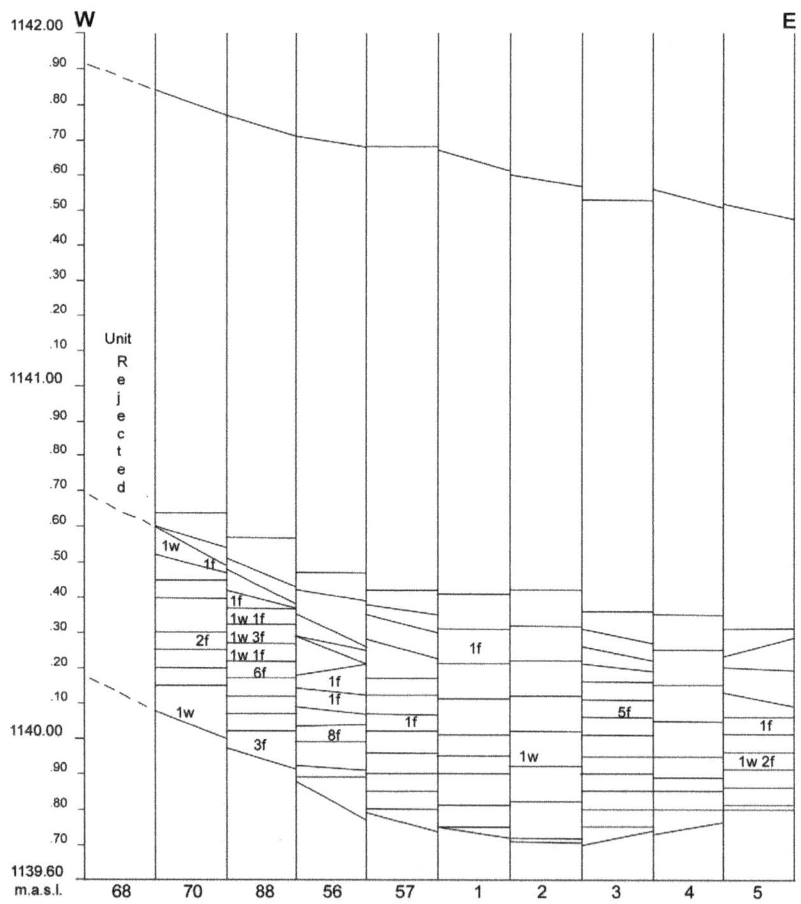

Figure 83: NORTHING 522 – The distribution of juniper seeds and seed fragments in Units 68 to 5. (w = whole seed; f = fragment)

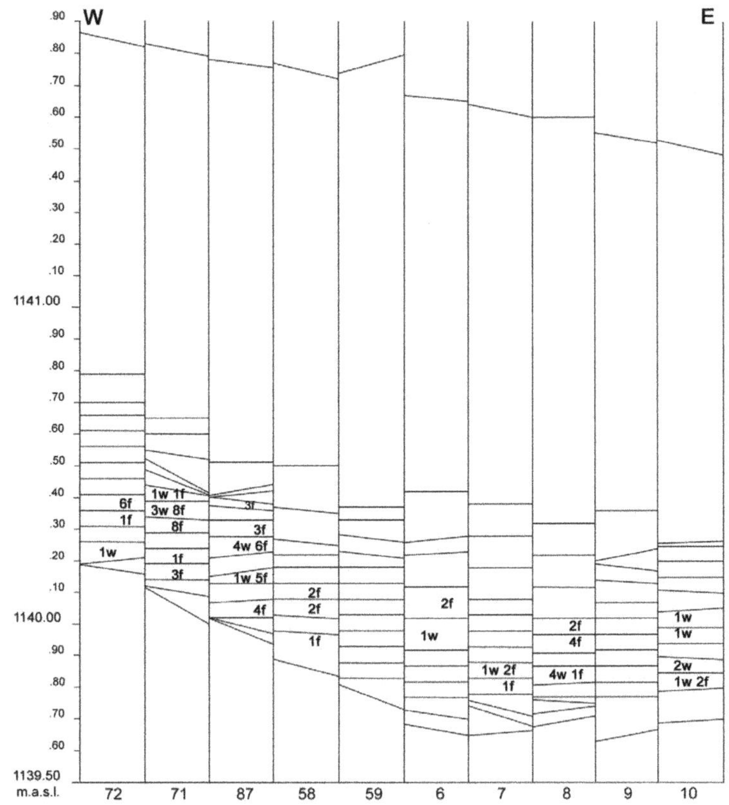

Figure 84: NORTHING 523 – The distribution of juniper seeds and seed fragments in Units 72 to 10. (w = whole seed; f = fragment)

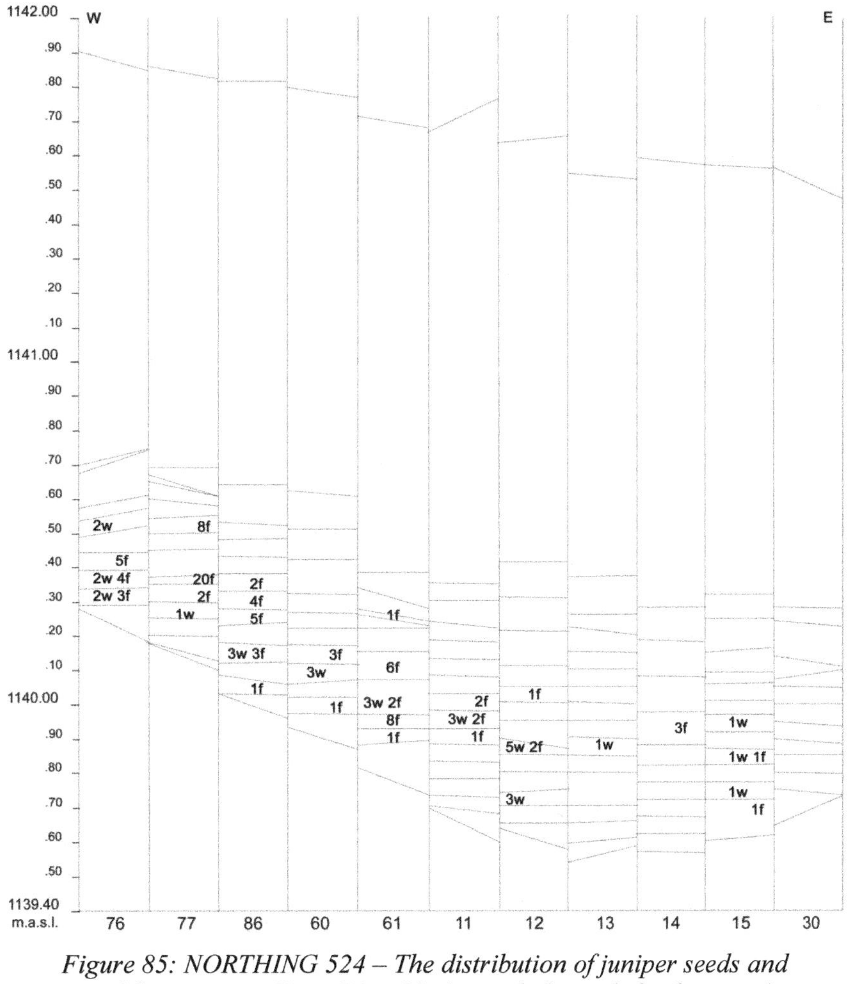

Figure 85: NORTHING 524 – The distribution of juniper seeds and seed fragments in Units 76 to 30. (w = whole seed; f = fragment)

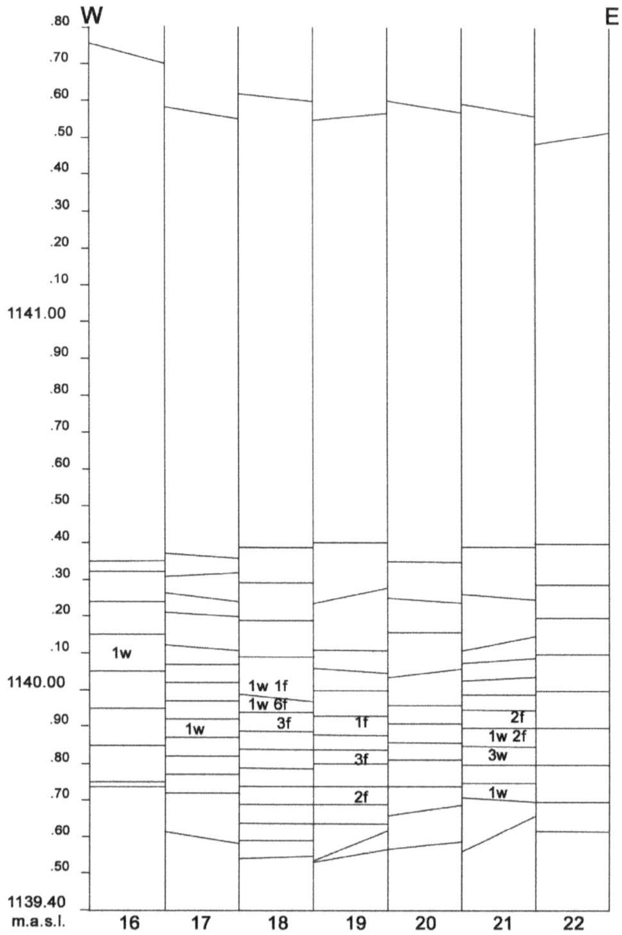

Figure 86: NORTHING 525 – The distribution of juniper seeds and seed fragments in Units 16 to 22. (w = whole seed; f = fragment)

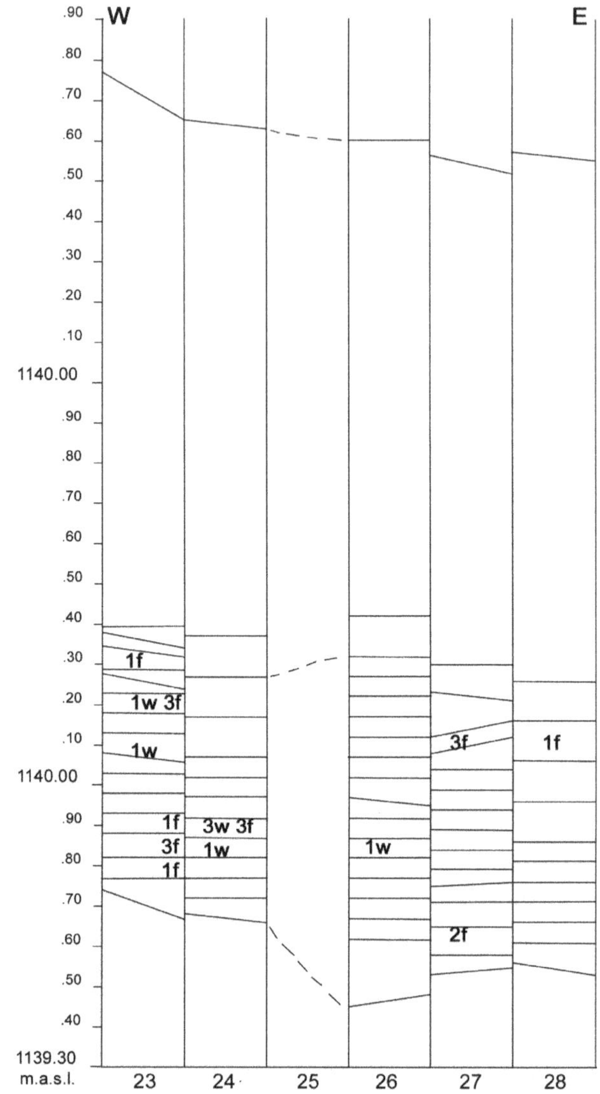

Figure 87: NORTHING 526 – The distribution of juniper seeds and seed fragments in Units 23 to 28. (w = whole seed; f = fragment)

duff cover, and areas where moisture and resilience helped the plant materials burn to the point of charring and being preserved through time.

Some experimental work on the effects of fire on materials that are recovered in archaeological contexts has been carried out by Picha *et al.* (1991). Cow rib bone, charcoal, lithic materials, ceramic remains and wood were among some of the items placed into selected plots with varying types of fuel loads of natural mixed prairie grasses and buckbrush in a North Dakota location (Picha *et al.* 1991). The items were placed as close to the ground surface as possible to mimic a natural deposition event. Bone appeared to be severely affected by the fire, becoming scorched and charred, however, charcoal was completely unaffected (Picha *et al.* 1991). This information at least suggests that the bone at the Tuscany site was more likely deposited after, or well before, any fires had burned in the area.

The information on the complexities of burning duff and the experimental archaeology in North Dakota reveal that we really do not understand what is involved in the burning of a site like the Tuscany depression. The hypotheses presented would require extensive research and experimentation to achieve some measure of understanding the variables of burning in combined natural and archaeological contexts.

The occupation zone at the Tuscany site appears more ephemeral when compared to that of the Vermilion Lakes site. It may be unfair to evaluate a site based on quantities of lithic debitage. Perhaps tool making took place at a different location for the people at the Tuscany site. Some of their lithic work was done outside the perimeter of the structure, as evidenced by the distribution of flakes, but this was not extensive. The material worked was silicified siltstone. The point found near the hearth is chalcedony that looks like it may be Knife River flint, a well-known material widely traded from its source in the Dakotas in the United States. Perhaps, the people who camped at the Tuscany habitation site had all the tools they needed and relied on curated tools brought from elsewhere. They could thus spend their time at the site doing other tasks like picking berries. This is not meant to imply that the kinnikinnick and juniper berries present at the site were the focus of berry picking endeavors. The site may have

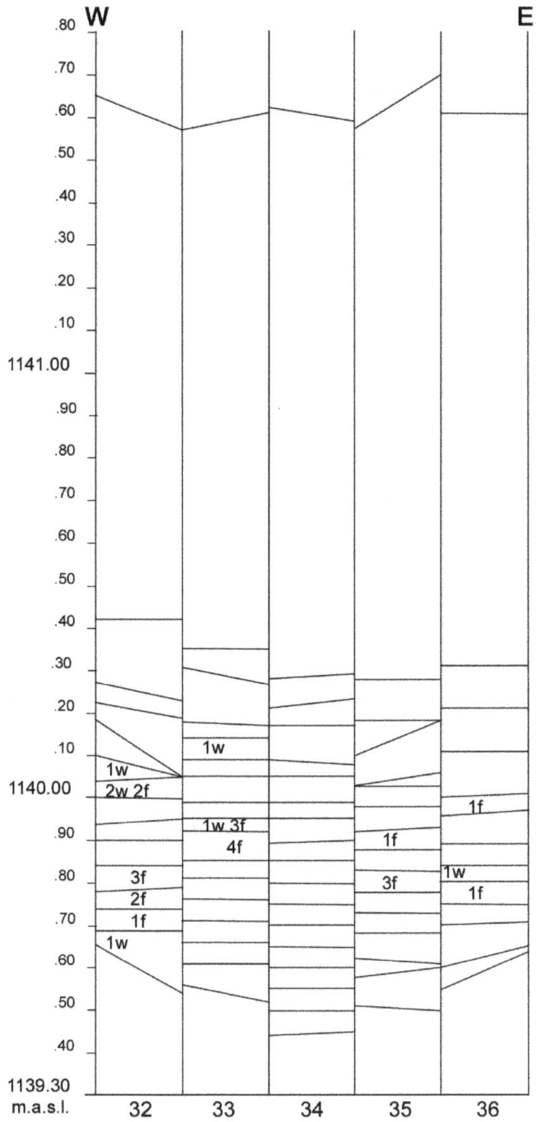

Figure 88: NORTHING 527 – The distribution of juniper seeds and seed fragments in Units 32 to 36. (w = whole seed; f = fragment)

served as a base camp from which the people went farther afield to gather other berries or plants. However, it must be stressed that there are no remains of other berries like chokecherries or saskatoons recovered from the flotation samples and this proposal is purely conjectural. Alternatively, the Tuscany site may have simply been a small base camp for a small nomadic hunting group and this is why there are few artifacts representing the occupation dating to about 7800 years BP.

Though the lithic scatter is not extensive at the Tuscany site, it does appear to be linked quite well to the projectile point in terms of the stratigraphy. The projectile point was found in Unit 37 level S (N521, E526) which was immediately overlying/within the upper concentration of charred plant remains (Layer J in the profiles) in the sediments. The flake concentration, 3 or 4 meters to the east of the hearth, was also found immediately overlying/within the upper concentration of charred plant remains (Layer J in the profiles). Part of the pattern observed between levels for the projectile point, the flakes and the concentrated burn layer may be a function of where the bulk soil sample was taken (generally the southwest corner) and the sloping surface of the site. The flakes were provenienced and thus reflect the true undulations of the ground surface compared to the projected flat surfaces of the levels from the arbitrary method of excavation. The profiles are based on the excavation levels, which yield a 'stepwise' view of the changing concentrations of materials recovered. As well, the original occupants of the site 7800 years ago may have trampled the flakes into the sediments, affecting the positions they were found in millennia later. A significant episode or series of episodes of burning of the vegetation in the locale appears to have taken place either just before the Lusk, Plains/Mountain complex occupation of the site or more or less simultaneously with it. If the burn event took place a year prior to the occupation, vegetation would have had some chance to recover and even attract animals to browse on the new growth, a phenomenon noted by Lewis (1982) to occur in burned patches in the boreal forest.

The combination of three sets of material remains at the Tuscany habitation site reveals a relatively short occupation by a small group of people. Three activity areas can be proposed from the distribution patterns of bone, lithic debitage and macrobotanical remains through the space of the site. For the following discussion refer to Figures 7 and 6 for the outlined activity areas and the location of units mentioned. First, the hearth is located in Unit 37, near the center of the east wall. The macrobotanical evidence delineates a proposed trample zone of 3 meters diameter around the hearth feature. The large bone pieces were mainly spread out to the northeast in excavation Units 58, 59, 6, 11, 16, 17, 19, 21, 27, 28, 35, and 36. Some larger scattered pieces of bone are also found in excavation Units 47, 41, 5, 8, and 10 to the east of the hearth. The lithic debitage scatter is found in excavation Units 48, 49, 50, 39, 40, 41, 2 and 3 to the east of the hearth. The lithic debitage scatter is not very large and appears to reflect a minor flintknapping activity area created from the reworking or manufacture of some small tools or points.

There are no tools that reflect possible plant processing. Implements like skin bags and containers from this time period would be manufactured from materials like hide and wood. Other tools that might reflect general plant processing strategies could be made from bone or antler, and stone. A mano and metate found in the Betty Greene site (Wyoming) appear to be associated with the terminal Paleoindian timeframe (Frison 1991), which the Tuscany site fits into. Frison (1991:339) further notes that the stationary grinding slabs of hard, fine-grained sandstone and associated manos began to appear in prototype form during the terminal PaleoIndian period and subsequently become part of tool assemblages until historic times. Chances of preservation for potential tool materials appears to be fair in the case of the Tuscany site as faunal remains have been recovered in association with the projectile point. Bone and antler often survive post

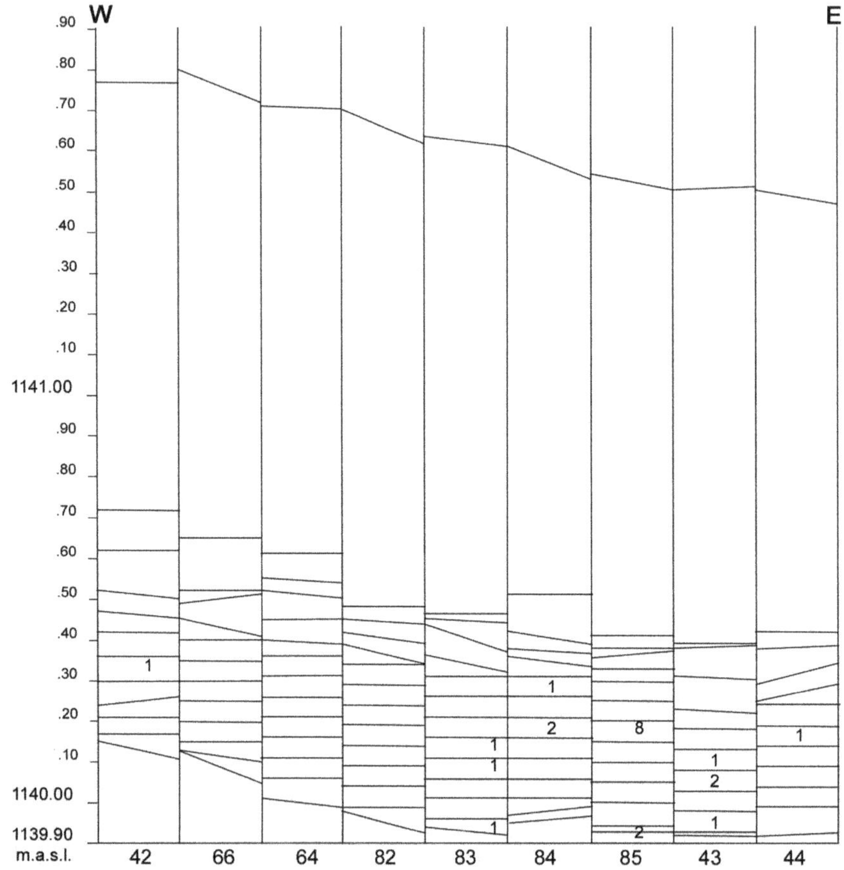

Figure 89: NORTHING 519 – The distribution of lithic remains in Units 42 to 44

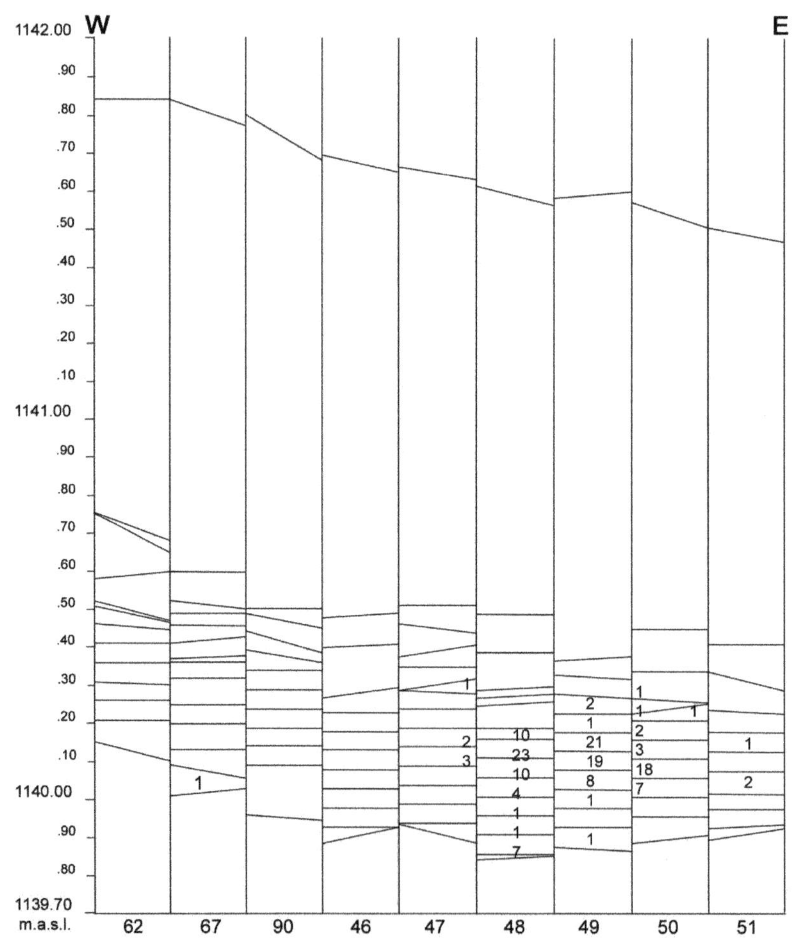

Figure 90: NORTHING 520 – The distribution of lithic remains in Units 62 to 51

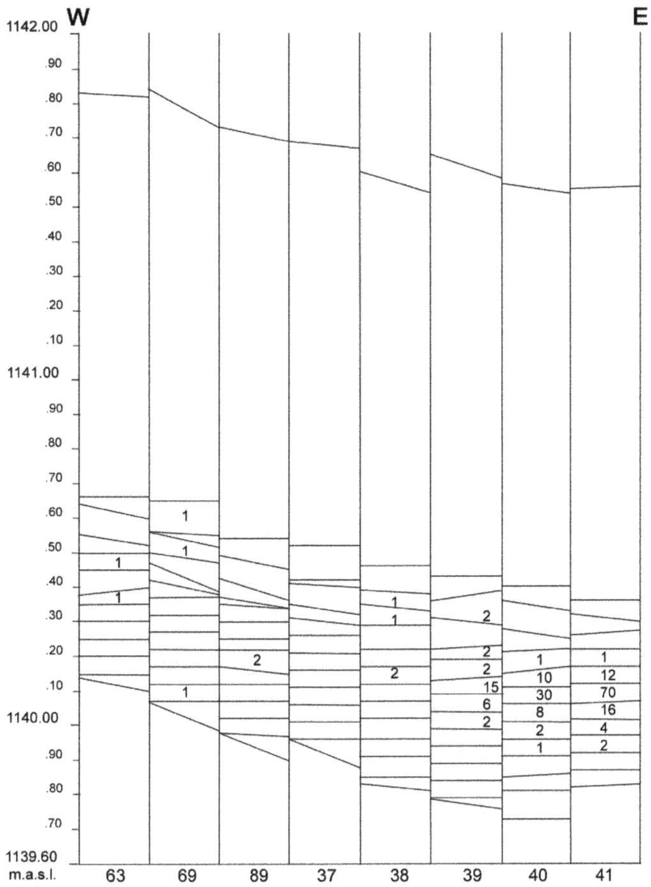

Figure 91: NORTHING 521 – The distribution of lithic remains in Units 63 to 41

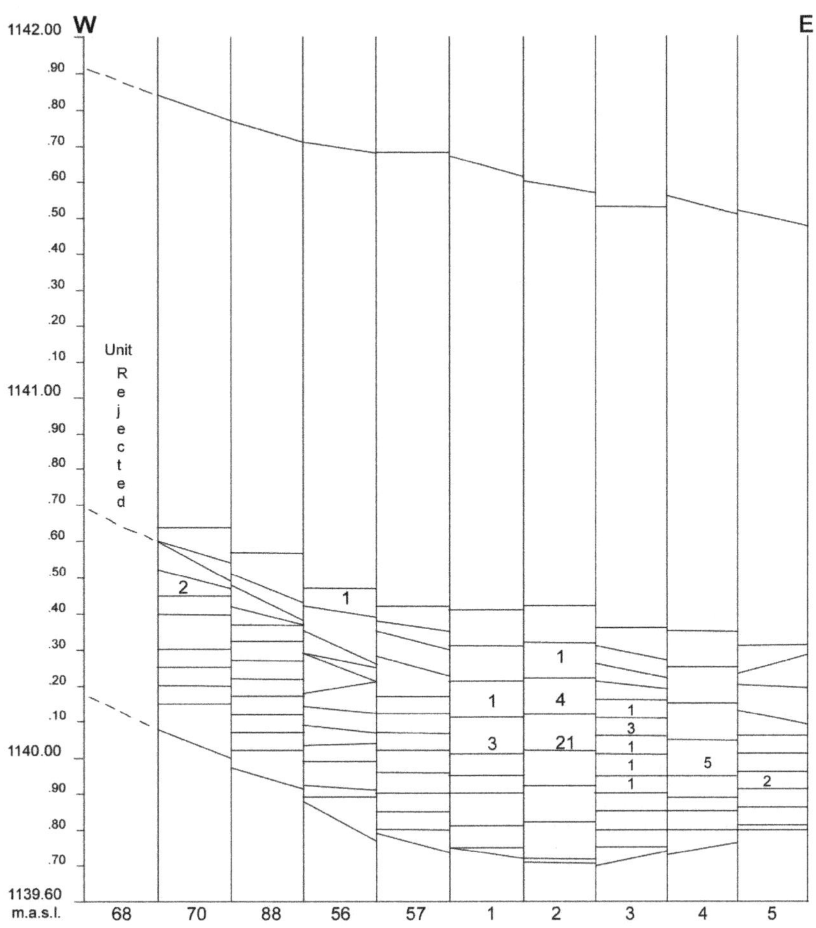

Figure 92: NORTHING 522 – The distribution of lithic remains in Units 68 to 5

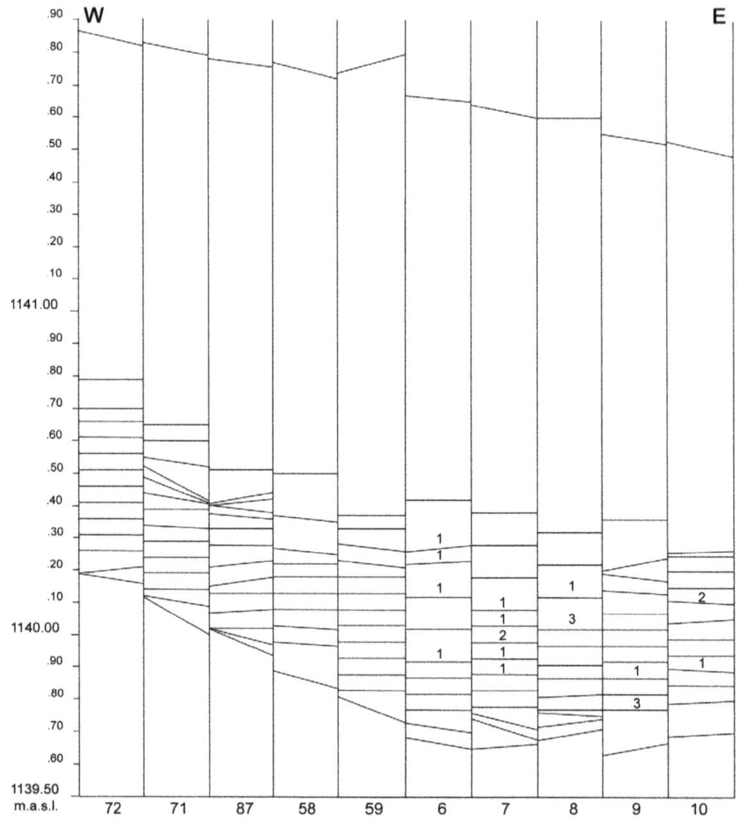

Figure 93: NORTHING 523 – The distribution of lithic remains in Units 72 to 10

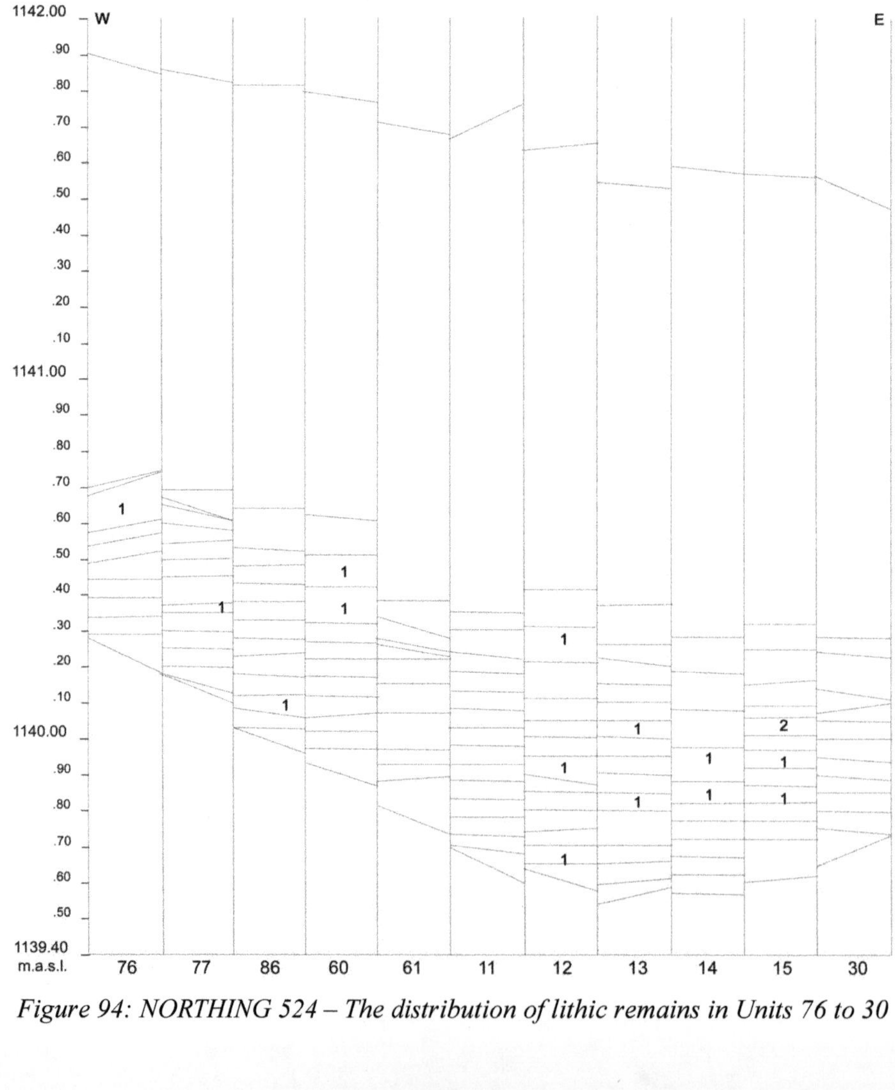

Figure 94: NORTHING 524 – The distribution of lithic remains in Units 76 to 30

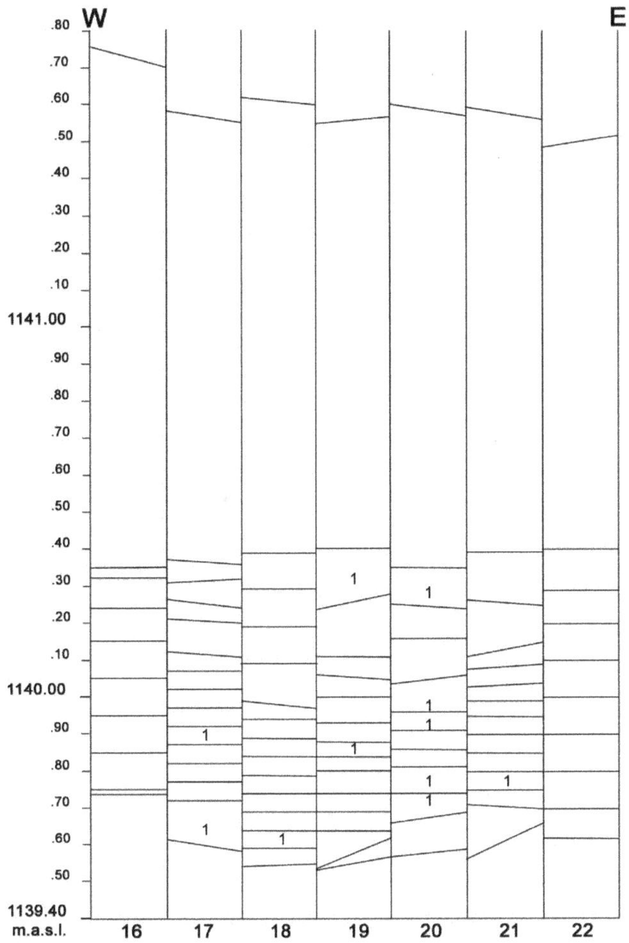

Figure 95: NORTHING 525 – The distribution of lithic remains in Units 16 to 22

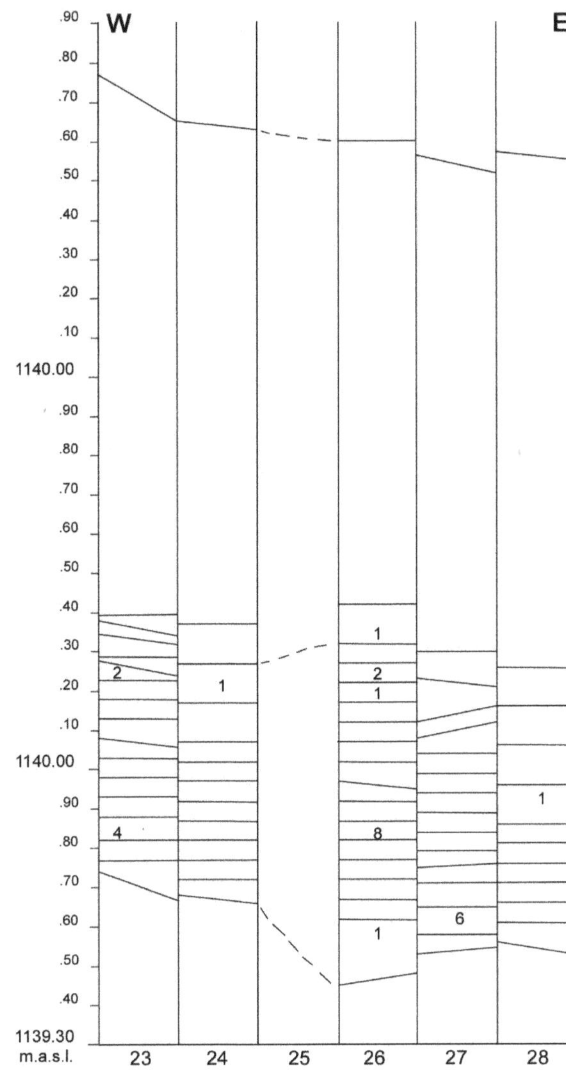

Figure 96: NORTHING 526 – The distribution of lithic remains in Units 23 to 28

depositional processes, however, wood rarely survives through time in buried contexts unless there are special conditions like water logging or extreme aridity. In the case of the Tuscany site, no tools made from antler, bone or wood have been recovered.

While there are curious amalgamations of seed remains their relationship to human activity is not completely decipherable. However, there are patterns of distribution that appear to be clearly correlated to a hearth feature and a distribution of cultural material remains. Thus, the distribution of paleobotanical remains is at least partially related to human activities at the Tuscany site at the time of occupation. The abundant kinnikinnick seed remains indicate a potentially carbohydrate-rich, though not necessarily very tasty, food source was present at the site. The next plant that might have provided an edible product is the rose, which produces rose hips that are nutritious and high in vitamin C. The *Chenopodium* species present at the site may also have represented a potentially rich food source from the leaves and seeds. The juniper berries present in abundance at the site may also have been a food source, though not extremely palatable. While there is not a lot of evidence for the presence of wild bergamot, it represents a plant that can be used to make a beverage tea along with the shrubby cinquefoil. Trees like spruce and poplar have potential uses for tools but also as food; the cambium is edible. Trees provided fuel for fires and materials for the erection of structures. Modern ethnobotanical studies (e.g. Leighton 1985; Marles *et al.* 2000; Peacock 1992; Siegfried 1994) for Alberta and Saskatchewan show that nearly every plant in an environment has the potential for some kind of use by human populations. Thus, we can certainly say that the potential existed for use of these different plants as foods, crafts/tools and medicines at the Tuscany site. Any specific uses of plants in the paleoenvironment of Tuscany are not discernable from this study, but there is the indication of a rich environment of great potential to human populations present in the area between 10,000 to 7800 years BP.

The spatial analysis of the distribution of charred botanical remains from both the planview perspective and the northing profiles perspective has demonstrated that except for the trample zone there is little discernable cultural patterning apparent for these remains. From an ethnobotanical perspective, there are conditions that would be expected from the interaction of people with a locale exploited for plants or used as a base for further

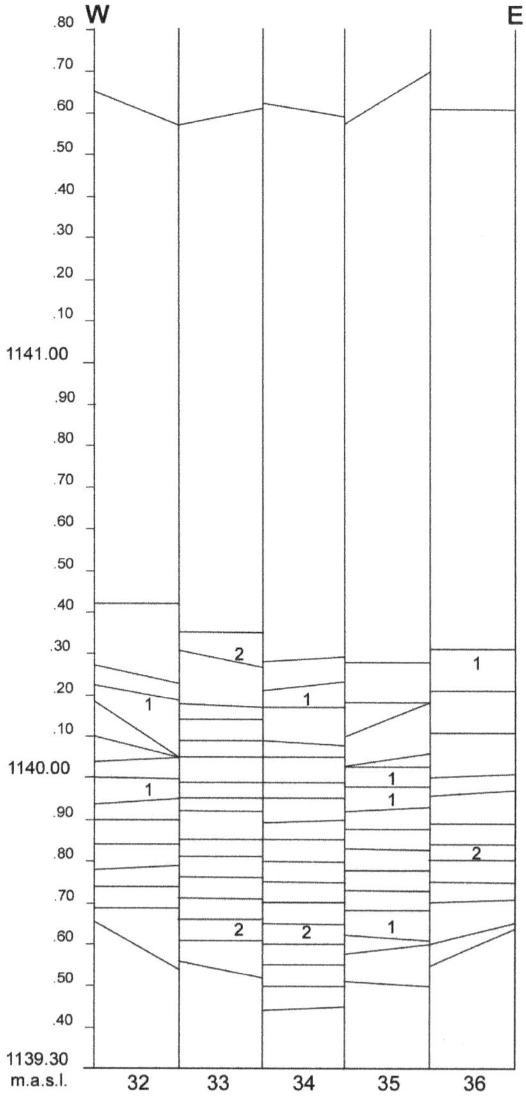

Figure 97: NORTHING 527 – The distribution of lithic remains in Units 32 to 36

processing of harvested plants or useful parts of plants. If we hypothesize that, the berries of the locale were exploited by people camping at the Tuscany site, the next question is, what kind of distribution patterns would we expect? Kinnikinnick and juniper have been carefully examined using the two spatial perspectives. Their charred seeds are distributed over the entire site grid in planview and in profile view. If these plants were exploited during the occupation of the site, a locus of preparation would be expected and required. The drying of berries for storage has been ethnographically demonstrated as a task that can take place on top of hides and mats spread out on the ground (eg Kuhnlein and Turner 1991; Peacock 1992). The distribution pattern that we might reasonably expect from this would be the loss of a few berries at the outer edges of a large blank area of hide/mat dimensions. However, the seeds contained by the berries would not be charred by this method of processing and the site would have had to burn, subsequent to the activity. As well, the only large blank area correlates with the hearth and the proposed hut structure.

The problems with a local exploitation of the two represented berry plants are even more complex. We must assume the berries were picked and then spread to dry or contained in bags or baskets of wood and skin. If a cache was forgotten at the site, it would have had to burn to preserve and that is what we could expect to find in a flotation sample. Given our 1% sampling of the site, chances of recovery of a burned cache is remote. No such feature was encountered during excavation. However, the problem with these two species of berry plants is further complicated by their propagation. Both kinnikinnick and juniper produce berries that will stay on the plant long after they ripen. The berries will dry out, while remaining attached to the plant. Many berries will fall to the ground as well, if they are not eaten by animals or collected by people. These old berries would not be an attractive resource and would then simply reflect the natural distribution of the plants in the landscape when a fire swept through the locale. Even if everything useful was picked during occupation of the site, it is unreasonable to expect that we could still discern one year of missing berry productivity, especially if a fire swept through and subsequently charred the remaining berries that were drying out on the plants, or decaying on the ground.

It is clear from the set of northing profiles that kinnikinnick and juniper plants occupied the locale long before people ever camped there. The distribution of the charred seed remains from them are scattered ubiquitously across the paleosols in the site and up through the paleosols of the site through time to the Mazama tephra boundary. The flotation samples from the upper part of the profile have been sorted and preliminary work indicates there are very few seeds of any kind above the Mazama tephra, though there are uncharred seeds present in samples closer to the sod surface. It is the ubiquity of the distribution of the charred kinnikinnick and juniper seeds that indicate their presence in the site is not the result of any human activity. When people harvest berries such as those identified here, they do not scatter them indiscriminately throughout the site. They collect them in containers or caches of some sort. What behavior would cause the berries to be scattered to all areas of the site? That is simply not a gathering or caching pattern. It is most probable that the berries became charred at the same time as the rest of the other charred botanical materials like the small wood twig/root fragments and the other charred seeds. However, the pattern of complete site coverage of all the charred botanical remains is also difficult to explain if we hypothesize that they originated from the hearth as an explanation for simultaneous charring. There was no concentration of these remains within the hearth and there was actually a drop-off in numbers immediately around the hearth. Further, how would the charred seeds and all the other charred plant fragments become incorporated into all parts of the paleosols prior to the level where the cultural material is found, if they were introduced to the site at the level where the hearth feature and projectile point appear? There is no evidence of such an incredible amount of disturbance within the paleosols to cause the degree of mixing within them that would spread all the charred plant remains from the top to the bottom. While there does appear to be some translocation of the lithic debitage, it spans three levels, or the equivalent of 15 centimeters of the paleosols and does corroborate a similar possible translocation pattern of the upper concentration of charred plant remains.

A number of arguments have been examined to test the hypothesis that the distribution of the charred kinnikinnick and juniper seeds could be attributed to cultural behavior. Reasoned explanation contradicts this idea. Rather, the ubiquity of the observed distributions is more attributable to the natural propagation of these plants across the landscape and the natural distribution of their seeds across the locale and through the time period represented by the entire set of paleosols. The botanical remains became charred when a fire or fires swept through the locale.

FIRE ECOLOGY

Fire appears to have played an enormous role in the landscape of the Tuscany habitation site. Evidence of fire impacting the site during the past has been recovered in the form of charred plant remains identified as small twigs/roots of willow or poplar trees, the seeds of several species of plants, and charred structures that have been identified as the cone scales of white spruce. These macrobotanical remains raise important questions about the role of fire in the paleoenvironment of southern Alberta and the relationship between people and the vegetation that burned.

The evidence of burning vegetation has been examined spatially within the confines of the site depression. A brief discussion on the characteristics of materials involved in burning landscapes has been addressed above in an effort to understand how the culturally deposited bone might have been affected by fire and subsequent deposition after human occupation of the Tuscany site. The spatial distribution of the charred botanical remains is proposed to have been the result of natural propagation and growth habits of the plant species identified as they grew within the confines of the site depression during the early Holocene. The distribution of the charred botanical remains, above and well below the level of the cultural component indicates that fires must have occurred at the depression locale at different time periods. This hypothesis appears to be verified by radiocarbon dating of charred botanical remains from two different locations and depths within the site, as discussed above. What is not completely clear from the data is whether or not human agency was involved in the burning of the landscape during the time of the main occupation of the depression.

The topic of fire ecology has generally been neglected until more recent decades by foresters and ecologists (Johnson and Miyanishi 2001). Fire ecology related to traditional knowledge of Indigenous populations throughout North America has also not been examined by researchers in anthropology and ethnobotany until more recent decades (Turner 1991). Studies by Lewis during the 1970's in northern Alberta represent some of the earlier work that began to examine the relationship that existed between Indigenous boreal forest populations (Cree, Slavey, Beaver and Chipewyan) and the vegetation landscapes they burned on a regular basis (Lewis 1982). It has long been assumed that fire is a destructive and disruptive phenomenon and control and suppression of fire has been the adopted policy for forestry departments all over North America for many decades. The recognition by ethnobotanists (e.g. Turner 1991) and other researchers (e.g. Lewis 1982 and Pyne 1982) of the incredibly complex nature of the relationship Indigenous people maintained with their fired vegetation landscapes and the amount of knowledge required to maintain them, has only surfaced with these recent studies. It has also become apparent that fire suppression as a blanket government policy at all levels, has resulted in management of forest and grassland resources that now appear as 'mismanagement', in retrospect (e.g. Collins and Wallace 1990 and Greenlee 1996). Fires are natural events caused by lightening strikes, but they are also events caused by people both purposefully and accidentally.

The question of interest to this dissertation is, whether we can extrapolate the intentional burning of vegetation landscapes by Indigenous populations during the early Holocene in what is now southern Alberta using data such as that recovered from the Tuscany site. Can information such as fire frequency records extrapolated backwards through time reveal the intentional burning of grasslands and forests by Indigenous people as a form of 'management' to promote better harvests of edible plants, better browse and grazing for ungulates, or for clearing brush to make travel easier? All of these purposes have been described for Aboriginal people in interior British Columbia (Turner 1991).

A study of fire frequency has been carried out for the Kananaskis River watershed located approximately 60 km west of Calgary. Johnson *et al.* (1987) developed a 398 year fire history record for the area, looking at it both spatially and temporally. The watershed was separated into the front range and the main range, which demonstrated variation in fire frequencies through time. During the first time period from, approximately 1582-1730 AD, the fire cycle was on average every 60 years in the main range and every 30 years in the front range (Johnson *et al.* 1987). Between 1730-1980 AD, the fire cycle was 132 years in the main range and 94 years in the front range (Johnson *et al.* 1987). What is important about these figures is that they demonstrate variability through time and space. During the first period the fire cycle was shorter in both the front and main ranges and became longer for both ranges during the second time period. Tree ring analysis shows a period from about 1520-1660 with narrower rings that are associated with warmer and drier weather and a period from about 1660-1860 with wider rings that are associated with cooler and moister weather (Johnson *et al.* 1987:36). The latter period is associated with a climatic cooling episode known as the Little Ice Age. The time between 1660 and 1730 represents part of a lag time that may be climatically driven or may be a product of the analytical technique employed (Johnson *et al.* 1987).

Aboriginal presence in the Kananaskis River watershed was maintained by the Kootenay before the 1780s. They spent their winters in the foothills where bison could be hunted (Johnson *et al.* 1987). However, a smallpox epidemic in 1781 killed 50-75% of the Kootenay and they retreated further west, forced out also by the Blackfoot in the 1790s and then the Stoney (Assiniboine) in the 1840s. The Stoney currently reside on a reserve at the mouth of the Kananaskis River (Johnson *et al.* 1987:37-8). Johnson *et al.* (1987) propose that a drop in fire frequencies between 1780 and 1830 could be related to the smallpox epidemic and the Kootenay drop in population and use of the valley. However, this is only a tentative hypothesis. Their overall conclusion was that regional climate was probably the primary agent affecting fire frequencies.

This fire frequency study provides inconclusive results concerning the relationship between the Aboriginal populations occupying the landscape and any impact on it by their use of fire. The difference in fire frequencies, where there were more fires in the warmer-drier period and fewer in the cooler-moister period, can be linked to short-term climatic trends apparent in the environmental record (i.e. the Little Ice Age). Tree ring data combined with pollen and charcoal records for different sites might be a useful method to clarify these kind of trends further back into the paleoenvironmental record, in the sense of positively correlating warm-dry and cool-moist periods with evidence of fires in pollen cores. This would require a dendrochronological record that extends back into the early Holocene. If records do not correlate positively, we may be able to infer human agency as a factor in high fire frequency rates.

Larsen (1996) carried out a study on fire and climate dynamics between 1850 and 1989 in Wood Buffalo National Park in northern Alberta. Historical records and dendrochronological forest stand-age data were combined to examine a period of annual variation between 1950-1989 compared to a period of semi-decadal variation between 1850-1989 (Larsen 1996). According to Larsen (1996), wildfires in Wood Buffalo National Park correlated to climatic variability observed in environmental records. When it was warmer and drier the tendency for the forest vegetation was to burn as a higher fuel load became available. Relationships between Aboriginal people of the region and burning practices were not addressed by this study.

Houston's (1973) discussions of fire frequencies in Yellowstone Park, Wyoming demonstrate that fire suppression was a clear goal for the park after it was established in 1872. Fire patrols began in 1886. According to Houston (1973), Aboriginal people in the park were removed. While historical records demonstrate that there are examples of fires being set deliberately by Aboriginal people and this would indicate an influence on fire frequencies, Houston (1973) does not believe this relationship can be quantitatively evaluated. Houston (1973) derived a mean interval of approximately 20-25 years between fires before the regime of fire suppression affected the park vegetation. Between 1931-1971 there were 99 fires started by lightning and then suppressed, in Houston's (1973) study area. This would appear to indicate that the fire frequency probably would have been high if suppression had not taken place. This does not help an argument for Aboriginal burning as a possible causal factor in high fire frequencies. Interestingly, Houston (1973) noted that the lack of fires led to changes in plant succession regimes within Yellowstone Park, visible with the increase in conifers and the general decline of deciduous aspen. This obviously played a role in the subsequent natural disastrous burning of Yellowstone forests during the summer of 1988.

According to Cannon (1996) fire frequencies vary depending on the dominant vegetation type. For example, a range of 20-25 years is cited for grassland fire frequency, 50-100 years for aspen and 100-200 years for lodgepole pine (Cannon 1996:3). Unfortunately, it is not completely clear how Cannon derived these figures. However, Cannon (1996) also recommends caution in the application of modern fire frequencies to past systems, noting that these intervals have very likely changed through time influenced by changes in the vegetation communities themselves and by changes in climate.

This discussion of fire ecology has not been exhaustive. It is clear, however, that there is great variability in fire frequencies from one region to the next because of the changing vegetation communities and the changing climates that play a role in influencing those changes. Modern fire frequencies do not represent very good comparable data because too many factors have disrupted the natural fire regimes. Aboriginal burning is difficult to establish in these records, since they can be interpreted to have changed as a result of climate changes rather than the demise of Indigenous people and suppression of their prior influence on the vegetation landscape.

Chapter 6
LOCAL ECOLOGICAL AND CULTURAL LANDSCAPES IDENTIFIED

ECOLOGICAL LANDSCAPES

The charred remains of juniper (*Juniperus* sp.) and kinnikinnick (*Arctostaphylus uva-ursi*) appear to have dominated the Tuscany landscape consistently through time. A conservative evaluation of the charred wood fragments viewed and photographed with the ESEM, identifies them to at least the deciduous level. The diffuse pore pattern observed indicates the wood is possibly from a *Poplar sp.* or *Salix* sp., both members of the Salicaceae or 'Willow Family'. A small number of *Rosa* sp. (rose) seeds, *Chenopodium* sp. (lamb's quarters) seeds, and tentatively *Potentilla fruiticosa* seeds (shrubby cinquefoil) are also identified. There are, also a number of *Picea* sp. seeds and cone scales that appear to resemble scales from *Picea glauca* (white spruce).

We can begin to describe a vegetation landscape with this small set of plant species. It is an area that must be moderately well drained. Both kinnikinnick and juniper thrive better in dryer areas. As noted by Archibald *et al.* (1996:9-12), kinnikinnick and juniper in an area or site indicate that conditions at that locality are dry. Junipers are intolerant of shade. However, *J. communis* was found growing in a sheltered area under poplar and spruce at the top of Twelve Mile Coulee, approximately 800 meters north of the excavation grid. It can be concluded that plants have a range of tolerance for varying conditions. While there are optimal conditions for them, they can adapt to soil types and moisture regimes along a continuum of variability.

With these problems in mind, we can use the data set from the Tuscany site to try and derive a possible interpretation of what the vegetation cover may have been like through time. The first part of this exploration entails examining studies of present-day landscapes to see if any parallels exist in the known botanical communities. A number of valuable ecological classification studies have been produced for the province of Alberta through the Canadian Forest Service. Thus far, Field Guides to Ecosites have been published for Northern Alberta, West-Central Alberta, and Southwestern Alberta. A thorough review of each guide was undertaken to look for plant communities that contain both kinnikinnick and juniper as part of the dominant vegetation. There are no such plant communities in the Boreal Forest Ecological Area, which is covered in the Field Guide for Northern Alberta (Beckingham and Archibald 1996). There are no kinnikinnick-juniper plant communities in the West-Central Region of Alberta either, which is the area covered in the Field Guide to Forest Ecosystems of West-Central Alberta (Corns and Annas 1986).

There are two major areas where kinnikinnick-juniper plant communities can be found in Alberta. One of these is located in the far northeastern corner of Alberta, known as the Canadian Shield Ecological Area (Beckingham and Archibald (1996). There are also several different ecosite and ecophase communities with varying percentages of both kinnikinnick and juniper, described in the Field Guide to Ecosites of Southwestern Alberta (Archibald *et al.* 1996). These ecosite and ecophase communities are found to the west of Calgary today. The relevant details of these plant communities will be summarized from Archibald *et al.* (1996) and Beckingham and Archibald (1996), to draw comparisons to the plant communities that may have existed at the Tuscany habitation site during post glacial times and up to the Mazama tephra deposition event.

This use of the information presented in the fieldguides to ecosites in Alberta is actually very different from their intended purpose which is to help understand ecological information and organize it in a format that provides structure for ecologically based land management (Archibald *et al.* 1996). The classification system is based upon the analysis of hundreds of plots where data were gathered for vegetation, soil and tree production.

Table 7 provides a vegetation summary of the dominant plants in two ecological areas: the Canadian Shield and

Table 7: Comparison of the vegetation in kinnikinnick-juniper ecophases in the different Alberta Subregions

		Canadian Shield Subregion	**Lower Foothills Subregion**	**Upper Foothills Subregion**	**Montane Subregion**	**Subalpine Subregion**
Tree	*Picea glauca*			5-9.99%		
	Pinus contorta			>20%	>20%	>20%
	Pinus banksiana	10-19.99%				
	Populus tremuloides		>20%			
	Populus balsamifera		5-9.99%			
	Pseudotsuga menziesii				2-4.99%	
	Picea engelmannii					2-4.99%
Shrub	*Arctostaphylos uva-ursi*	2-4.99%	10-19.99%	*<2%	>20%	10-19.99%
	Linnaea borealis			5-9.99%	5-9.99%	2-4.99%
	Juniperus spp.	5-9.99%				
	Juniperus communis		5-9.99%	*<2%	5-9.99%	5-9.99%
	Picea engelmannii					2-4.99%
	Picea glauca			2-4.99%		
	Pinus banksiana	5-9.99%				
	Populus tremuloides		5-9.99%			
	Rosa acicularis		2-4.99%	2-4.99%	2-4.99%	*<2%
	Shepherdia canadensis		2-4.99%	2-4.99%	5-9.99%	2-4.99%
	Pinus contorta				*<2%	2-4.99%
	Vaccinium myrtilloides	2-4.99%				
Forb	*Frageria virginiana*		2-4.99%		*2-4.99%	*<2%
	Achillea millefolium		*<2%			
	Anemone multifida					*<2%
	*Aster ciliolatus**		*<2%			
	Epilobium angustifolium		*<2%			
	Galium boreale		*<2%			
	Lathyrus ochroleucus		*<2%			
	Aster conspicuus			2-4.99%	*<2%	
Grass	*Agrostis scabra*	2-4.99%				
	Calamogrostis rubescens		5-9.99%		5-9.99%	
	Elymus innovatus		5-9.99%	10-19.99%	2-4.99%	10-19.99%
Moisture Regimes **		Subxeric Xeric submesic	Submesic	Submesic Mesic	Submesic Subxeric Mesic	Subxeric Mesic
Nutrient Regimes **		Poor Very poor	Medium Rich	Medium Poor	Poor Medium	Poor Medium Rich
Drainage**		Rapid Well	Well	Well Mod. Well Rapid	Well Rapid	Rapid Well Mod. Well
Soil type**		Very dry/sandy; Dry/sandy	Dry/Fine Loamy-Clayey	Moist/Fine Loamy-Clayey; Dry/Fine Loamy-Clayey; Dry/Silty-Loamy	Dry/Silty-Loamy; Very Dry/Coarse Loamy; Moist/Fine Loamy-Clayey; Dry/Coarse Loamy; Dry/Sandy	Very Dry/Fine Loamy-Clayey; Very Dry/Sandy; Dry/Fine Loamy-Clayey; Moist/Fine Loamy-Clayey; Very Dry/Coarse Loamy; Very Dry/Silty-Loamy

* Characteristic species with a prominence value <20% for all plant community types.

** Moisture and nutrient regimes, drainage and soil type are listed in order of dominance in landscapes – i.e. first one is most common.

Source: Adapted from Beckingham and Archibald (1996), Archibald *et al.* (1996)

Southwestern Alberta. The summaries present the characteristic plant species found in ecosites and their representative cover on the landscape. In each case, the four dominant classes of vegetation (tree, shrub, forb and grass) are subdivided. There are a limited number of places where the combination of kinnikinnick and juniper will be found. Interestingly, the Canadian Shield subregion is different from the other subregions of southwestern Alberta in all four classes of vegetation cover. The common feature between the far northeastern subregion and the southwestern set is the shrub cover combination of kinnikinnick and juniper with a pine species as the dominant tree cover. However, *Pinus banksiana* dominates the Canadian Shield landscape while *Pinus contorta* dominates the southwestern Alberta landscape.

Clearly, there have been very different sets of ecosystem dynamics operating through time in these two major regions. This is an important point. These areas are widely separated in geographic space. However, similar conditions are available for plant growth in terms of moisture, nutrient and drainage conditions. Kinnikinnick and juniper today, appear to do well in areas with poorer soils that are well drained and have drier moisture regimes. In the northeastern shield area, though, juniper dominates over kinnikinnick. In the southwestern areas, kinnikinnick clearly dominates over juniper by large percentages except in the Upper Foothills subregion where they are both found in amounts of less than two percent. The two plants appear to thrive best as a combination in the Montane subregion which is found in the Bow Valley Corridor and along the foothills bordering the eastern flanks of the Rocky Mountains from approximately 80 kilometers (fifty miles) southwest of Calgary, and down into the United States (see Figure 98: Map of the Southwestern subregions in the greater Calgary region).

The tree cover is different for each of the southwestern subregions, demonstrating clearly the influence of changing elevation. The Lower Foothills subregion is dominated by deciduous tree and shrub cover whereas the remaining three subregions are dominated by combinations of conifers, with the primary species being *Pinus contorta.*

Table 8 examines the climate data summarized for the five subregions in this discussion. Unfortunately, the data for the Canadian Shield area is not complete, however, there is enough information to make some general observations. The data show that the Canadian Shield subregion fits in with the Southwestern subregions in all aspects. There is nothing dramatically different between the temperatures or precipitation records for these areas. There appears to be one variable that differentiates the Canadian Shield from the Southwestern subregions—it is the soil type. The soils are Very Dry/Sandy and Dry/Sandy in the Canadian Shield subregion. This appears to make a difference to the growth of kinnikinnick, which does not grow as abundantly in the region.

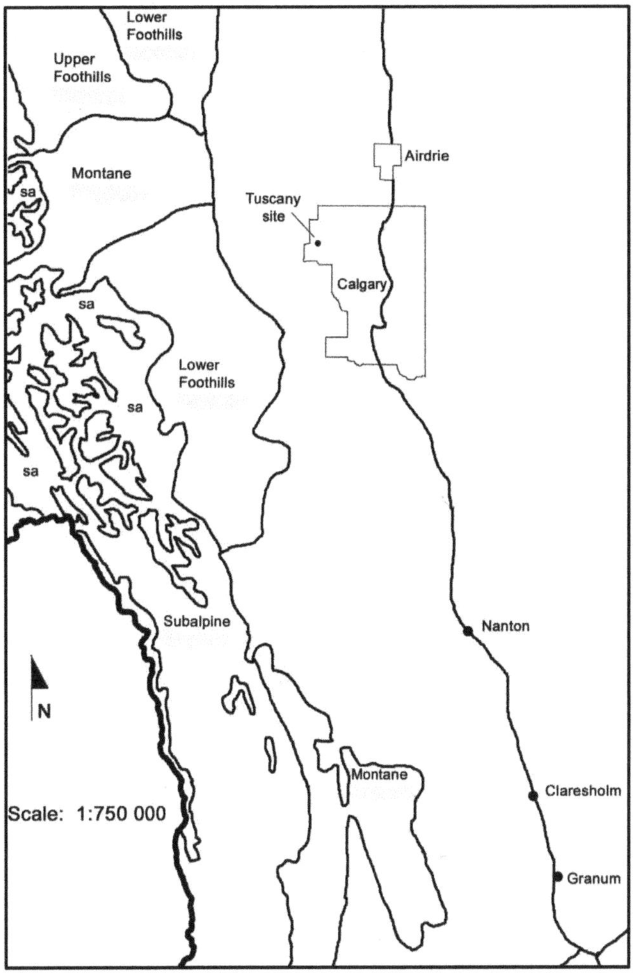

Figure 98: Map of the Southwestern Alberta Subregions located in the greater region west of the City of Calgary. Source: (adapted from Archibald et al. *1996: Map: Natural Subregions of Southwestern Alberta)*

There is a means of viewing how well these plants fit together in the same locale. Archibald *et al.* (1996) provides a series of summary graphs for the major plant species found in the Southwestern Alberta ecosites. These diagrams summarize the maximum, minimum, and optimal, moisture and nutrient regimes against the range and mean elevation data for each plant. A number of these diagrams have been put together here, to view their relationships to each other. Figure 99 presents the first set of summary data diagrams for five shrub and one forb species. Figure 100 presents a second set of diagrams for the six main tree species, and Figure 101 presents a final shrub, forb and grass set. There is an underlying order for these diagrams. The first figure (Figure 99) compares the shrub and forb species found in the flotation samples from the Tuscany habitation site, for which there was an available moisture and nutrient diagram. All the species included are associated with the kinnikinnick-juniper ecosites found in Southwestern Alberta today.

The trends in these diagrams are important. In Figure 99, the plant species have very similar ranges of tolerance for moisture and nutrient regimes and means of elevations. Their present-day values indicate that there should not be

Table 8: Comparison of climate data from Northeastern and Southwestern Alberta

	Canadian Shield Subregion	Lower Foothills Subregion	Upper Foothills Subregion	Montane Subregion	Subalpine Subregion
Summer- May, June, July, August					
Mean typical temperatures (°C)	12.0	12.8	11.5	11.9	9.4
Minimum typical temperatures (°C)	6.9	6.9	5.9	5.1	3.8
Maximum typical temperatures (°C)	16.9	18.3	16.7	18.9	14.7
Total precipitation (mm)	266	295	340	210	263
Growing degree days	883	1008	752	773	558
Number of days <0°C	12	8	13	16	28
Winter- November, December, January, February					
Mean typical temperatures (°C)	--- (no data)	-7.8	-6.0	-5.5	-8.9
Minimum typical temperatures (°C)	---	-14.3	-12.5	-11.3	-14.5
Maximum typical temperatures (°C)	---	-2.1	0.5	-0.2	2.8
Total precipitation (mm)	---	60	60	93	146
Annual					
Total precipitation (mm)	---	464	538	440	568
Mean temperature (°C)	---	3.0	3.0	3.5	1.3

Source: Adapted from Beckingham and Archibald (1996), Archibald *et al.* (1996)

any reason (based on these variables) why they could not be found together in a local area of the landscape. The assumption here, based on the macrobotanicals recovered in the flotation analysis, is that these six plant species once cohabited the Tuscany habitation site depression. This combination of plants requires a mesic to submesic moisture regime. If poplar and white spruce represent the upper canopy, the site area vegetation at various times during the past, would probably have appeared similar to the Parkland of today.

Figure 100, summarizes the tree species found in Kinnikinnick-Juniper ecosites and demonstrates the interesting relationship between elevation changes and optimal sites for the different species to thrive. We have seen that *Picea engelmannii* and *Pseudotsuga menziesii* are never prolific, but they are in the landscapes at certain higher elevations. *Picea glauca* appears to prefer slightly lower elevations than the previous two species and from previous data presented, also enjoys a more productive existence in the Upper Foothills subregion. All three higher elevation subregions are dominated by *Pinus contorta* in percentages greater than twenty.

This discussion has relied on the modern distributions of plants in landscapes of today. The primary purpose of this argument has been to establish what the vegetation trends are today and whether or not they compare to the information that has been derived from the flotation analysis. The conclusion here is that there is correlation visible through time. Following deglaciation, the landscape was recolonized by vegetation. The question of where the colonizing plants came from is a question of debate but it seems most likely that vegetation that existed throughout the continental glaciation period, moved northward from refugia south of the glacial maximum margins. Logically, plants in that area would have had the best and easiest chance, in terms of distance, to move into denuded landscapes and establish themselves. Pine appears to be a colonizing tree species along the eastern flanks of the Rocky Mountains, according to fossil pollen studies. In the southern regions of Alberta the species that dominates is *P. contorta* (lodgepole pine). The dominance of this tree in the landscape appears to have continued through time in areas of higher elevation.

If pine did extend its boundaries as a colonizer species, it did not maintain a presence in the more easterly landscapes, away from the Lower Foothills regions. There is certainly no evidence of lodgepole pine in the Tuscany paleosol sediments. There is, however, evidence of *Picea glauca* (white spruce) in the Tuscany paleosols. This indicates the presence of more trees in the landscape above the level of Twelve Mile Coulee, where a Parkland vegetation assemblage can be found today. Twelve Mile Coulee is functioning as a kind of refugium today, harboring plants that cannot survive the conditions available in the open grasslands vegetation of the Tuscany landscape today.

According to MacDonald and Cwynar (1985:1042), modern geographical distribution data of vegetation are used to reconstruct historical biogeographies. This approach clearly must be augmented with the use of fossil data. In the case of the Tuscany site, the modern distribution of plants in the landscape certainly does not yield the same vegetation scenario that is evaluated from the charred fossil macrobotanical data that has been recovered through flotation analysis. Though many of the same plants are present in the landscape today as in the past, they are not in the Tuscany site area itself. They are

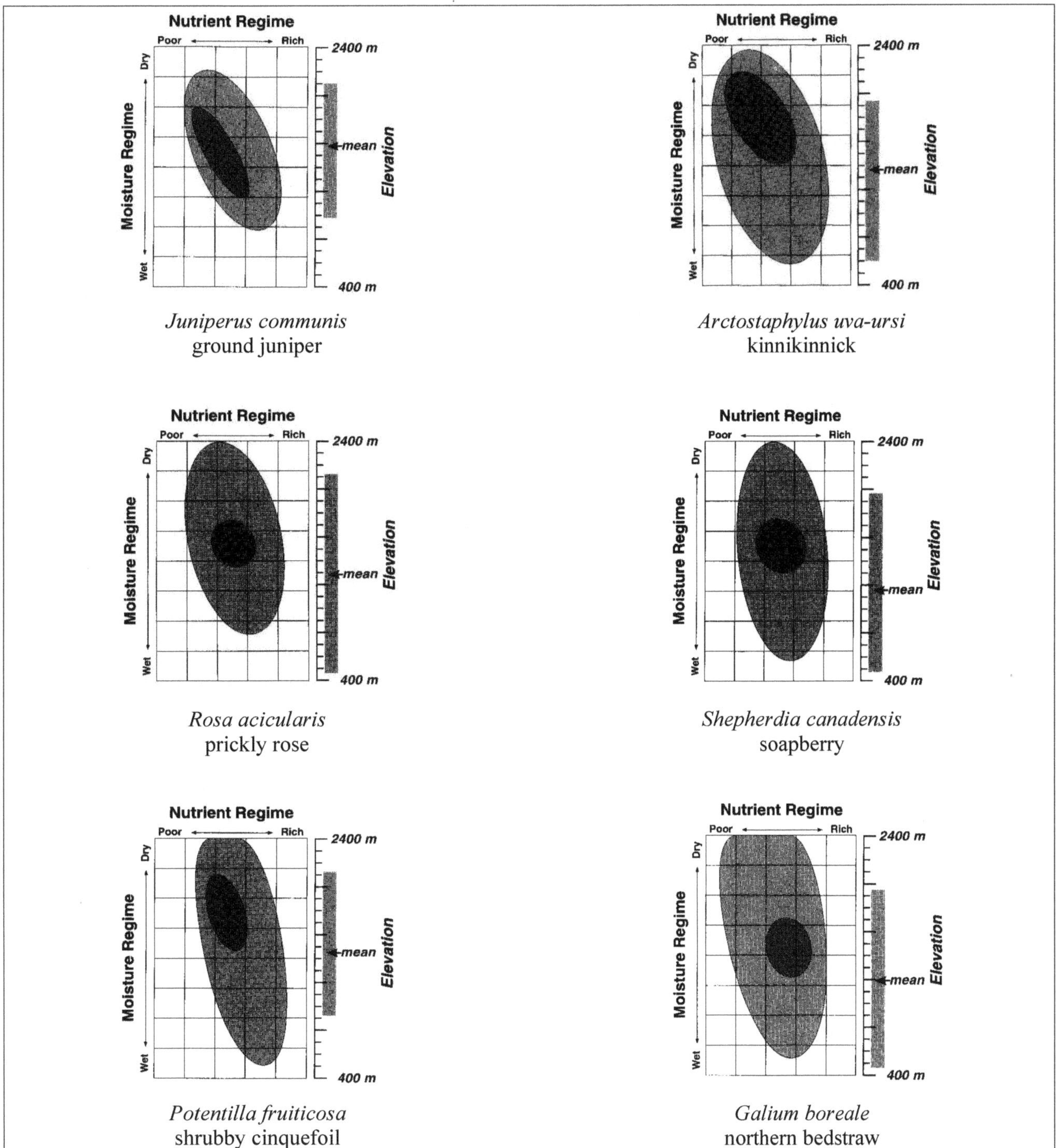

Figure 99: Comparison of the moisture and nutrient regimes and range of elevation of six species of shrubs and forbs associated with ecosites of Southwestern Alberta. The data are from modern specimens found in ecotsites today. Source: Archibald et al. *(1996:13-24,18,37,41,33,69)*

hundreds of meters away at the edge and on the slopes going down into Twelve Mile Coulee. The differences between the Tuscany grid site and the coulee slopes appear to be related to soil and moisture regimes.

It is the combination of kinnikinnick and juniper that appears to have played a major role in the understory of the Tuscany habitation site in the past and has been the focus of this discussion because of the quantities of charred seeds recovered for these two species. There are also a number of other plants that made an appearance at the Tuscany site, though they are not as well represented in the number of seeds recovered through the flotation analysis. While making assumptions about biogeographic histories of plants based upon present day landscapes is not recommended, the combination of plants represented from the flotation analysis has been shown to have a history in this area that has carried through time. Species that left charred seed remains in the paleosols can be found in the Tuscany landscape today.

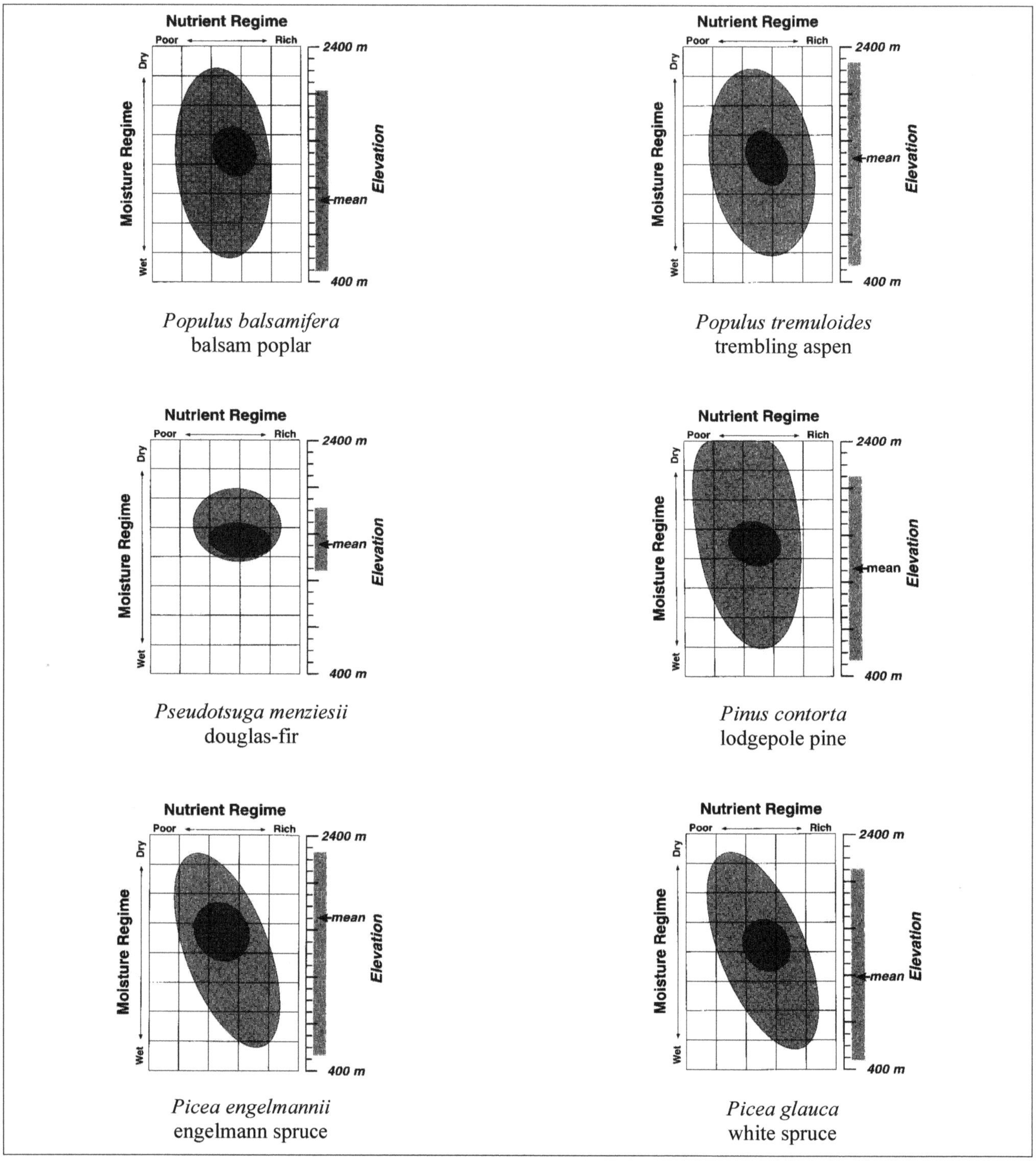

Figure 100: Comparison of moisture and nutrient regimes and range of elevation of six species of trees associated with ecosites of Southwestern Alberta today. Source: Archibald et al. *(1996:13-12,13,14, 10,6,7)*

USING FOSSIL POLLEN STUDIES TO DERIVE VEGETATION INFLUENCES

Kinnikinnick and juniper obviously played an important role in the vegetation cover of the immediate site area during the formation of the paleosols. Do we find kinnikinnick and juniper growing today at the Tuscany site? No. Only grasses, forbs and snowberry shrubs are found in the immediate site location. However, we do find kinnikinnick and juniper growing along the Bow River escarpment to the south of the site and on the southeast facing slopes of Twelve Mile Coulee to the east of the site. Both plants prefer sandier, well drained, mesic to submesic soils.

Kinnikinnick and juniper as combinations of understory vegetation are also found to the west of Calgary today in the Lower and Upper Foothills, the Montane and

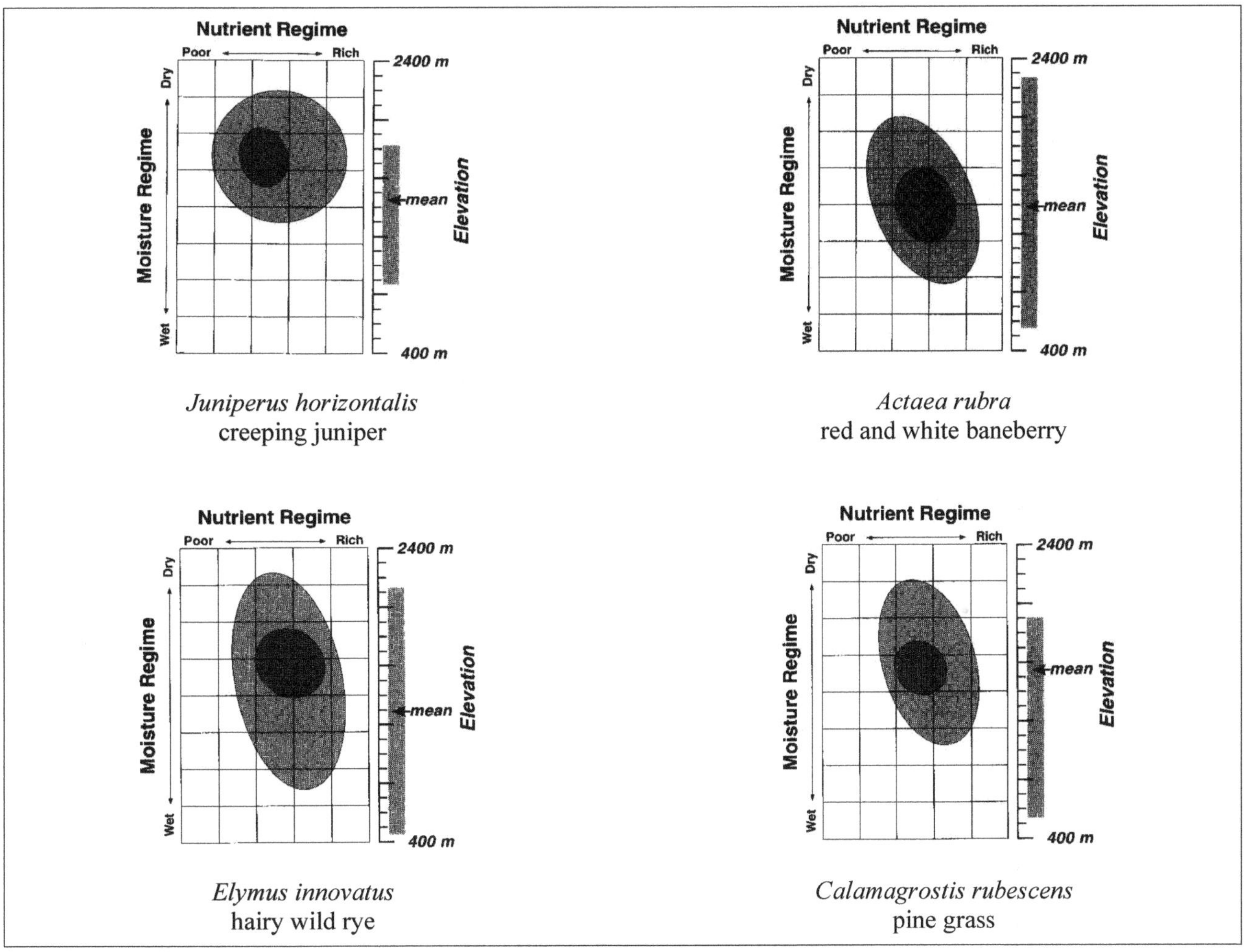

Figure 101: Comparison of the moisture and nutrient regimes and range of elevation of shrubs, forbs and grasses associated with ecosites of Southwestern Alberta today. Data are from modern specimens. Source: Archibald et al. *(1996:13-25,52,99,97)*

Subalpine vegetation areas (Archibald *et al.* 1996). They appear to thrive best in the Montane areas based on percentage cover figures. Thus, these two shrubs are not "strange bedfellows" occurring in the Tuscany sediments. They appear to have an association that is "writ large" in the landscape to the west of Calgary.

The question of association through time for these two shrubs can be examined through pollen studies that have been done on sites through the Foothills and Rocky Mountains to the West of Calgary. A number of key sites have been examined for the patterns of vegetation change that have taken place following deglaciation at the end of the Pleistocene and through the early Holocene to the Mazama tephra boundary. Unfortunately, the macrofossil studies that have been examined for comparison to this research are all from former pond or slough environments, which produce quite a different inventory of seed remains. There is no other macrofossil assemblage from a site like Tuscany that I am presently aware of. Thus, I am left with the fossil pollen studies as a baseline for comparison of plant species, genera and families present in the past environment.

Following deglaciation the denuded landscapes that had been scoured by ice began to be populated by vegetation that survived at the peripheries of the major ice sheets. The southwestern Alberta Rocky Mountains appear to have been revegetated from refugia that thrived in the mountain flanks further south during the late Pleistocene, in what is now Montana. MacDonald and Cwynar (1985) have demonstrated a migration pattern for lodgepole pine (*Pinus contorta* ssp. *latifolia*) moving northward from southerly refugia during the Holocene. The continental ice extended south to positions in what is now the continental United States. While it has been noted that pollen studies are lacking for the Great Plains region in general, work by Barnosky (1989) at Lost Lake and Guardipee Lake in Montana has provided two records that closely bordered the maximum extent of the Laurentide ice. In areas bordering the eastern ice-sheet in Montana, there appears to have been virtually no deciduous or coniferous tree cover on the landscape. The Continental Divide 60 km to the west of Guardipee Lake was clear of Cordilleran ice by 12,400 years BP and coniferous pollen accumulations in the lake indicate that re-vegetation of the western mountainous landscape must have been taking place. Guardipee Lake itself appears to have been located in an ecoregion that was primarily grassland already at 12,400 years BP, during the time when revegetation of the landscape would have been taking place.

Juniper, aspen, willow and rosaceous shrubs are proposed to have been thriving along the mountain flanks—their pollen was blown east to Guardipee Lake (Barnosky 1989). To the east of Guardipee Lake in the Dakotas, spruce advanced across the landscape before 12,000 years BP (Ritchie 1976). Deciduous trees followed the spruce advance and then slowly disappeared to be replaced by the prairie landscape that remained. At The Andrews Site in south-central Saskatchewan, spruce trees existed at 10,230 ± 140 ^{14}C years BP and were replaced by deciduous poplar trees which continued to thrive until some time after 8800 years BP (Yansa and Basinger 1999). The landscape eventually turned to prairie and the area dried out thereafter.

At Harris Lake, at the base of the north side of the Cypress Hills the Cypress family Cupressaceae, of which junipers are a member, is represented throughout the pollen core from 9120 ± 250 ^{14}C years BP onward (Sauchyn and Sauchyn 1991). There is no sign of the Heath family, Ericaceae, of which kinnikinnick is a member.

At Toboggan Lake, just 25 kilometers southwest of Calgary, juniper is found in small amounts of 1-3% in the pollen core dating from 9400-7600 years BP (MacDonald 1989). Again, there is no sign of any Ericaceae. Approximately 50 kilometers to the northwest, the Yamnuska Bog core contained *Juniperus* fossil pollen in amounts of 24-12% between about 10,400 and 9400 years BP (MacDonald 1982). Interestingly, at Wedge Lake only a few kilometers south but higher in elevation, there is no sign of *Juniperus* in the pollen core.

Moving up into the Bow Valley corridor to Copper Lake, Juniperus makes an appearance in zone 1 at 3% at >10,500 years BP but drops to <1% by 10,300 years BP (White 1987). At Kingfisher Pond Juniperus appears at 6% in zone 1 at >10,000 years BP but drops to 2% in zone 2 at 10,000 years BP and does not appear at all in zone 3 (White 1987). At higher elevations like Lake O'Hara Juniperus is found at <1% representation from 10,100-3000 years BP (Reasoner and Hickman 1989). Close by and only 200 meters higher, Juniperus is found at <1-2% from 8530-3000 years BP. At Wilcox Pass, the most northern and highest site reviewed, Juniperus is represented at 5-7% in zone 1 dated at >9600 years BP and drops down to 1-2% from 9600-6350 years BP (Beaudoin and King 1990).

It is clear that *Juniperus* species have played a role in the migrating vegetation cover that moved into the denuded landscapes of southern Alberta following deglaciation. The pollen studies reviewed show *Juniperus* occurring throughout the eastern Rocky Mountain flanks from the earliest dated cores from Montana, up through Alberta and over the provincial border into British Columbia. Distribution maps of *Juniper horizontalis* and *J. communis* show that these shrubs have moved into nearly every region of Canada, east and west and in the northern regions (e.g. Farrar 1995; Porsild and Cody 1980). Interestingly, junipers do not appear to be important in the shrub canopy of the boreal forest in Alberta.

In the Rocky Mountains *J. scopulorum* is also a species that has to be considered in the inventory but it is restricted to the higher elevations today, according to Farrar (1995). It is probably not present in the charred seeds recovered from the Tuscany site. However, given the similarity of the seed morphologies this assumption must remain open to correction if required. The distribution of *J. horizontalis* and *J. communis* in the present-day landscape of west Calgary is definite and assumed to have existed through time. The pollen studies summarized do not distinguish the juniper species.

The relationship between kinnikinnick and juniper that is clearly visible with the charred seeds recovered from the Tuscany site is not seen in the pollen studies. The macrofossil studies reviewed contain seeds from plants that thrive in ponds and standing water and are quite dissimilar to the Tuscany seed inventory. Thus, the relationship between these two plants is not demonstrated with either of these paleoenvironmental investigation methods.

Modern vegetation inventories for the regions to the west of Calgary contain plant community types that include kinnikinnick and juniper as part of the combined cover. Kinnikinnick grows at elevations ranging between 600-1950 meters above sea level (masl) with a mean of 1400 masl. *J. horizontalis* (creeping juniper) thrives best between 900 and 1800 meters masl with a mean of 1400 meters masl and *J. communis* (common juniper) thrives between 1000 and 2100 meters masl with a mean of 1600 masl (Archibald *et al.* 1996). Clearly kinnikinnick and creeping juniper have the greatest overlap in elevation preferences. Of the two junipers present in the Tuscany landscape, *J. horizontalis* or creeping juniper was by far the most common occurring, especially along the south facing terrace edge of the Bow River valley. The Bow River enters the west edge of Calgary at an elevation of 1130 meters elevation. The Tuscany site sits on a terrace at an elevation of 1142 meters masl overlooking the Bow River valley, which is well within the range of elevation preferred by kinnikinnick and creeping juniper. This elevation is closer to the lower boundary of elevation preferred by common juniper and may be part of the reason why it is less common in the landscape. Other factors or variables must be involved as well, of course. Interestingly, in terms of moisture and nutrient regimes, common juniper has a greater range of variability than creeping juniper (see Figures 99 and 101). Creeping juniper is by far the more prolific plant in the Tuscany landscape today and is found growing together with kinnikinnick along the terrace escarpment edge.

Comparison of the climatic information for Calgary (see Environmental Background) to that of the Natural

Subregions (see Figure 98) outlined by Archibald *et al.* (1996) show small differences between the average mean temperatures of the different regions. It is in the breakdown of the landscape into Ecoregions and Ecodistricts (see Chapter 3) where changes can be seen in the vegetation that grows east and west of Calgary today. To the east of Calgary is Ecoregion number three known as the Fescue Grasslands part of the physiographic component known as the Western Alberta Plain. The division between the third ecoregion and the fourth ecoregion known as Aspen Parkland runs north to south, through the center of the City of Calgary. The Tuscany site is located near the edge of this division between the two ecoregions, in the physiographic component known as the Southern Alberta Uplands (see Figure 5).

The list of plants identified in the landscape around the Tuscany habitation site reveals the mixture of plant species that show the transition from Fescue Grasslands to Aspen Parkland, something not very discernable to the naked eye. The amounts of charred twig remains found in the Tuscany flotation analysis along with the similar set of remains found at the Bowfort Road site (see Chapter 7) indicate the possibility that this division between the Aspen Parkland and the Fescue Grasslands was more pronounced in the area during the early Holocene. Fire ecology has also played an important role in the landscape of the Tuscany habitation site during the past and certainly contributed to clearing the area of shrubby and possibly larger tree taxa through time. Following the Mazama tephra deposition boundary the aeolian deposited sediments formed a chernozem. There is very little evidence of charred plant remains in this chernozem, certainly nothing like the quantities of small charred twigs present in the paleosols. Possible explanations for this include: 1) the possibility that all burned remains were blown away after fires swept through, 2) there were far fewer fires from 6730 years BP to the present and thus few burned remains, or 3) the nature of the vegetation cover was different leaving behind fewer remains once burned.

The vegetation cover at the site today is indeed quite different from that evidenced by the charred seed remains recovered through flotation analysis. The third explanation is preferred here. A thorough analysis of the sediments through the site profile may clarify differences observed between the upper and lower soil zones.

There is a distinct possibility that a parkland-like tree and shrub dominated vegetation cover may have extended further east during the early Holocene. Strong (W.L. Strong, Ecological Land Surveys Ltd. 1992) has shown that the Fescue Grass Ecoregion to the east of Calgary parallels the Cordilleran climate more than Grassland because it sits at higher elevations where there are cooler summers, warmer winters, and increased precipitation.

The general pattern of change through time that has taken place at the Tuscany habitation site is presented in a series of figures that illustrate the changing sediment deposition combined with vegetation cover proposed for the site locale. In Figure 102, the denuded landscape is represented for the period of time following the draining of Glacial Lake Calgary at 13,900 calendar years ago (Fisher 1999). Glacial Lake Calgary is proposed to have drained (Fisher 1999), exposing the sediments at the Tuscany site that were deposited during the existence of the lake. This event preceded the developments that then took place in the Tuscany depression locale.

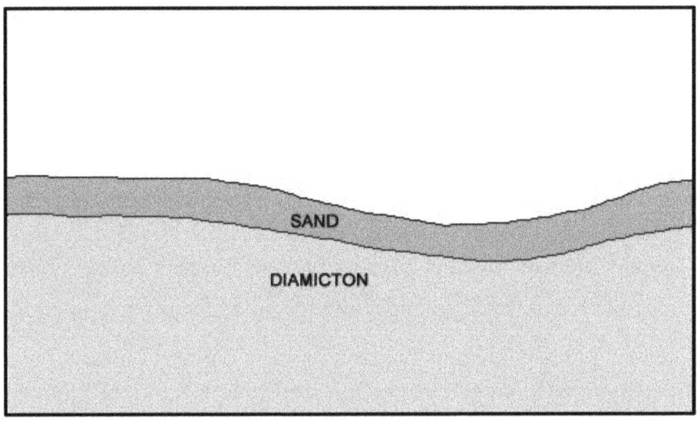

Figure 102: Diagram depicting the sediment profile at the Tuscany habitation site during the period immediately following the draining of Glacial Lake Calgary. The timeframe for this event ranges around 13,900 years BP. The scale of the drawings presented here is intended to depict the area covered by the excavation grid, which was 12 meters east to west

There must have been forbs, grasses and some shrubs present in the area, and possibly forests developing to the west of Calgary at what is now Cochrane. This vegetation must have existed to support some of the megafauna present in the landscape whose fossil remains have been recovered from deposits in Calgary and Cochrane at approximately 11,400 years BP. The deposition and development of the fifth pedogenic zone identified in the Tuscany sediment profile (Layer M) began to take place. Since charred plant remains have been recovered at this level, there must have been fires in the region starting at an early point in the colonizing vegetation. The radiocarbon date of 9950 ± 120 years BP derived from charred plant remains is from Layer M. Figure 103 depicts a landscape that developed after 10,000 years BP. At approximately 10,000 years BP deciduous and coniferous trees appear to have thrived in various areas of the landscape that are part of the Great Plains region today (e.g. the Andrews site (Yansa 1998) and the Fletcher site (Vickers and Beaudoin 1989). It is not certain how extensive the tree cover may have been in the Tuscany area, but there appears to be evidence of at least some shrub cover and perhaps some trees during this time frame. Figures 49 to 57 provide profile views of the densities of charred woody plant remains recovered from the bulk sediment samples from each level within the paleosols of each unit. These profiles provide part of the evidence used to make the inference that there was probably some shrub cover in the landscape of the site during the deposition of sediment that eventually

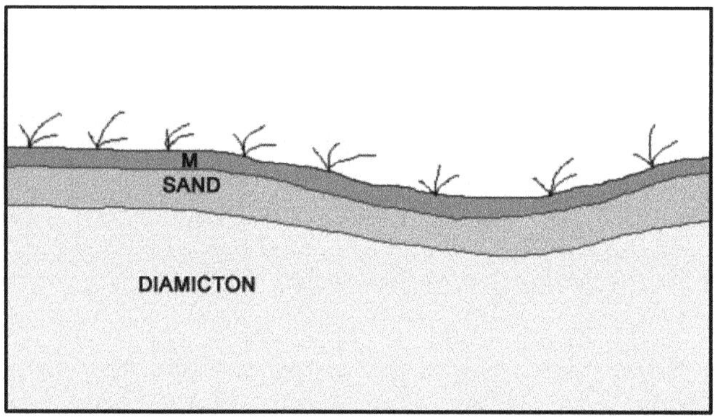

Figure 103: The initial vegetation colonizing the landscape were grasses and forbs. Layer M (the fifth pedogenic unit described in Chapter 5) was formed, initial soil development at the site. The timeframe is just prior to 10,000 years BP. Fires must have already begun to burn the landscape. A few charred plant remains were recovered from sediments comprising Layer M

cell structure that looks like *Salix* sp. (willow) or *Populus* sp. (poplar). This identification is very tentative as there are thousands of pieces of charred woody material in the Tuscany flotation samples and identification of one example is not statistically significant. However, the presence of *Picea glauca* (white spruce) seeds and cone scales indicates that there was possibly a mixture of deciduous trees, shrubs and occasional conifers covering the landscape of the site during this time period. Kinnikinnick and juniper comprise a significant part of the low vegetation cover. Again, the evidence for this inference is provided in the profile views in Figures 71 through 88 for the kinnikinnick and juniper seed remains from the site. Nearly all the macrobotanical materials recovered through flotation are charred and certainly all the remains from the lower levels of the excavation are charred. This indication of periodic fires sweeping through the landscape fits in with the evidence of charcoal remains quantified in the fossil pollen studies reviewed.

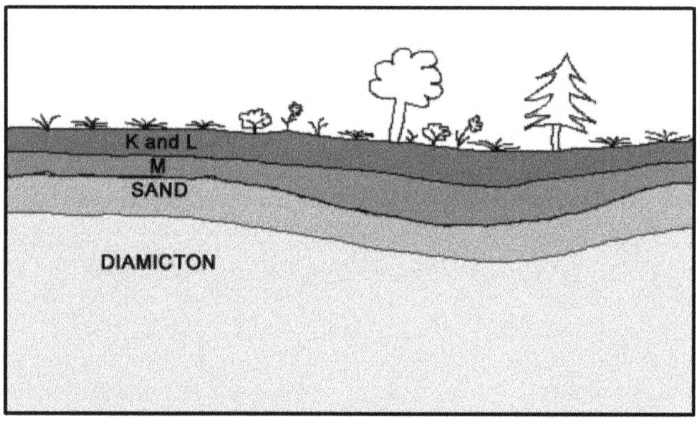

Figure 104: Kinnikinnick and juniper begin to fill in the landscape along with willows, deciduous and conifer trees. Layers K and L are formed. This initiates the development of the fourth pedogenic unit. This vegetation burned at some point at least once, possibly more than once. The timeframe depicted here spans 10,000 to 9000/8500 years BP

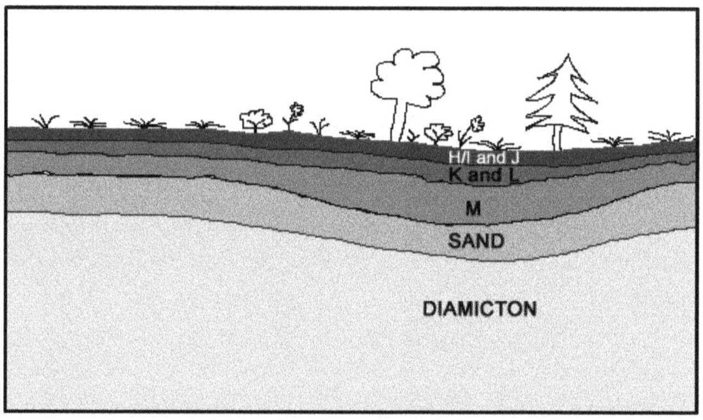

Figure 105: Kinnikinnick and juniper fill the land surface again, along with willows, deciduous and conifer trees and herbaceous plants as undergrowth. Layers H/I and J are formed as the third pedogenic unit. This vegetation burns off again at some point, possibly more than once. People came to camp and hunt. The timeframe depicted here is ~8000 years BP

comprised the formation of Layer M. Closer examination of the woody charred plant remains is required to clarify the identification of the plant species represented by these materials. Charred seed remains do not appear in the bulk sediment samples until points 15 to 20 centimeters above the sediments identified as Glacial Lake Calgary sands. This pattern of the first occurrence of charred kinnikinnick and juniper seeds in layers above Layer M is visible in Figures 71 through 88.

Following the deposition and development of Layer M, the landscape in the Tuscany site area appears to have become covered with a significant amount of shrub and tree cover. This is depicted in Figure 104, which also shows the deposition and development of the fourth pedogenic zone (Layers K and L) during the time period following events that took place in Layer M. The examination of a piece of woody material from excavation Unit 14 Level U reveals a pore pattern and

In Figure 105, the timeframe of approximately 7800 years BP is represented. Eventually, Layers H/I and J of the third pedogenic zone developed, probably over the time period following the occupation of the locale by a nomadic hunting group. During the time of occupation of the site kinnikinnick and juniper covered the landscape again along with willows, deciduous and conifer trees and herbaceous plants as undergrowth. The entire series of profiles including Figures 49 to 57 for the charred plant fragment densities and Figures 71 through 88 for the kinnikinnick and juniper seeds in the profiles provide evidence for this inference. This suite of vegetation is susceptible to fire, which appears to have occurred at least once and possibly more often in the area. The charred botanical remains are concentrated in Layer J, which is designated as a B horizon of this pedogenic zone. The cultural lithic remains are concentrated in this layer along with the charred plant material. The Lusk projectile point is located in excavation Unit 37 Level S (N521, E526), which is in between Layer H/I and J of the pedogenic zone. The soil appears to have developed further for some time after people had camped in the area, differentiating into an A and B horizon with the already insitu cultural materials incorporated well within them.

In Figure 106, the time period of 7800-6730 years BP is depicted. Kinnikinnick and juniper remain in the landscape as low ground cover, however, their growth may have been more dispersed and in lower amounts than in the previous time periods. Refer to Figures 71 through 88 for the observed decrease in the occurrence of these charred seed remains in the upper parts of the paleosols. Grasses, forbs and shrubs continue to occupy the landscape. This inference is, again based on the lower densities of charred plant remains visible in Figures 49 to 57 and the decrease in kinnikinnick and juniper seed remains in the upper paleosols observed in Figures 71 through 88. Fires sweep through the area possibly more than one time but very few remains of a shrubby/woody nature are left behind compared to the previous densities observed in the flotation samples. Thus there appears to be far less cover that is made up of shrubs or trees, hence few substantial wood fragments are produced and the insubstantial charcoal of grasses and forbs are not preserved. No spruce seeds or cone scales are found in this part of the profile or in any part above. The sediment constituting Layers F and G is deposited and the formation of the last pedogenic zone takes place over time, differentiating this sediment into an A and B horizon eventually. This is the second pedogenic zone. It is the uppermost of the four pedogenic zones (zones 2 to 5) within the paleosols.

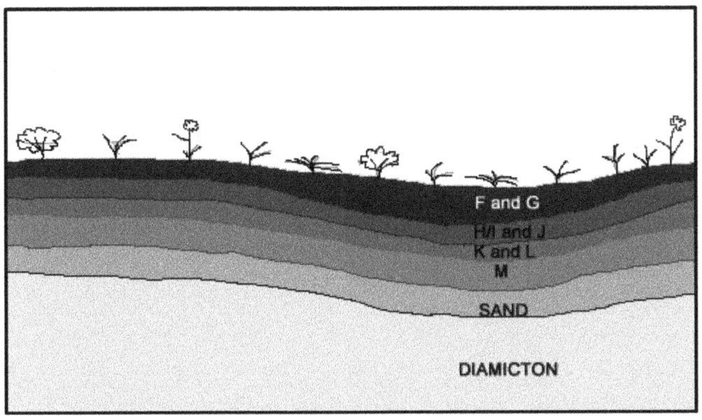

Figure 106: Kinnikinnick and juniper remain in the landscape but in low numbers along with grasses, shrubs and forbs. Fires sweep through but very few remains are left behind because most plants are not woody. Layers F and G are formed as the second pedogenic unit. They represent the last buried paleosol formed at the site. The timeframe depicted here is from 8000-6730 years ago BP

In Figure 107, the deposition of the Mazama tephra is depicted, which capped the paleosols. Following this event dated to 6730 ± 40 years BP, the sediment composition appears to be quite different. It is aeolian in origin and deposited to depths of 1.2 meters in parts of the depression. The first pedogenic zone is a chernozem that developed in this last sediment deposition.

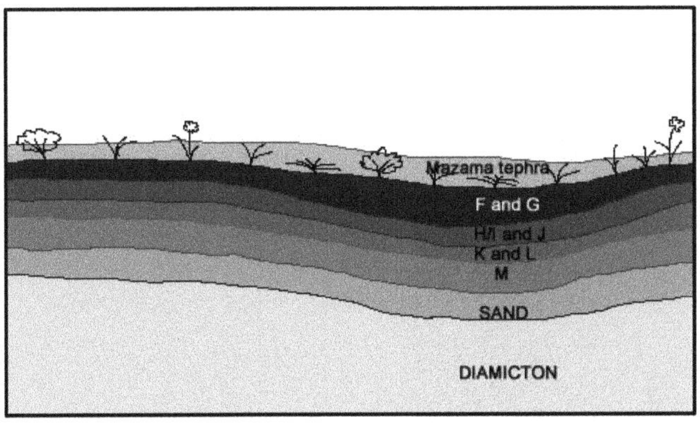

Figure 107 The timeframe is now 6730 ± 40 years BP. Mazama tephra covers the landscape

Figure 108 depicts the depositional profile of the site as a general scheme. At the bottom are the glacially deposited clays and sands. The paleosols lie above followed by the Mazama tephra deposition. The developed chernozem of today is the uppermost pedogenic zone and is covered by grasses and forbs with only occasional, very low *Symphoricarpus occidentalis* (buckbrush) shrubs. The difference between the amount of botanical materials found in the upper sediments above the Mazama tephra and the paleosols below is marked. There are only very occasional pieces of charred plant in the upper part of the site. Thus, the vegetation appears to be similar to that visible today which would leave only a few remains behind if a fire swept over the landscape. Any charred plant material produced from grasses and forbs would not be robust and hence would not be preserved as a record of this vegetation.

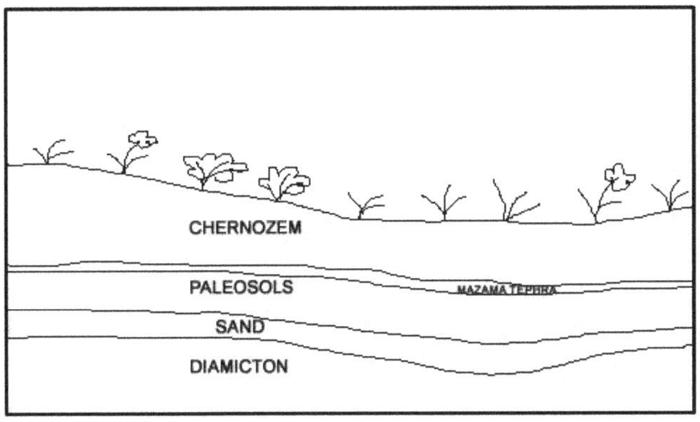

Figure 108: This diagram represents the present day landscape, which is covered with buckbrush, grasses and forbs

CULTURAL LANDSCAPES

One of the problems for this paleoethnobotanical analysis of the Tuscany paleosols is the associated cultural material and its placement or fit within the cultural chronological scheme that has been derived for the Great Plains. Generally poor preservation of wood and bone have led to the primary use of diagnostic lithic material remains such as projectile points as the evidence to fit sites into the accepted chronology. In the case of the Tuscany site, there was a small lithic assemblage recovered that is available for analysis. The most diagnostic item recovered from the paleosols was the small lanceolate projectile point shown in Figure 109. A line drawing of the point (Figure 110) is useful to view the flaking scar pattern that is the result of its manufacture. One of the important characteristics revealed in the line drawing is that the point appears to have been broken and reworked. The sides have been ground laterally from the base for a length of approximately 1 cm. The base has not been ground and appears to also have been broken partially. The flaking pattern used to create the point was collateral and parallel. The flaking scars from reworking the upper half of the point appear to be more expedient, to fix the point so that it could be useful again.

Figure 109: Two views of the Lusk point found in the Tuscany habitation site excavation in Unit 37 Level S (N521, E526)

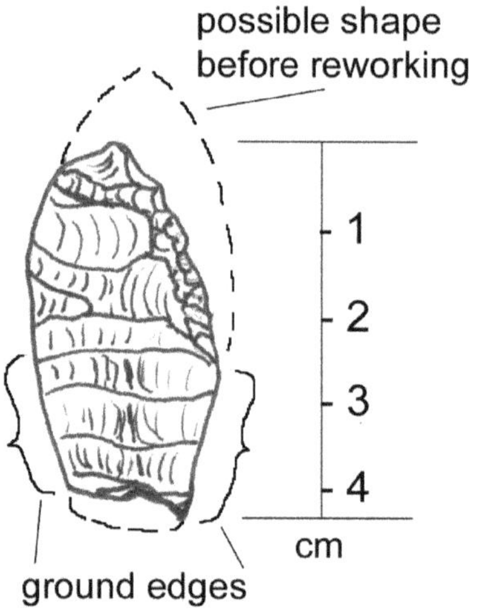

Figure 110: A line drawing of the Lusk projectile point. The collateral and parallel flaking pattern is clearly visible on the bottom half, from the base up to 2 cm. The dotted line shows the possible size of the point before possible breakage and indicates the point where the flaking pattern ceases to be collateral and parallel. The upper half of the point is far more crudely flaked and has been resharpened along the right edge

The projectile point was initially identified as a reworked Agate Basin point. As noted by Kooyman (2000), reworking of points is a common feature of the Agate Basin Complex. In the chronological cultural scheme for the Great Plains region, the dates for the Agate Basin Complex range from 10,500-10,000 years BP, in the United States. As noted by Vickers (1986), few Agate Basin points have been found in Alberta and the established chronology is considered acceptable as an extension of the scheme derived from eastern Wyoming. In terms of appearance, the point did fit the reworked Agate Basin designation. However, subsequent dating of charred botanical remains and bison bone in close association with the point, yielded two uncalibrated radiocarbon dates that placed the projectile point near the 8,000 year-old timeframe. This resulted in a re-evaluation of the point, which appears to be part of the Plains/Mountain complex as defined by Reeves (1975). As part of the Plains/Mountain complex, the point is classified as a Lusk point, which would place it within the general timeframe that has been derived for Lusk of 7,950 to 7,450 (Vickers 1986:37). The following discussion evaluates available information on the relationship between the Agate Basin and Plains/ Mountain complexes in the northwestern Plains to assess the associated foraging pattern observed for the Tuscany habitation site and the conclusions that may be drawn.

According to Hofman and Graham (1998), there are problems associated with our understanding of the origins and distribution of the Agate Basin Complex. Projectile points that resemble the Agate Basin type appear to be distributed over a large area that extends outside the Plains including Canada (Hofman and Graham 1998:105). However, technological relationships between the materials from these extended areas are not clear and the dates are also problematic because they are often <10,000 years BP (Hofman and Graham 1998:105).

The following discussion examines some of the ideas that have been advanced with regard to Agate Basin complex materials. It has been suggested by Arnold (1985) that there should be a range of dates for Agate Basin complex materials because of the movement of populations across a number of regions through time. The possibility of a time-transgressive northward occurrence of the Agate Basin complex has been proposed by Ebell (1982) in southern Manitoba and is linked to the observation of a similar time-transgressive movement of environmental change. Ebell (1982) has proposed that as the grasslands migrated to the north and east following the continental glacial retreat, a concurrent migration took place of a population that manufactured and used Agate Basin complex tools. This migration of population and tools eventually ended up in the boreal forest environment as far north as the Acasta Lake and Grant Lake sites in the Northwest Territories where the remains date to less than 7000 years BP (Ebell 1982).

Arnold (1985) conducted a statistical analysis on the lithic assemblages of three separate Plano complexes that had been proposed to be related including the Lakehead (northwest Ontario), Caribou Lake (southeastern Manitoba) and the Beverly Unit (Keewatin District, NWT). The analysis showed the assemblages were actually different from each other and that the Lakehead and Caribou Lake complexes were not related or derived from the Agate Basin complex. However, the Beverly Unit materials did appear to be related to Agate Basin materials from the Wasden site in Idaho. Materials from the Grant Lake site, which is part of the Beverly Unit, have been dated between 9000-8000 years BP by Wright (1976:91). This interpretation would appear to change the focus of possible migrations from eastern to western origins.

Wilson and Burns (1999:228) note that a "Northern Plano" series of points from the Northwest Territories, dating between 9000-6000 years BP and looking similar to Agate Basin have been used to try and establish links between southern and northern areas. However, Wilson and Burns (1999:228) further note the morphology of these points persists through to the later times of Acasta and Taltheilei tradition in the Northwest Territories. According to Wilson and Burns (1999:228) the morphology of this point style is simple and could be easily reinvented—the south-north migration linkage is extremely tenuous.

Agate Basin projectile points have been fitted into the Plano complex which includes the unfluted series of points that follow Clovis and Folsom fluted projectile points (Hofman and Graham 1998). Importantly, the Plano complex ties together a series of projectile points that date between 10,200 and 8000 years BP and are generally associated with a near year-round focus on bison hunting (Hoffman and Graham 1998:103). In Alberta the Plano points series includes Agate Basin/Hell Gap, Alberta/Cody, and Frederick/Lusk, dating between 10,450 to 7450 years BP, which is essentially the same as the outline from the United States (Vickers 1986:36). Agate Basin points have been found throughout the province of Alberta. Interestingly, Vickers (1986:36) notes "...most other Early Prehistoric lanceolate projectile points in Alberta are from surface contexts." However, these kinds of finds are not very useful to the problem of understanding cultural historical models (Vickers 1986:39).

The problem with the present cultural historical model for the Plano sequence in this part of the northern Plains lies with the evidence that is available to derive the conclusions that have been advocated. Frison (1991:164) has noted that systematic communal bison hunting is associated with the Agate Basin timeframe. However Frison (1991:164) also notes that only one bone bed has actually been found for the Agate Basin complex—the Agate Basin component from the Agate Basin site.

It is at this point that the material evidence available from the Tuscany site becomes problematic. The first problem arises with the dates that have been derived from two different sources associated with the reworked point and the hearth. The dates of 7840 ± 100 ^{14}C years BP from charred plant remains from the hearth and 7610 ± 70 ^{14}C years BP from bone collagen, are much later than expected if the point is from an Agate Basin component. The oldest date derived for the Tuscany site, 9950 ± 120 ^{14}C years BP, is problematic because it is also much more recent than expected, as noted above. However, a date from collagen from a bison third phalanx at 6940 ± 70 ^{14}C years BP, is very well positioned for the date of 6730 ± 40 ^{14}C years BP for the Mazama tephra deposit immediately above the bone. The question remaining is whether or not these dates should be accepted. There is no legitimate reason apparent yet, to reject the dates for the Tuscany site. If accepted, they indicate that the point is more likely to be Lusk, rather than Agate Basin.

A problem of a different nature arises if we place the Tuscany habitation site into the Agate Basin complex with its associated systematic communal bison hunting. The lowest/oldest cultural component of the Tuscany site appears to have a broad scale foraging pattern associated with it, though it is still not completely clear if all the faunal remains are the result of successful hunting endeavors by the past occupants of the site. One method to examine the general 'fit' of the Tuscany site within the cultural model is to compare the evidence available from other sites that are regionally associated.

Excavations at DlPo-20, a site "located in the Front Range of the Rocky Mountains in southwestern Alberta" (Reeves and Dormaar 1972:325), recovered an archaeological assemblage including an Agate Basin projectile point that was assigned to the Mountains-Foothills cultural complex dating between 10,000 to 7500 years BP (Reeves and Dormaar 1972:333-334). While a specific faunal assemblage was not itemized, the site was identified as part of a:

> ....larger land/resource utilization pattern which included the use of base camps, other hunting camps, fishing stations, and bison kills, for the seasonal exploitation of a variety of environmental resources in various life zones and habitats in the southern Alberta Rocky Mountains. (Reeves and Dormaar 1972:333-334).

By 1975, extensive research had been completed (between 1967-1971) in Waterton Lakes National Park in the southwestern corner of Alberta (Reeves 1975). Assemblages from 12 archaeological sites dating between 10,000 to 7500 years BP, solidified the interpretation of a resource utilization pattern identified now as the Plains/Mountain cultural complex (Reeves 1975). This complex is identified as unique from both the Cody complex found to the east on the Plains and the Cascade phase from the Plateau area to the west (Reeves 1975:245). Reeves (1975:245) notes that the basis for the Plains/Mountain complex stems from a model called the Mountain-Plains culture where Swanson (1962:154) proposed:

> For deductive, hypothetical purposes, I would view this culture as adapted to parkland forests of the

Rocky Mountain Range and adjacent grassy plains which intergrade with mountain valleys and which form part of the natural setting.

Reeves (1975:245), agreed with Swanson (1962) that the Agate Basin phase, dating to 10,000 years BP, was antecedent to the Plains/Mountain complex.

Frison (1991) outlines an interesting dichotomy that appears in the archaeological record along the eastern flanks of the Rocky Mountains in the United States, which appears to parallel the model outlined by Reeves (1975), for the equivalent topographical area located further north in Canada. According to Frison (1991), from 10,000 years BP and later, a Foothill-Mountain Paleoindian occupation chronology is found with diversified subsistence strategies that appear similar to later groups from the Archaic period. On the open Plains, the Paleoindian occupations demonstrate a subsistence strategy oriented to the part-time hunting of bison (Frison 1991). As noted by Frison (1991), numerous sites in Montana and Wyoming have aided in defining the Foothill-Mountain grouping.

A site that appears to fit the Plains/Mountains model within the greater region to the west of Calgary and the Tuscany site, is the Vermilion Lakes site located in the Bow Valley corridor near the town of Banff, approximately 130 km west of Calgary. The deepest/lowest cultural component of this site contains evidence of structures that may be similar to the proposed Tuscany trample area outlining an activity zone. Unfortunately there are no diagnostic points associated with the structures and their material remains dated to 10,770 ± 175 years BP. Nevertheless, tools and lithic debitage from the occupation are tentatively proposed to be most representative of Clovis and Folsom style artifact remains from other sites in Alberta (Fedje *et al.* 1995). The materials that have been identified as Agate Basin complex at the Vermilion Lakes site date between 10,310 ± 190 (RIDDL 528) and 9570 ± 150 (RIDDL 75) ^{14}C years BP (Fedje *et al.* 1995). The combined cultural record including stemmed projectile points, tools and bones have been used to assign the cultural associations in both cases. The foraging pattern associated with the early cultural components of the Vermilion Lakes site appears to be broad spectrum though there is a focus on bighorn sheep (*Ovis canadensis*). Other faunal remains include bison (*Bison* sp.), caribou (*Rangifer tarandus*), beaver (*Castor candensis*), and snowshoe hare (*Lepus americanus*) (Fedje *et al.* 1995). This foraging pattern covers both the Clovis-Folsom and Agate Basin components.

Study of the lithic materials from Vermilion Lakes clearly differentiates two groups. As noted by Fedje *et al.* (1995), the first group of lithic materials are comparable to fluted point traditions like Charlie Lake Cave in northeastern British Columbia (Fladmark 1996, Fladmark *et al.* 1988), which contained a Clovis component dated to 10,550 years BP. The faunal assemblage for the Clovis component at Charlie Lake Cave includes a variety of 30 different species, predominantly small mammals and birds. The smaller animals (e.g. 163 snowshoe hare) are considered to represent prey brought into the site by owls and raptors (Driver 1996:22-25). Only the bison bones associated with the cultural materials are considered the product of human predation activity (Driver 1996, Fladmark 1996, Fladmark *et al.* 1988).

The second group of lithic materials from Vermilion Lakes is noted (Fedje *et al.* 1995) to be comparable to the Stemmed Point tradition of the Columbia-Great Basin region that has been proposed to have replaced the Fluted Point tradition by 10,500 years BP from Wyoming to southern Alberta.

The paleoenvironment of the Vermilion Lakes site has been derived from a combination of pollen and macrobotanical analyses from collected soil samples and comparison to studies from other sites. The initial vegetation cover appears to have been low and herbaceous to shrubby in nature with sage (*Artemisia* sp.), plantain (*Plantago* sp.), and wolf willow (*Elaeagnus commutata*), plants that would have thrived on bare, recently deglaciated terrain. The presence of a macrofossil needle of *Abies* (fir) indicates that conifer trees were possibly in the areas close by. By ca. 10,300 years BP the landscape appears to have been covered by low shrubs, herbs, and grasses changing to a tree-dominated landscape by 9,900 years BP, based on the Copper Lake pollen analysis (Fedje *et al.* 1995).

Gryba (1983) recovered Agate Basin points from the excavations at Sibbald Flats (EgPr-2). The stratigraphy of the site was very compressed and interpretation was problematic. There were no faunal remains that could provide any meaningful interpretation of a foraging pattern to associate with the Agate Basin materials.

The Crestmont Estates site (EgPn-428) was located on an area known as Paskapoo Slopes on the south side of the Bow river, which is south and slightly east of the Tuscany site (EgPn-377) on the north side of the Bow river. The stratigraphy of EgPn-428 was very similar to that of EgPn-377, containing a paleosol in the sediments beneath a Mazama tephra layer. A date of 6840 ± 70 (Beta-125408) derived from collagen extracted from four bison cervical vertebrae was considered unacceptable (Vivian and Reeves 1999). According to Vivian and Reeves (1999), an acceptable date would have been in the range ca. 9000-7750 years BP, based on the stylistic attributes of a lanceolate projectile point found within the paleosols and determined to be Lusk, from the Plains/Mountain complex. The faunal evidence fits into the subsistence pattern established for the complex, though some of it may be considered tentative. Actual faunal remains of bear and a dog-like species are present, however, bison, sheep deer and rabbit are indicated only by blood residue analysis acquired from lithics.

Frison (1991) notes that there are a number of problems with the Lusk cultural complex. It was first proposed based on a small number of artifacts from one level of the

Hell Gap site and from the Betty Greene site, 64 km north of Hell Gap (Frison 1991:66). While there is a date of about 7900 years BP from the Betty Greene site, its provenience requires clarification, according to Frison (1991:67). Vivian and Reeves (1999:24) note that the Lusk complex is "poorly documented" as yet.

One site of particular interest to paleoethnobotanical research along the eastern flanks of the Rocky Mountains is the Barton Gulch site in southwestern Montana. This site is classified into the Foothill-Mountain grouping by Frison (1991:67). Projectile points recovered are lanceolate with a parallel-oblique flaking pattern (Frison 1991). The Alder complex of the Barton Gulch site has been dated at 9410 ± 140 years BP (no lab number listed). Armstrong (1993:3) notes that the faunal assemblage of the Alder complex indicates the use of local resources as were available in the environment at the time of occupation of the site. The list of faunal remains is noted to be incomplete but noteworthy is the fact that "no bison remains were recovered from the Alder complex stratum" (Armstrong 1993:3). Armstrong lists utilized faunal taxa, which include cottontail rabbit (*Sylvilagus* sp.), indeterminate rabbit (Leporidae sp.), mink (*Mustela vison*), and deer (*Odocoileus* sp.).

Far more important information related to plant and animal foodstuff preparation has been recovered from the Barton Gulch site. A total of 18 hearths, 57 roasting pits, and 125 post molds were excavated. Sediment samples were taken for further examination (Armstrong 1993:11). Macrofossil remains included charred seeds from 28 taxa and 5 non-seed remains of 3 other taxa for a total of 31 identified plant taxa remains from the Alder complex component of the site (Armstrong 1993). In contrast to the samples and charred plant remains recovered from the Tuscany site, the Alder complex samples and charred plant remains were all recovered from feature fill and thus represent a very different set of data for analysis. The Alder complex remains are clearly associated with cultural activity.

For comparative purposes, the characteristics of a broad scale foraging pattern can also be found with the Jurgens site located approximately 90 km north of Denver and 1.6 km from the town of Kersey in Colorado (Wheat 1979). The oldest cultural component of the site dated to 9070 ± 90 years BP (SI-3726) (Wheat 1979:151). The points pictured by Wheat (1979) appear to be Scottsbluff, though the site is described as being Agate Basin. The timeframe is appropriate for Scottsbluff materials. Area 1 contained the remains of at least 31 butchered bison. Interestingly, there were also the remains of white-tail and mule deer, moose, antelope, elk, rabbit, beaver, muskrat, canid, birds, turtle, and fish (Wheat 1979; Dixon 1999). Reworking of broken points appeared to be a common practice at the Jurgens site as well. Of great relevance to a paleoethnobotanical perspective, is the presence of stone slabs for the processing of plant materials (Dixon 1999), an important indication of a diversified diet spectrum. The potential use of plants hints at an alternative, expanded and more dynamic view of the landscape by these early populations, if we consider that hunting was not the only focus of their interaction with the environment.

SEASONALITY

Season of occupation for the sites examined here has not been specifically discussed. Clearly, any site containing seed remains that have ended up in a feature like a roasting pit, hearth, or storage pit, could have been occupied during any season. Items such as grinding stones, manos and metates indicate probable occupation of a site during the latter part of the summer when most edible plants produce seeds, a fruiting body containing seeds, or mature roots/tubers or other plant part that can be harvested and processed. There are some plants that are better harvested when immature, like the center pith of cattail stems (*Typha latifolia*) (Siegfried 1994). However, as a general rule, seeds, berries and roots require maturity, which comes with the late summer. If storage is employed as a harvesting strategy, a site could be occupied during the winter and spring continuously, long past the time of harvest. A form of storage can also take place if roots/tubers are left in the ground and harvested the following spring. Thus, other means must be employed to ascertain which season(s) may represent the time of site occupation. Faunal remains can provide alternate seasonal information based on the maturity of the skeletal material; i.e. the presence of fetal bones indicates an animal was killed in the winter or spring.

The season of occupation of the Tuscany site is not clear. There were no tools or grinding implements recovered that might have been used in plant processing. This statement is not meant to imply any expectation that there should be such tools present, it is a simple observation of the facts. It is clear that such tools have been found at other eastern slopes sites in Paleoindian timeframes (discussed above), however, this is not observed for the Tuscany site. One of the main reasons for ambiguity with the seasonality of the plant remains lies in the nature of the sampling regime employed at the Tuscany site. There were no features to sample except for the single hearth. The bulk sediment samples represent random examples of the sediments within the entire site context. The charred seeds that have been recovered do not indicate a season of occupation for the Tuscany site, as discussed above, due to the nature of seed propagation for the plant species identified. Unfortunately, the faunal remains recovered from the site do not yield any evidence of possible season of occupation for the site either. None of the bone remains examined reveal any diagnostic characteristics that can be used to determine possible age at death, which can be used with general life cycles to indicate a possible time of year when an animal died (G. Oetelaar, pers. comm.).

The possible season of occupation of the Tuscany site remains a matter of conjecture. The occupation does not appear to have been extensive. Given the amount of faunal remains recovered and the animals represented, it is estimated that the occupation represents no more than approximately three weeks of habitation in the locale by a small group of 3-8 people. It is hypothesized that the weather must have been warmer at the time of occupation. The activity area of flintknapping represented by the lithic remains is located two to three meters away from the proposed shelter surrounding the small hearth feature. Since these two activity areas are separate, flintknapping at the site appears to have taken place outside the shelter. The seasons when the weather would permit this activity without shelter extend from the late spring, through summer and late fall. If plant collection and processing was one of the activities carried out by the group, occupation during the latter part of summer and early fall would be expected, when plants matured and berries ripened for gathering.

THE SITE LOCATION

This discussion on cultural landscapes has examined what kind of settlement the Tuscany site represents and how it might fit within a larger settlement pattern for a population living within the greater region. The site appears to represent a relatively short occupation of a locale by a small, nomadic hunting and foraging group of people. The diverse inventory of animal species recovered from the site indicates that a broad-scale foraging strategy was being practiced by the small group of nomadic hunters.

The question of why this particular site locality was chosen for occupation can only be answered hypothetically. It is clear that hunting endeavors were successful, and it can be suggested that the site was chosen for the range of fauna available for the small group to hunt. The macrobotanical data provide another perspective to our perception of the locale. It appears to have been a relatively sheltered spot on the landscape with shrubs and possibly a few trees that would have lent some cover for a small encampment. There is evidence that at least two berries, kinnikinnick and juniper, were available perhaps in quantities suitable for harvesting and a number of other plants represented in the inventory indicate diversity of vegetation in the local environment. Thus, it can be suggested that vegetation itself may have been a factor in the choice of location. However, it is suggested that more than likely both the faunal inventory and the vegetation inventory played fairly equal roles leading to the choice of the site as a place to camp.

The Tuscany habitation site appears to fit into the model outlined for a broad-scale foraging pattern of landscape occupation, similar to that proposed for the Plains/Mountain complex. In this model, the environment must be widely explored and known so that all available resources of potential use to a population can be taken advantage of when required. Intimate knowledge of the resources available at different locales, governs the choice of camp locations. The scale of resource choice determines the size, intensity, and length of occupation of these locales selected by populations living within the known landscape.

A model for subsistence strategies within the the Crowsnest Pass area of southwestern Alberta has been outlined by Driver (1978, 1993). The Crowsnest Pass represents a unique geographical and environmental area within the southern Alberta eastern slopes region. There is a significant extension of the Plains environment, westward into the Crowsnest Pass, which is part of the Old Man River floodplain. Driver (1978, 1993) has proposed a seasonal subsistence pattern which is broadscale during winter in the westernmost reaches of the Crowsnest Pass, changing to a focus on bison found in the eastward grazing areas during the summer. This pattern appears to be corroborated with information from places like the Welsch site (Van Dyke and Unfreed 1992), located on a terrace above the Old Man River in the open, grazing areas east of the Crowsnest Pass.

This model presents a different scenario when compared to observations for the Tuscany site faunal assemblage and evidence of the site seasonality. The Tuscany site appears to have been part of a transitional Parkland landscape during the early Holocene. The human occupation of the site is proposed to have taken place during the early, middle or late summer. This does not fit into the seasonally oriented model for the Crowsnest Pass. As well, if the Plains/Mountain complex is identified as having a strictly seasonally adapted foraging strategy similar to that observed for the Crowsnest Pass area, the Tuscany site would not appear to fit completely well within that model either. The foraging pattern for the Tuscany site appears to be broadscale and related to the surrounding environment and animals found within it; the people hunted what was available in the landscape. Thus, season of occupation does not appear to be a factor related to the foraging pattern observed at the Tuscany site. The question that remains here is whether this site represents an anomalous pattern within the paleo-southwestern Alberta landscape or whether it can be demonstrated to be similar to other sites within the greater region. These ideas remain to be tested against sites with comparable stratigraphy and cultural and environmental remains.

Chapter 7 POLLEN AND MACROBOTANICAL STUDIES IN SOUTHERN ALBERTA

INTRODUCTION

One of the research objectives of the Tuscany Archaeological Project is to derive an understanding and description of the change in vegetation and climatic trends at the site during the early Holocene time period (10,000-6800 years BP). The present study has recovered charred macrobotanical remains which have provided a partial vegetation inventory. Four radiocarbon dates derived from materials deposited within the paleosols indicate that deposition and soil formation processes fall within the 10,000-6800 years BP time frame.

Most paleoenvironmental reconstruction for southern Alberta and adjacent regions has been derived from pollen analyses and more recently by a number of small-scale macrobotanical studies. A small number of studies have yielded information on buried paleosols from both geographic and archaeological perspectives. Both sets of information are examined here for comparative purposes to test the value and fit of information derived from the Tuscany site study. Table 9 provides a list of the pollen and macrobotanical study sites reviewed here. Locations of the study sites are provided in Figure 111.

REVIEW OF HOLOCENE VEGETATION ZONES FROM POLLEN PROFILES IN THE ROCKY MOUNTAINS: POSTGLACIAL TO 6730 ± 40 YEARS BP (MAZAMA TEPHRA EVENT)

The Wilcox Pass pollen profile demonstrates a definite pattern of changing vegetation through time. The interpretation of the earliest assemblage zones proposes an initial establishment of tundra-type cover which is characteristic of recently denuded post-glacial environments. According to Osborn and Luckman (1988), by the beginning of the early Holocene the limits of glaciation in the mountains were close to the present day boundaries. Vegetation cover has usually been interpreted as a colonizing response. However, it is apparent that vegetation cover continues to change through time. Initially, tundra-type vegetation is proposed to have colonized the exposed land surfaces. This was followed by a trend to conifer tree cover prior to 9600 years BP, which began with *Pinus* species which appear to be arboreal colonizers that later give way to *Picea* through time.

Beaudoin and King (1990) interpret this trend as the migration of vegetation and successional regimes rather than change due to climatic oscillations. The pattern appears to be an oft-repeated one, recurring in many pollen profiles throughout the Alberta montane and foothills regions (Beaudoin and King 1990). During the early to mid Holocene time period, conditions are interpreted to have been warmer and drier than during post-glacial or present times. The increase in *Picea* percentages is interpreted to be suggestive of warmer conditions. As well, the predominance of haploxylon pine (*Pinus albicaulis* and/or *P. flexilis*) is probably related to migrations and competition within the vegetation landscape (Beaudoin and King 1990).

The Crowfoot Lake core provides the oldest date, at 11,330 ± 220 years BP, for any of the mountain sites summarized here. It also provides a long clear record of deposition from the immediate postglacial period to the Mazama tephra boundary (and beyond). The record shows the same gross trends as the Wilcox Pass core with an initial period of low grass-tundra-shrub vegetation on the landscape, followed by an arboreal stage. Unfortunately, the Wilcox Pass core does not have the same chronological control that has been obtained for the Crowfoot Lake core. The Lake O'Hara core (see below) identifies the boundary at the beginning of the warming trend in the climate where trees quickly colonized the landscape.

Importantly, the Crowfoot Lake pollen record yields an indication of why there was an open grass-tundra-shrub stage phase prior to the arboreal phase of colonizing

Table 9: Pollen and macrobotanical studies of the Rocky Mountains, Foothils and Plains in eastern British Columbia, Alberta and Montana

Name of Site	Latitude N (approx.)	Longtitude W (approx.)	Elevation Masl (approx.)	Radiocarbon Dates in Years BP and Lab No.	References
1. Wilcox Pass	52°15'	117°13'	2355	9600 ± 305 (GX-8785)	Beaudoin & King (1990)
2. Crowfoot Lake				11,330 ± 220 (CAMS3065)	Reasoner and Huber (1999)
3. Lake O'Hara	51°21'	116°20'	2015	10,100 ± 200 (RIDDL-433)	Reasoner & Hickman (1989)
4. Opabin Lake	51°21'	116°20'	2280	8350 ± 190 (RIDDL-510	Reasoner & Hickman (1989)
5. Copper Lake	51°15'	115°55'	1426.7	10,490 ± 160 (RIDDL-664	White (1987)
6. Kingfisher Pond	51°24'	116°09'	1595	11,370 ± 250 (BGS-1091)	White (1987)
7. Wedge Lake	50°52'	115°10'	1500	10,400 ± 110 (GSC-2965)	MacDonald (1982)
8. Yamnuska Bog	51°04'	115°03'	1300	*>10,400 estimate	MacDonald (1982)
9. Toboggan Lake	50°46'	114°36'	1480	10,400 ± 70 (TO-149)	MacDonald (1989)
10. Chalmers Bog	50°37'	114°25'	1370	**18,300±380 (GSC-2668) **18,400 ± 1090 (GSC-2670)	Mott & Jackson (1982)
11. Beauvais Lake	49°25'	114°06	1368	Ca. 10,700 est. from sedimentation rates	Powers (1994)
12. Guardipee Lake	48°33'	112°43'	1233	Ca. 11,400 based on Mt. St. Helens Jy tephra. Bottom est. to 12,200	Barnosky (1989)
13. Lost Lake	47°38	110°20'	1019	9325 ± 150 (Beta-21578, ETH-3108)	Barnosky (1989)
14. Harris Lake	49°40'	109°53'	1234	9120 ± 250 (S-2908)	Sauchyn and Sauchyn (1991)
15. The Andrews site	50°20'	105°52'	720	10,230 ± 140 (AECV-2047C)	Yansa (1998); Yansa and Basinger (1999)
16. The Fletcher site	49°35'30"	111°49'20"	--	9380 ± 110 (TO-1097)	Vickers and Beaudoin (1989)
17. The Webb Dugout	51°08'	111°05'	--	10,150 ± 140 (AECV-1596C) 9890 ± 120 (AECV-1597C)	Beaudoin (1992); Yansa and Basinger (1999)
18. The Jenner Dugout	50°44'	110°47'	--	10,050 ± 110 (AECV-1666C)	Beaudoin (1992); Yansa and Basinger (1999)
19. Hampton's Dugout	52°45'	112°45'	--	10,200 ± 130 (AECV-1396C)	Beaudoin (1992); Yansa and Basinger (1999)

* Date is estimated by comparison of stratigraphy to a similar site (Wedge Lake)
** These dates are problematic and will be discussed below

vegetation. In this scenario the recession of the continental glaciers would have taken place prior to the 11,330 years BP date derived from a *Salix* (twig) in the core. Following the recession of the glaciers there must have been a lag time in which the denuded landscape lay barren and exposed for an unknown period of time, perhaps several hundred years. The grass-tundra-shrub cover was the first vegetation colonization of that landscape which was still cool, especially in the high elevations. At 10,000 years BP, a warming trend appears to have taken place over a very short period of time according to Reasoner and Huber (1999) and tree colonization happened quickly.

Reasoner and Hickman (1989:306) note that the inferred pioneering herbaceous community of postglacial times observed in zone 1 (dated to >10,000 years BP) of the Lake O'Hara pollen profile is similar to a number of late-glacial tundra-like vegetation assemblages that have been examined from the northwestern United States, as well as some records from Alberta. Zone 1 represents a period of low pollen productivity and thus a sparse, shrubby, colonizing herbaceous vegetation community is inferred. The plant species found in this early assemblage include taxa like *Artemisia* and the nitrogen fixers including *Alnus* like *A. sinuata*, *Shepherdia canadensis*, and Gramineae. Nitrogen fixing plants seem to invade quite readily into newly exposed mineral materials found in recently deglaciated areas (Reasoner and Hickman 1989:305-6).

A lack of macrobotanical remains like conifer needles in zone 1 at Lake O'Hara suggests the pollen presence of

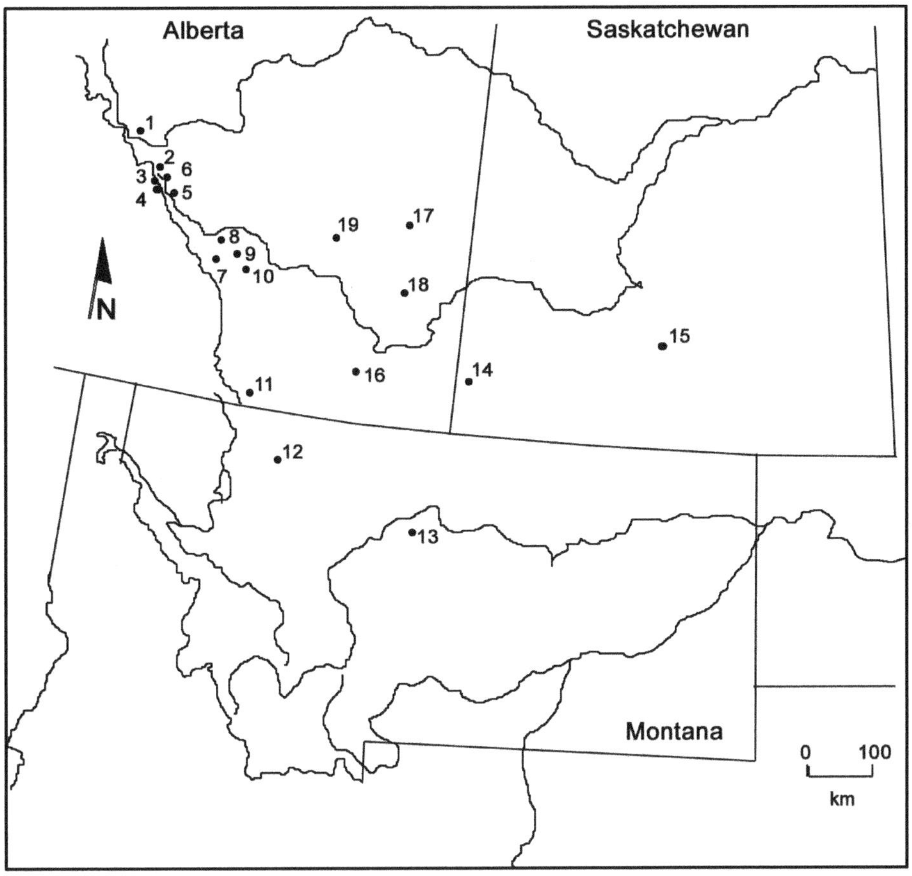

Figure 111: Locations of the pollen and macrobotanical study sites discussed in this chapter. The locations are only gross approximations based on information from Beaudoin (1993), Yansa and Basinger (1999) and Barnosky (1989). 1. Wilcox Pass; 2. Crowfoot Lake; 3. Lake O'Hara; 4. Opabin Lake; 5. Copper Lake; 6. Kingfisher Pond; 7. Wedge Lake; 8. Yamnuska Bog; 9. Toboggan Lake; 10. Chalmers Bog; 11. Beauvais Lake; 12. Guardipee Lake; 13. Lost Lake; 14. Harris Lake; 15. The Andrews Site; 16. The Fletcher Site; 17. The Webb Dugout; 18. The Jenner Dugout; 19. Hampton's Dugout

these trees is not local, but rather of long distance origin. There is an increase of conifer needles in zone 2 and this is accompanied by an abrupt increase in the amounts of pollen from these tree species. Changes between haploxylon and diploxylon pine pollen representation are attributed to absence of competition between tree species, rather than warmer conditions in the early Holocene. The timberline is proposed to have been higher than it is today (Reasoner and Hickman 1989). Conditions changed very gradually between 7000 and 3000 years BP as the warm conditions cooled. Conditions around the time of the Mazama tephra deposition at 6730 ± 40 ^{14}C years BP appear to have been much the same as in the early part of zone 2.

One of the first visible differences between the Opabin Lake and Lake O'Hara pollen profiles, is the time span covered by them. While the Lake O'Hara core extends back prior to 10,100 years BP, the Opabin Lake core begins at >8530 years BP (Reasoner and Hickman 1989). Thus the Opabin Lake pollen OPB1 zone 1 appears to begin during the middle of Lake O'Hara's LOH25 zone 2. It is not clear how far back >8530 years BP actually refers to.

The key contrast that can be made between Lake O'Hara and Opabin Lake throughout zones 1, 2 and 3 is diversification. Colonizing taxa invade the landscape, initiating an enrichment process that contributes to the migration of the less tenacious plant species into the landscape after some time. Opabin Lake appears to have been denuded of ice at a much later date than Lake O'Hara but benefited quickly by proximity once the ice was gone and plants could move into the freshly scoured land surface. Initially a tundra-like environment appears to have been established which quickly evolved into a complete forest-type environment. *Pinus* is the colonizing conifer, but once *Picea* is established it steadily increases in coverage through time, being more shade tolerant.

The two major differences between Kingfisher Pond, Lake O'Hara and Wilcox Pass are in the percentage representations for *Pinus* and *Artemisia.* At the sites located at higher elevations – Lake O'Hara and Wilcox Pass – *Pinus* initially dominates the landscape with a tall arboreal canopy, which would not be conducive to the growth of low-growing herbaceous species. The landscape at Kingfisher Pond appears to have been far more open and we see a subsequent rise in low-lying plants, with *Artemisia* at 38% in zone KP1 dated to >10,000 years BP. The relationship between *Pinus* and *Artemisia* percentages appears to be the same between Opabin Lake and Kingfisher Pond zones 1, though they cover very different time periods. The crucial plant in the landscape is *Pinus*. When *Pinus* is abundant, *Artemisia* is not. In time, the *Pinus* cover in the higher elevation sites gradually diminishes whereas it appears to increase dramatically at the lower elevations and more southerly latitudes during the same timeframe, ca. 10,000-6800 years BP. However, spruce also appears to thrive well through time at the lower elevations, at representations of 20-26%. *Populus* is present in smaller values (6-2%) which probably translates to it being under-represented in the samples. *Populus* pollen does not preserve well and we can probably infer a greater presence than the percentage values indicate. Given pine's overproduction of pollen in general, it appears that the arboreal cover is becoming much more of a mixed conifer forest rather than single canopy

Radiocarbon dating of the Copper Lake core appears to have been a little problematic (White 1987). Nevertheless, in terms of vegetation, the relationship between *Pinus* and other tree taxa appears more complicated in the Copper Lake pollen profile. The percentage proportions of arboreal versus non-arboreal plants at Kingfisher Pond appear to be very similar to the pollen percentages for Copper Lake in zones 1 (dated to >10,500 years BP) and 2 (dated to >10,500 to 10,300 years BP). However, Kingfisher Pond KP3 (10,000 – >6850 BP) exhibits very similar final percentages for *Pinus, Picea*, *Abies* and *Betula* to Copper Lake zone CL4b which is dated between 7000-4600 years BP. The difference in the timeframes coupled with the calibration problem for any radiocarbon dates derived from analyses makes this comparison interesting, but problematic. According to White (1987), the Copper Lake pollen profile compares most readily with that of Yamnuska Bog at Morley Flats (>10,400 years BP) (MacDonald 1982). This is interesting because the next closest site to Copper Lake that is at a similar elevation is the Wedge Lake site in the Kananaskis Valley.

Interestingly, a series of 15 species of mollusks were found within the matrix of the pollen cores of both Wedge Lake and Yamnuska Bog. At Wedge Lake, fewer mollusks in zone 1 (dated to 10,400 ± 110 years BP) are interpreted to indicate cooler climatic conditions during that time period (MacDonald 1982). An overall slight increase in the diversity of southerly ranging species in zone 2 (dated to 9935 ± 215 years BP) is interpreted to indicate climatic amelioration of the mid-Holocene, especially since it is accompanied by high *Pinus* pollen concentrations in the core up to the Mazama ash level (MacDonald 1982). Thereafter, both phenomena decline.

One of the most noticeable differences between the pollen diagram for Wedge Lake and others examined is the overall lack of species diversity in terms of what has been recognized and tallied in the study. It is not clear whether this is attributable to identifier bias, poor pollen preservation, low numbers of airborne pollen types, or simply the lack of diversity itself. Only re-examination of the area could provide an understanding of trends within the data.

The relationship between *Pinus* and *Artemisia* that has been observed in other pollen profiles does not exist in the Wedge Lake data. At lower elevation sites, *Pinus* appears to continue to increase in its percentage representations through the Holocene, as opposed to a decreased representation through time at the higher elevation sites. However *Picea* increases substantially as well and *Populus* is represented in lower values. Given the overproduction of *Pinus* pollen and the poor preservation of *Populus* pollen, the forest cover is probably becoming much more of a mixed canopy than the percentage values on the pollen diagram indicate.

REVIEW OF HOLOCENE VEGETATION ZONES FROM POLLEN PROFILES AND MACROBOTANICAL ANALYSES EAST OF THE ROCKY MOUNTAINS: POSTGLACIAL TO 6730 ± 40 BP (MAZAMA TEPHRA EVENT)

The trend toward arboreal domination of the landscape appears to be slowed at the lower elevations. As well, proximity to the mountains and foothills proper may be a factor to consider in initial trends toward a more open landscape. MacDonald (1982) interprets the early non-arboreal taxa dominance as indicative of aggressive colonizing of pioneer species. Reference is made to similarities between taxa represented in zone 1 in the diagram for Yamnuska Bog (MacDonald 1982) and tundra communities found on the Arctic Slope today. However, it is noted that the taxa of zone 1 (dated to >10,400 years BP) are common to a great variety of plant communities of different latitudes throughout western Canada (MacDonald 1982). In zone 2 (dated to approximately 9400 years BP), *Pinus* and *Picea* dominance is interpreted to represent the establishment of montane and subalpine forests (MacDonald 1982). This landscape appears to have remained relatively stable since its establishment. Two taxa, *Picea* and Gramineae, both appear to fair somewhat better away from the higher elevations and mountainous landscape. The high initial representation of *Populus* in the pollen profile is interesting given the problems that have been noted for its preservation. Deposition environment may be a factor in this unusual recovery rate.

The change in forest mixture becomes more pronounced as we move further south to Toboggan Lake and Chalmers Bog. Toboggan Lake (1480 masl) is not much lower in elevation than Wedge Lake (1500 masl) and is actually higher than Yamnuska Bog (1300 masl). Toboggan Lake is located only 35 kilometers southwest of Calgary in an area classified as Lower Boreal Cordilleran Ecoregion (W. L. Strong Ecological Land Surveys Ltd.: Map 1992). The reference vegetation for the area today is listed as Aspen-Balsam Poplar-Lodgepole Pine (succession to white and black spruce and balsam fir) (W. L. Strong Ecological Land Surveys Ltd.: Map 1992). The arboreal canopy of Toboggan Lake zone 1, dated at 10,400 ± 70 years BP, provides some interesting contrasts in species representation when compared to zone 1 at Wedge Lake dated at 10,400 ± 110 years BP. Wedge Lake is also only a few miles further north and west of Toboggan Lake in the Kananaskis valley. *Pinus* at Toboggan Lake (24-32%) does not climb anywhere near the proportions seen at Wedge Lake (40-50%) during the same time period. *Picea,* as well, makes a significant jump from 0% to 25% at Toboggan Lake, while remaining fairly steady (21-28%) at Wedge Lake during this timeframe. *Abies*, *Populus* and *Betula* are found in similar percentage representations at both sites during this time, as well. The contrast that can be seen is that, through time, the Toboggan Lake vegetation assemblage changes from being herb and shrub dominated to being arboreal dominated. The pollen profile does resemble those of Wedge Lake and Yamnuska Bog to a very large degree, though they are in different localities and at slightly different elevations. The common connection between these sites is that they are on the edge of the subalpine-Parkland transition zone in terms of the major vegetation zone classifications for the region today. Obviously, their positions in the overall landscape have rendered common changes through time that are reflected in their pollen profiles.

The difference that is apparent between the three sites is the timing of events. Toboggan Lake takes on the vegetation of Wedge Lake, but at a slower pace. Initially, the Toboggan Lake pollen profile resembles Yamnuska Bog, but with time both profiles show a dramatic change in vegetation from non-arboreal to arboreal. These forests also appear to be well mixed in terms of numbers of the different tree species present in the landscape. There is more diversity and the appearance of a conifer-deciduous cover more typical of areas classified as Parkland today.

The dynamic between *Pinus* and *Picea* is noticeably different between sites at higher versus lower elevations, but there are problems in the pollen records and in the timing of events relative to the southern areas recognized as glacial refugia for *Picea.* MacDonald (1989:165) notes, "...marked increases in *Picea* pollen percentages provide evidence that spruce forests began to develop at the northern foothill sites by 10,000 yr B.P." but "...delayed migration cannot explain the late development of *Picea* forests in the south...". Part of the problem in these observations has to do with the dating of limnic sediments. MacDonald (1989:165), points out that "...none of the core chronologies from the eastern slope, except for Toboggan Lake is based on radiocarbon dates from terrestrial plant macrofossils". However, this statement is somewhat confusing in light of his other work at Wedge Lake (MacDonald 1982) where a date was derived from a piece of *Populus* wood from the base of zone 1 of the pollen core analyzed. These problems aside, MacDonald (1989) proposes the development of a forest at Toboggan Lake, represented by the rise in *Picea* and *Pinus* pollen percentages over time. *Betula,* as a tree rather than a shrub, appears to make a distinct impact on the forest appearance between 9000-8000 years BP (MacDonald 1989:165-166). As time passed, the forest around Toboggan Lake became more Parkland-like for a while. However, at around 5000 years BP, the trend reversed when *Picea* and *Pinus* dominated the landscape once again (MacDonald 1989:166).

The pollen profile presented for the Chalmers Bog site has been argued to provide evidence for a very early opening of the classical Wisconsinan "ice-free corridor" at approximately 18,000 years ago (Mott and Jackson 1982). Two radiocarbon dates were derived for CB3, the oldest zone (18,300 ± 380 years BP GSC-2668; 18,400 ± 1090 years BP GSC-2670), on moss fragments from clay near the bottom of the pollen core (Moss and Jackson 1982). More recent studies of terrestrial plant macrofossils and aquatic moss from the Toboggan Lake site have demonstrated serious potential problems with the reliability of any ^{14}C dates derived from these materials (MacDonald *et al.* 1987). For this reason, the two early dates for zone CB3 at Chalmers Bog have been purposely ignored in this summary of information from the site. Instead, comparison can be made between this site and others, with the raw information derived from the oldest portion of the pollen core.

At Chalmers Bog the early initial landscape after glaciation would appear to have been sparsely covered by vegetation as has been seen with some interpretations of other pollen diagrams. This trend is revealed in the percentage representations for *Artemisia*, Chenopodiaceae and Gramineae. Cyperaceae pollen percentages may represent the more immediate area local to the bog. Local conditions could have been moist or even wet. *Typha* (cattail) does make a brief appearance in CB2 and in the early portion of CB1 (the youngest zone) which indicates wet conditions existed, probably shallow standing water.

Otherwise, the percentages for *Pinus* and *Picea* in zone CB3 have been seen before in the pollen diagram for Yamnuska bog zone 1 (est. date >10,400). They are also similar to some measure to the percentages seen in Toboggan Lake and Wedge lake, though both taxa appear to be slightly better represented at an earlier point at these sites. The overall diversity of taxa represented in the pollen diagram at Chalmers Bog is interesting. Many of the taxa represent plants (not necessarily the same species) that can be found today on the western edge of the City of Calgary, which is situated right at the Plains-Parkland transition zone boundary. Many of the taxa can also be found in Parkland areas and the Boreal Forest environments throughout Alberta today. Obviously, these plants are very adaptable, overall. The Chalmers Bog site has been proposed/presented as a very early tundra environment that remained essentially the same for several thousand years until the migration of arboreal taxa to the region (Mott and Jackson 1982). For various reasons this scenario does not fit with the information derived from other sites examined here. Rather, the pattern of vegetation of the landscape follows the trends seen in other sites which date to the >10,000 years BP timeframe. The other sites examined here, more or less, agree with one another overall.

Three substantial reasons can now be listed for a reconsideration of the age and interpretations derived for the Chalmers Bog pollen core. 1) Given the problem with dating aquatic mosses at Toboggan Lake the 18,000 year-old date from Chalmers Bog zone CB3 should probably be discounted. 2) The original date derived for the Chalmers Bog site led to an interpretation of the vegetation pattern that fitted the older date. Mott and Jackson (1982) interpreted the pattern of vegetation to represent a tundra-like setting, established 18,000 years ago and existing for a very extended period of time. They noted that Chenopodiaceae-type pollen is found in small amounts in zone CB3 and that it is usually co-dominant with *Artemisia*, Gramineae and Cyperaceae in grassland assemblages (Mott and Jackson 1982:510). However, Chenopodiaceae can also be found in high arctic-alpine tundra assemblages, but is not a prominent part of the group. Low influx values combined with the very old radiocarbon dates encouraged an interpretation of the area as a tundra-like environment rather than grassland. If we disregard the old dates and reconsider the more usual co-dominance of Chenopodiaceae with the other three taxa mentioned, in grassland assemblages rather than tundra, we can call for a reinterpretation of the pollen profile that fits in with other studies presented here. 3) If we then compare the arboreal assemblage of Toboggan Lake zone 1 (oldest zone) to Chalmers Bog CB3 (oldest zone), there is some measure of similarity in the relatively low representations of *Pinus, Picea* and *Populus* in both of these pollen profiles.

Upon reconsideration it is proposed that the lower percentages for arboreal pollen at Chalmers Bog may be attributed to the area being characterized as grassland rather than as a tundra type of environment. However, a more fit description would be to classify the area as 'transitional Parkland-grassland'. The other low-growing herbs, represented in very small percentages in the Chalmers Bog pollen diagram, are plants that are found today in transitional Parkland-grassland areas. Given the proximity of Chalmers Bog to Toboggan Lake (18 km) and the fact that "...few indicators of either tundra or grassland vegetation occur in the dominant taxa of the palynoflora..." (MacDonald 1989:163) at Toboggan Lake, it is logical to disregard the tundra characterization of Chalmers Bog zone CB3. Further to this argument, every plant species noted in the pollen diagram for Toboggan Lake is found on the Chalmers Bog pollen diagram. The exception is *Pseudotsuga,* which is rare in the vegetation cover today in Parkland-grassland transition zones-but does occur at present in the Calgary City transition zone.

The Beauvais Lake pollen record begins at approximately 10,700 years BP (uncorrected date inferred from sedimentation rates). *Pinus* at 15% and *Picea* at 10% in zone 1 indicate the presence of taller arboreal cover at an early point in the pollen record. However, it is noted that low numbers of *Pinus contorta* in a region can yield pollen frequencies of 15-20% in terrestrial pollen sums (Powers 1994). Long distance transport of *Pinus* pollen can also contribute to over-representation of this species in a pollen sum. Thus, a frequency of 15% may simply reflect the presence of *Pinus* at the regional level, masking the true local representation of pine in the canopy. The low representations of *Pinus* and *Picea* may also reflect poor soil development and warmer climatic conditions following deglaciation, according to Powers (1994).

In summary, early, low frequencies of *Pinus* and *Picea* pollen are problematic in pollen analyses such as Beauvais Lake. The trees may be present locally, but if so are they are few in number and are perhaps more representative of a colonizing presence, given the dramatic changes that take place in Zone 2a. Zone 2 (including 2a and 2b) covers the time period between 10,500 BP and 7,800 BP (using the uncorrected dates for the chronology). Powers (1994) notes that the high frequencies for grass and herb species during this period may indicate there was more extensive grassland cover than exists in the area today. Powers (1994) proposes that the presence of *Alnus* and *Betula* in the pollen assemblage is more likely attributable to the small, shrubby varieties of these plants, *Alnus crispa* (Ait.) and *Betula glandulosa* (Michx.).

Interestingly, the changes in pollen frequencies appear to parallel changes in soil textures in the depositional environment of the lake. Zone 1 has been labeled as consisting of silts whereas Zone 2a consists of sandy clayey silts and Zone 2b consists of clayey silts. This is an indication of changes taking place in the climate of the region. These changes were significant enough to cause moisture regimes to alter which is reflected in the changing vegetation visible from the pollen profile. This change in moisture regimes gave rise to drier conditions, which allowed movement of sediments of larger grain size proportions (sand) along with the clay and silt fractions (Powers 1994).

Indeed, this change in the depositional environment parallels a dramatic shift in vegetation. There is an abrupt and significant decrease in grass (Gramineae) and sage (*Artemisia*) accompanied by a less dramatic decrease in *Betula, Alnus* and Chenopodiaceae. This is congruent with an abrupt and significant increase in the conifer species, *Picea* and especially *Pinus.* Zone 3 contains Mazama tephra throughout and represents a hiatus in the pollen diagram where pollen slides have not been assessed.

The transition between Zones 2a and 2b is significant because of the drop in arboreal pollen representation in the pollen profile. This is accompanied by a rise in the representation of the lower shrub cover (*Alnus* and *Betula*) and the low-growing herbaceous plants present in the pollen profile. *Pinus* appears to recover slowly from 15% to 40% for a good length of time and then fluctuates between 60% and 25% to the Mazama tephra boundary. As noted by Powers (1994), the fluctuating reductions in

Pinus representation correspond with abrupt or gradual increases in *Picea* at the same time. This changing pattern suggests a local fire regime where the fire-successional lodgepole pine (*P. contorta*) recolonized quickly after a fire and *Picea* followed at a slower rate, gradually moving the landscape toward a successional climax vegetation canopy (Powers 1994). Fluctuations in the arboreal canopy are mirrored by corresponding changes in the lower shrubs and the low-growing herbaceous plants represented in the pollen profile. When the trees proliferate, the grasses, sages and Chenopodiaceae do not flourish and vice versa.

Guardipee and Lost Lake in northwestern Montana in the northwestern Great Plains region contain deep sediments that have provided pollen records extending back to 12,200 years BP and 9400 years BP, respectively. These two pollen studies compliment each other because of the time periods that are best represented in their cores. In the case of Guardipee Lake the core covers the timeframe 12,200 to 9300 years BP but becomes problematic after that because of breaks in the deposition record (Barnosky 1989). The Lost Lake core covers the timeframe 9400 years BP to the present (Barnosky 1989).

The interest here is limited to the lower/older portions of the analyses up to 6600-6800 years BP. According to Barnosky (1989) the Marias Pass area on the Continental Divide to the west of both sites, was free of ice by 11,400 years BP. The Guardipee Lake core begins at 12,200 years BP with representative pollen from local and distant windblown vegetation sources that appear to be well established already by that time. Pine (lodgepole and whitebark), spruce and fir made up a landscape of open parkland (Barnosky 1989).

Barnosky (1989) notes that shrubby plants like Juniper, Aspen, Willow and the Rose family found represented in the cores probably came from farther afield in the forest/grassland transition areas along the flanks of the Rocky Mountains to the west. Grasses and forbs covered the local landscape around both lakes. In the early part of the record there are indications that a cooler and wetter climate regime was in place. This began to change around 9400 years BP at Guardipee Lake with the onset of aridity which appears to have lasted till ca. 6000 years BP in this part of the West.

Barnosky (1989) points out that the vegetation history from these two lakes differs from the patterns observed in the eastern Great Plains and the Midwest. According to Ritchie (1976), a *Picea* forest covered the entire interior of North America from ice margins in Manitoba, Saskatchewan and Alberta to northeastern Kansas, east through central Illinois and the Great Lakes during the timeframe of 14,000-11,000 years BP. There were areas where there appears to have been an open landscape between the ice margins and the forest, which have been interpreted as 'tundra' (Ritchie 1976). However, in the case of Guardipee and Lost Lake the landscape appears to have simply been treeless through time similar to the Columbia Basin further west, the Snake River Plain in Idaho and the Powder River Basin to the south (Barnosky 1989).

At Harris Lake in the Cypress Hills, Saskatchewan, the pollen core extended to the early Holocene with a basal radiocarbon date of 9120 ± 250 years BP, however, it did not go as far back as the Guardipee Lake core from Montana. Thus, the pollen record begins with evidence of a period of aridity starting at the bottom of the core. This coincides with the interpretation for the beginning of the warmer, drier period at 9300 years BP at Guardipee and Lost Lakes a few hundred miles to the southwest of the Cypress Hills. Sauchyn and Sauchyn (1991) interpret zone II as the period of maximum aridity, especially between 7700 to 6800 years BP.

The *Picea* representation indicates that the temporary spruce-dominated period, discussed above, had already occurred. The low percentage representations of pine and spruce are problematic for Sauchyn and Sauchyn (1991). Their appearance in the landscape could be the result of the spruce forest that spread across the interior during the late Pleistocene. The really low representation of their pollen at <15% in a core indicates they may not have even be present in the landscape—their pollen grains may have blown in from elsewhere. However, it is proposed that some trees survived the arid period that is evidenced, surviving in small, mixed stands of poplar (Sauchyn and Sauchyn 1991). This is very possible. Given the right aspect and slope, there are areas where the moisture regime could have been maintained to support a small tree population through time.

The Andrews site (50°20' N, 105°52' W) at 720 masl, is a small, rimmed kettle with closed drainage, located on the Missouri Coteau upland in southwest Saskatchewan (Yansa 1998, Yansa and Basinger 1999). A basal date of 10,230 ± 140 ^{14}C years BP was derived from a piece of *Picea* wood (root) from the bottom sediments overlying glacial diamicton (Yansa 1998). The material recovered from this site is very different from the studies summarized thus far in this chapter. Postglacial age sediment samples were collected from a section of a profile (5.8-3.1 m) that was exposed when the kettle was excavated to make a livestock watering hole (Yansa 1998). Paleoenvironmental reconstruction is based on autochthonous plant and invertebrate macrofossils recovered from processing 50 cm^3 subsamples of the sediments (Yansa 1998). According to Yansa (1998), upper sediments (above 3.1 meters) in the site did not contain preserved remains, bracketing the dateable material between 10,200 and 5800 years BP.

Macrofossil recovery was productive with "...42 taxa of fossil plants representing 20 families of vascular plants and 2 families of non-vascular plants..." (Yansa 1998:432) identified from the site. Five zones were identified in the sediment profile based on macrofossil assemblages associated with stratigraphic changes observed. Zone I, the basal diamicton contains macrofossils that were poorly preserved and probably transported with deposition.

The Andrews site appears to fit into a general scheme that has been proposed for the evolution of the landscape as noted above in the discussion on Guardipee and Lost Lakes. There is evidence of a large spruce forest that seems to have existed between 14,000-11,000 years BP in the interior of the North American continent (Ritchie 1976). At the Andrews site, an open spruce forest was on the landscape at 10,200 years BP and was gradually replaced by a deciduous parkland cover consisting of balsam and aspen poplar and river birch, which continued till at least 8800 years BP (Yansa and Basinger 1999). A burning event or series of events appear to have occurred in the landscape at this time though there does not appear to have been a lot of charred material produced at the site (Yansa and Basinger 1999). The fires may have been due to aridity in the landscape overall, however, the pond did not dry up at this time. At 8790 ± 140 ^{14}C years BP (TO-5019) there are indications that the pond was drying out (Yansa and Basinger 1999). At 7670 ± 80 ^{14}C years BP (TO-4780) there is evidence of fire playing a role in the landscape (Yansa and Basinger 1999). After 5800 years BP conditions at the Andrews site changed to the point that the preservation of macrobotanicals no longer took place. Aridity during this time may have been a factor, infilling of the kettle with sediments may have been a factor, or fluctuating groundwater regimes may have affected the local water table (Yansa and Basinger 1999). The pond eventually dried up completely.

The Fletcher site (49°35' 30" N; 111°49'20" W) has been examined several times since its discovery in 1963 when creation of a dugout for watering cattle exposed deposits of interest to archaeologists (Forbis 1968, Vickers and Beaudoin 1989). Subsequent excavations yielded four Alberta and seven Scottsbluff points in association which gave the deposits a timeframe of 11,000-7000 years BP (Vickers and Beaudoin 1989). Dates derived from bone fragments recovered from the site were problematic, yielding dates between 4000-6000 years BP (Vickers and Beaudoin 1989). However, dating of plant macrofossils recovered by wet-sieving of sediment samples taken from a small excavation in 1988/89 yielded an AMS date of 9380 ± 110 years BP (TO-1097) (Vickers and Beaudoin 1989).

The dated macrofossils were located 35 cm below the Cody Complex occupation, which means the cultural materials must be younger. According to Hofman and Graham (1998:113), the Cody Complex consists of Scottsbluff and Eden points and dates to around 9000 years BP. The Alberta/Cody materials from the Hudson-Meng site in northwestern Nebraska have been dated by associated charcoal to 9820 ± 160 years BP (SMU-224), associated bone apatite at 8990 ± 190 years BP (SMU-52), and associated bone collagen at 9380 ± 100 years BP (SMU-102) (Agenbroad 1978).

Seeds of aquatic and wetland plants have been recovered from the Fletcher site and include Cyperaceae, *Ranunculus, Typha, Potamogeton* and *Mentha* (Beaudoin 1996). While there were no macrofossil indications of trees in the site location during its formation in the post glacial time period, the remains of beaver (*Castor canadensis*) and deer (probably mule deer (*Odocoileus hemionus*)) indicate the presence of woody vegetation (Beaudoin 1996). The site is located within an area known as the Palliser Triangle. Today, this is the driest area in the Canadian prairie and the most climatically sensitive in the country (Lemmen and Vance 1999). The question is how does the paleo record of the Fletcher site compare to other sites in the greater region? Is this sensitivity a product of time and changed landscapes?

During the terminal and post glacial period a number of other sites similar to the Fletcher and Andrews sites were also being formed in the region. Some of these other sites, all in Alberta, have been sampled and dates have been derived from plant macroremains recovered.

At the Webb Dugout, east of Coronation, two dates were derived from *Populus* wood, some of which had evidence of beaver gnawing present (Beaudoin 1992). Dates of 10,150 ± 140 years BP (AECV-1596C) and 9890 ± 120 years BP (AECV-1597C) (Beaudoin 1992) give some proof of the presence of deciduous trees in the same time period when the Andrews site appears to have been changing from a spruce dominated forest to deciduous dominated bush.

At another dugout site close to the town of Jenner, east of Dinosaur Provincial Park, a piece of unidentified wood was recovered from sediments at the bottom of the dugout (Beaudoin 1992). A date of 10,050 ± 110 years BP (AECV-1666C) (Beaudoin 1992) is again comparable to the Andrews site, though the type of wood would make the information even more useful. Macrofossils of other plant taxa were present in the sediments including aquatics like *Ceratophyllum, Typha* and *Zannichellia* and plants of moister areas like *Ranunculus, Mentha* and *Chenopodium* (Beaudoin 1992). Except for *Ceratophyllum*, these species compare to those from zone II and III at 10,200-8800 years BP at the Andrews site.

Finally, at another site called Hampton's Dugout near Rowley, a piece of *Populus* wood was collected by the owner from four meters down in the sediments. A date of 10,200 ± 130 years BP (AECV-1396C) was derived, which again compares to the Andrews site and the others summarized here (Beaudoin 1992).

Thus, the paleo-record of the Fletcher site does appear to compare to other sites in the greater region. The changes seen in the Palliser Triangle, which is so climatically sensitive, do appear to be a product of time and changed landscapes since deglaciation and its position within the topography of the northern Plains is an important part of the whole scenario.

SUMMARY DISCUSSION

It appears that a transition of vegetation from wooded to grassland took place in the northern Plains beginning somewhere between 49 and 50 degrees latitude during the

10,200 years BP timeframe. It appears that the occupation of the landscape by arboreal vegetation was related to the changing climatic regime following deglaciation. Pine, spruce and poplar appear to have migrated north along the Rocky Mountain Flanks into southwestern Alberta, while spruce and poplar migrated into southeastern Alberta from areas to the south and east, in what is now the interior of the United States. As the interior high Plains region became warmer, the moisture and temperature regimes changed to the point that the spruce and poplar forests could not survive. Along the foothills, local fire regimes may have encouraged the migration of pines and at the same time encouraged the proliferation of spruce, which appears to have followed and filled in behind pine, taking over the landscape in many areas. White spruce appears to be more generally adapted to the lower elevations moving out of the foothills toward the Plains and Fescue grasslands near Calgary. Given the right slope and aspect, White spruce along with poplar, will retain a foothold in coulees and along the major drainage channels of the northern Plains in Alberta today.

In the mountain and foothill sites, where there is a pollen record that shows a transition of the landscape from glaciated to being covered with vegetation, there appears to be an initial grass-tundra-shrub phase followed by an arboreal phase. This phase was most clearly demonstrated in the Crowfoot Lake pollen analysis, which shows the transition at 10,000 years BP to the warming up of the early Holocene. That grass-tundra-shrub transition is not visible yet in the Plains sites. It does not appear at the Andrews site at latitude 50°20' in southwest Saskatchewan. However, this does not mean that it did not take place. The type of research adopted for a site can clearly make a difference in the results and interpretation of events. The Andrews site used macrobotanical analysis to reconstruct the paleoenvironment. Phytolith studies of the sediments might constitute a future research approach that may also be used in conjunction with macrobotanical analyses to try and reconstruct the vegetation transitions that took place over time in the northern Plains following deglaciation.

The most recent pollen analysis included in this discussion was the study of the Crowfoot Lake sediments from the upper Bow valley corridor in the Rocky Mountains. The sequence of vegetation cover that migrated into the landscape along with the dates derived from macrofossils from core CRW6 clarified the series of cooling and warming events quite well. One of these events, known as the Younger Dryas, 11,000-10,000 years BP, is recognized as a period of climatic cooling that was global in scale (Reasoner and Huber 1999). At the Crowfoot Lake site it is proposed that the first suite of vegetation that colonized the landscape was a grass-tundra-shrub cover that moved in following deglaciation. It appears that the Younger Dryas cold period caused a lag in the migration of arboreal plants into the higher elevations through this timeframe. According to Reasoner and Huber (1999) the Younger Dryas ended quite suddenly at 10,000 years BP and the climate became warmer. As soon as the climate warmed up, *Pinus, Picea* and *Populus* appear to have invaded the mountains and foothills from refugia to the south.

At Wilcox Pass, the site at highest elevation, the arboreal invasion does not begin until after 10,000 years BP at least—trees were well established by 9600 years BP. At Crowfoot Lake the arboreal invasion began in the latter part of the Younger Dryas and trees were well established by 10,100 years BP. At the next high elevation site, Lake O'Hara, the arboreal invasion was established by 10,100 years BP. The timing of events is not quite as clear at Opabin Lake, which is 200 meters higher than Lake O'Hara. As we drop down further in elevation to Kingfisher Pond near Lake Louise, the arboreal invasion appears to have been established by 10,000 years ago. The same pattern holds true for Copper Lake, which is lower and further south of Kingfisher Pond. The Wedge Lake core dates only as far back as 9395 ± 215 years BP, however, the arboreal invasion appears well established by that time. Yamnuska Bog has problematic dating, but the pattern appears to resemble that of Wedge Lake, which is close by. At Toboggan Lake, which is at a lower elevation and further south, the arboreal invasion appears to have established between 10,000 and 9600 years BP. Events at this site, which is away from the Mountains proper, appear to be lagging behind the higher elevation sites. The same pattern appears to be the case for Chalmers Bog, though there are issues with the dating of this site as well. There really is no lag time in the vegetation trends. What is different here is the canopy mixture at the lower elevation sites compared to the higher sites. The lower sites appear to trend toward a canopy that is much more a mixture of conifer and deciduous trees rather than single canopy or just conifer cover as seen in the mountain sites. The lack of *Pinus* overpowering everything else in the pollen profile yields the illusion that the arboreal cover is not expanding. It is expanding, but it is *Picea, Abies* and *Populus* that are more adaptable to the lower elevations than *Pinus*. As well, there is also an increase in shrubs, forbs and grasses at the lower elevation sites, which trend toward a Parkland vegetation cover.

It is clear that there were no trees or very few trees through time, 12,200 years BP to the present, only a short distance from the mountains in the Montana Plains and areas further south, as noted by Barnosky (1989). The pollen records from the Montana sites of Guardipee and Lost Lake demonstrate that the different tree species were present in the mountains south of the limits of the continental glaciation. A few hundred kilometers to the northeast, just north of the Cypress Hills in Alberta in what is now part of the northern Plains region, there were *Picea* and *Populus* trees in the landscape at 10,200 years BP. However, the *Picea* trees disappeared quickly and the *Populus* eventually followed by 8800 years BP. These trees appear to have migrated into the Plains region from the east following the retreat of the Laurentide continental ice.

This pattern of changing arboreal vegetation appears to be related to the changes in climate following deglaciation. As noted by Campbell and Campbell (1997) two phases appear in the North American paleoenvironmental records. The first phase takes place and peaks in the early Holocene when it becomes extremely dry, which is then followed by a reversal toward an increase in effective moisture (Campbell and Campbell 1997). The cause of these climatic trends has been the orbital perturbations of the earth, which have induced variations in insolation with a maximum peak of seasonality at 10,000 years BP (Campbell and Campbell 1997; Schweger and Hickman 1989). As noted by Campbell and Campbell (1997), this means that at 10,000 years BP, winters were very cold and dry and summers were very hot and dry compared to the present day conditions of cool moist winters and warm moist summers.

The trees that were able to survive on the lower elevation northern Plains north of the Cypress Hills region during the colder climate experienced at the end of the Pleistocene, could no longer survive when the climate warmed up. Thus the trees on the northern Plains vanished and the trees in the mountains expanded their niche. As the early Holocene became warmer, the trees at the lower mountain elevation sites began to diminish as well, visible in the Toboggan Lake and Beauvais Lake pollen records. Fire appears to have played an important role throughout the Mountains and Foothills and Plains between 10,000 and 8400 years BP and perhaps even slightly longer, to the 8000 years BP timeframe. Evidence of fire regimes is present in nearly every pollen analysis and there was plenty of fuel available with the migration of mixed conifer and deciduous forests along the eastern flanks of the Rocky Mountains.

Chapter 8
REGIONAL PALEOSOL STUDIES AND THE TUSCANY SITE

DATED PALEOSOLS FROM SITES IN ALBERTA

The Tuscany habitation site (EgPn-377) is one of a number of unique and interesting sites in Alberta that contain sediments identified as paleosols. The term 'paleosol' is problematic because, as Fenwick (1985:3) notes, "...the criteria by which such a unit is identified are not clearly defined." Ruhe (1956) provided a broad definition for paleosols as soils formed in landscapes of the past. Ruhe (1967) subsequently refined this definition, recognizing three types of paleosols. The first type includes soils that formed the landscape surface at some time during the past and were subsequently buried. The second type includes paleosols that were buried but then exhumed. The final type includes paleosols that have always remained at the surface of the landscape but display relict features that can be attributed to different environmental conditions existing during the past.

The present study is *not* a soil analysis focused on the paleosols of the Tuscany site. However, the macrobotanical remains, which this dissertation examines, were recovered through flotation analysis of bulk samples from the paleosols identified at the Tuscany habitation site. These sediments present unique characteristics that have been observed at other sites located in southern Alberta. Unfortunately, the Tuscany paleosols have not been subjected to many of the detailed analytical techniques utilized in some of the paleosol studies included in the comparative discussion. On the other hand, none of the other sites discussed appear to have been studied utilizing flotation on the scale of this dissertation. Therefore, macrobotanical remains have not been recovered from the other sites. Comparison to other site locations may reveal trends that will allow some interpretation of paleoclimatic factors at work in past landscapes in southern Alberta.

Table 10 presents a number of sites that contain pre-Mazama paleosols in Alberta and Saskatchewan. Many of the sites have had radiocarbon dating completed upon materials found in the soil profiles. Most of the sites have been excavated for archaeological recovery of cultural material remains. Some sites have been explored for geomorphological or soil studies/interpretation.

Four criteria were important for inclusion of a site in the table. First, the site had to contain one or more recognizable paleosols. Second, recognition of a Mazama tephra layer, either continuous or discontinuous within the stratigraphy, was very important. Third, the paleosols had to be located beneath the Mazama tephra layer. Their relative position beneath the tephra indicates they must be older than the material overlying. The fourth criterion for inclusion of a site in Table 10 was that a radiocarbon date had been derived from material from the paleosol(s) to give a chronological reference to the sequence of deposition and soil development.

The series of sites with dated paleosols listed in Table 10, cluster generally between 10,000 and 7000 years BP. Not all the paleosols listed are closely related to the Tuscany sediments. The Hawkwood site and EgPn-428 most closely resemble the observed Tuscany stratigraphy. What is important about the sites listed is the fact that paleosols were being formed in a series of different landscapes during the same time period. In areas like The Gap with its degraded alpine eutric brunisol it is obvious that different variables influenced the soil forming processes in the landscape. Moisture regimes must have been different and different parent materials were being deposited, which led to the formation of different soils and different vegetation was covering those landscapes.

The Fish Creek Paleosols

The Fish Creek site was examined by Kingston (1982), from a soil studies perspective, with the goal of developing methodology for the study of paleosols. The site is located in Fish Creek Park, in southwest Calgary. Three profiles were investigated along a cutbank overlooking Fish Creek. As Kingston (1982) notes, this site, with its buried paleosols, had been well documented

Table 10: Paleosol sites with radiocarbon dates from southern Alberta

Laboratory	Radiocarbon Date (Yrs. Bp)	Location	Collector	Material Dated
BGS-619*	13,900 ± 700	Fish Creek	Kingston (1982)	Carbonates
Beta-125408*	6840 ± 70	EgPn-428	Vivian and Reeves (1999)	Bison cervical vertebrae
Beta-107315	7030 ± 70	EgPn-230	Vivian *et al.* (1998)	Bison bones
GSC-1819	8400 ± 150	Bowfort Road 85th St. NW Pit	J. F. Harrison (1973)	Carbonized twig fragments
RL-905	8710 ± 230	Bowfort Road 85th St. NW Pit	M. Wilson (1983)	Carbonized twig fragments
GX-6397-A	8300 ± 280	Mona Lisa: Loc. C Lower paleosol	M. Wilson (1983)	Burned bone
GSC-1209	8080 ± 150	Mona Lisa: Loc. A	M. Wilson (1983)	Bison bone
RL-1276	7030 ± 210	Hawkwood – Component 2	Van Dyke and Stewart (1985)	Bone appetite
RL-1277	6820 ± 280	Hawkwood – Component 2	Van Dyke and Stewart (1985)	?
RL-1554	8250 ± 330	Hawkwood – Component 1	Van Dyke and Stewart (1985)	Bone collagen
GX-0956	9520 ± 240	The Gap	Reeves and Dormaar (1972)	Charred bison cervical vertebrae
GSC-1158	8000 ± 150	The Gap	Reeves and Dormaar (1972)	Charcoal
GSC-1298	6720 ± 140	The Gap	Reeves and Dormaar (1972)	Charcoal
GSC-1944	8030 ± 200	Rough Creek	B. Reeves (ref. in Waters 1979)	Charcoal
GSC-236	9290 ± 260	Willow Creek	A. M. Stalker (ref. in Waters 1979)	Freshwater gastropod shells
? not given	9860 ± 320	Bellevue-Crowsnest Pass Ah horizon	Reeves (ref. in Valentine *et al.* 1987)	Bison skull

* Date not accepted – see discussion for details

and correlated within a regional framework of sites exhibiting similar stratigraphy resulting from similar pedogenic processes.

Research by Kingston (1982) at Fish Creek Park in south Calgary, utilized Ruhe's (1956; 1967) definitions to describe and identify a paleosol. The paleosol discovered by Kingston (1982) was determined to be of the first type, described above, and had thus been buried over time. Kingston's (1982) work determined that the paleosol was underlying a Mazama tephra deposit. At Kingston's (1982) Fish Creek site, there are three important sedimentary sections to the sediment profile. The first sedimentary section extends down from the landscape surface to the Mazama ash deposit, which is the second sedimentary section of the profile. The paleosol found underlying the ash deposit is the third sedimentary section. The Tuscany habitation site contains the same three sedimentary sections in the upper half of the wall profiles that contained evidence of human occupation through time at the site locale. It is very clear that the paleosols have been buried over time and are of Ruhe's (1967) first type.

The Fish Creek paleosols provided a radiometric date on carbonates from the 'A' horizon of 13,900 ± 700 years BP. Kingston (1982) was faced with the dilemma of a problematic date that does not fit into the general framework visible from other dated paleosols in the greater region. Comparisons of the sediments were made to those at the Bowfort Road site (Harrison 1973) and Fish Creek was placed into the timeframe of about 8400 ± 150 (GSC-1819) by Kingston (1982). Interestingly, while two of the Fish Creek profiles contained chernozemic paleosols, the third profile was designated a Brunisolic paleosol.

Waters (1979) also examined a sediment profile from Fish Creek. The profile studied was also along a cutbank and appears to be close to the location examined later by Kingston (1982). The paleosol was identified as an orthic dark brown chernozem of colluvial origin made up of an Ah, a Bm, and a Ck horizon (Waters 1979). Phytolith materials of grassland environments were found in the Ah horizon. The Bm horizon was described as having a red hue and had undergone some illuviation of clays (Waters 1979). No date for the paleosols was provided in this earlier study. While the red hue of the Bm horizon is very similar to the Tuscany paleosol profile, there is no mention of any charcoal or charred plant material contained in the sediments.

The Crestmont Estates Site (EgPn-428)

The Crestmont Estates site (EgPn-428) was located on an area known as Paskapoo Slopes on the south side of the Bow River, which is south and slightly east of the Tuscany site (EgPn-377) location (on the north side of the Bow River). The stratigraphy of EgPn-428 was very similar to that of EgPn-377, containing a paleosol in the sediments beneath a Mazama tephra layer. A date of 6840 ± 70 years BP (Beta-125408) derived for the paleosol from collagen extracted from four bison cervical vertebrae was considered unacceptable (Vivian and Reeves 1999). According to Vivian and Reeves (1999) an acceptable date would have been in the range ca. 9000-7750 years BP, based on a lanceolate projectile point found within the paleosols that was determined to be Lusk, from the Plains/Mountain complex.

As previously discussed, there are a number of problems with the Lusk cultural complex. The date from the vertebrae at the Crestmont Estates site relative to the sediments they were deposited in is open to interpretation. However, the proposed extension of dates ranging between 9000 and 7750 years BP for the cultural material is not problematic.

Valley Ridge (EgPn-230)

EgPn-230 was located in a low swale on the western edge of the City of Calgary, on bench lands above the south banks of the Bow River (Vivian *et al.* 1998). The stratigraphy is comprised of a chernozem topsoil 75 cm thick with occasional pockets of Mazama tephra underlying and also mixed in with the C horizon, followed by a buried paleosol consisting of an A horizon underlain by a Bf horizon, which was then underlain by yellowish clays (Vivian *et al.* 1998). The buried paleosol is proposed to contain characteristics that indicate "...it developed under an open coniferous parkland" (Vivian *et al.* 1998:4). A radiocarbon date of 7030 ± 70 years BP on bone associated with the paleosol, is very near the Mazama tephra date boundary of 6730 ± 40 years BP. The date is comparable to the uppermost paleosol at the Tuscany site, which dated to 6940 ± 70 years BP and was also very near the Mazama tephra boundary. This paleosol is, however, more recent than the Tuscany paleosol that contains the oldest cultural component with two dates of 7840 ± 100 and 7610 ± 70 years BP.

Bowfort Road/85th Street Pit

The Bowfort Road, or 85th Street pit, exhibited similar stratigraphy to the Tuscany habitation site. From the surface down, the profile began with a massive silt, followed by a Mazama tephra, then a massive sand and silt, a paleosol, another massive sand grading to a massive silt and then crossbedded sand. The paleosol had a 10-15 cm 'A' horizon underlain by a red brown 'B' horizon. Within the paleosol, both Harrison (1973) and Wilson (1983) found charred twigs of 5mm dimension. Importantly, Harrison derived a date of 8400 ± 150 years BP on charred twig fragments from the paleosol. These charred twigs are very similar in size to the material from the Tuscany habitation site. There are four buried soil sequences present in the paleosols at the Tuscany site. The third pedogenic zone (128-161 cm in the excavation Unit 14 description) at Tuscany was made up of an 'A' horizon followed by a distinctly reddish brown 'B' horizon. This 'B' horizon contained the greatest evidence of burning vegetation at Tuscany with the number of small charred twig fragments recovered via flotation. Though these two sites have at least two similar parts (Mazama tephra and paleosol), the rest of the sequence is quite different and this is no doubt due to location, differing deposition processes and possibly erosion. Wilson (1983) designates the paleosol at Bowfort Road a Chernozem; this would represent a soil which formed in a predominantly grassland environment but that clearly contained a few small shrubby plants that left behind the small burned twigs.

Mona Lisa Site (EgPm-3)

The Mona Lisa site (EgPm-3) was located in the south Calgary City core in 1968 (Wilson 1983). Two profiles from two separate areas were described for the site. The general stratigraphy for Area 'A' from the surface down, includes a post-Mazama tephra alluvium, followed by a clear Mazama tephra layer, a silty clay layer, a bone bed with a single Bison kill, followed by silty clays, an organic silt loam with charred flecks present in it and finally, a bottom unit of coarse alluvium (Wilson 1983). The Area 'C' profile differed slightly having pre-Mazama sediments beneath a Mazama tephra, followed by a clay layer, then two organic layers, a sand layer, a cobbly silty clay layer, gravels, and bedrock (Wilson 1983).

This site was located on the Bow River floodplain and the deposition is clearly attributable to flooding episodes in the valley bottom. Thus there is no reddish brown 'B' horizon present in these profiles. Nevertheless, the bone beds have been dated and fall within the general range that has been noted for other locations with buried paleosols. In Area 'C' a date of 8300 ± 280 years BP was obtained on burned bone from the lower organic layer (Wilson 1983). Another date of 6580 ± 205 years BP (GX-6396-A), on unburned bone from the upper organic layer is a bit problematic as the date postdates the Mazama tephra yet is located beneath it (Wilson 1983). Wilson (1983) notes it is within the range if the standard deviation is invoked, but it is not included in the table since it is problematic.

The Hawkwood Site (EgPm-179)

The Hawkwood site, excavated in 1981, was located on the southwestern edge of the Nose Hill Uplands in northwest Calgary. It was located approximately 3 km north and east of the Tuscany habitation site, which is on a lower bench just above the Bow River floodplain. The Nose Hill Uplands rise about 150 meters above the valley floor and contain a very high density of archaeological sites, likely due to the location itself and the habitat and resources it presented.

Two large depressions, designated the East and West basins, were excavated. Three archaeological components out of six were located below a Mazama tephra horizon. Component 1, the lowest component in the East basin, was assigned to the Lusk Plains/Mountain Complex based on an unnotched atlatl lanceolate projectile point that was 4.0 cm long and 2.3 cm wide (Van Dyke and Stewart 1985). A date of 8250 ± 330 years BP was derived from bone collagen from Component 1 (Van Dyke and Stewart 1985). Component 1 is the most relevant because it is contained in the oldest stratigraphic sediments that appear to compare, by position below the Mazama tephra horizon, to the lowest/oldest part of the paleosols at the Tuscany site, Layer M. As well, the projectile point in Component 1 does have some similarities in size and shape to the point from the Tuscany habitation site, which has now been designated as Lusk as well.. The sediments in which the Hawkwood lanceolate point was found were a series of paleosols separated by clay-rich pond deposits (Van Dyke and Stewart 1985). While the topographic setting was very similar to the Tuscany site, the paleosols do not appear to have been similar based on the short description provided. There was no reddish 'B' horizon described for the Hawkwood sediments. Their stratigraphy was attributed to the effects of periodic ponding in the basin topographies.

Another archaeological component was present at the Hawkwood site that is comparable to the Tuscany site. In the east basin at Hawkwood, a point determined to be Bitterroot was found in Component 4, which was located immediately above the Mazama tephra layer (Van Dyke and Stewart 1985). A point determined to be Bitterroot was found at the Tuscany habitation site in the upper, paleosol-Mazama tephra boundary. Bitterroot atlatl points have been found in radiocarbon dated contexts ranging from 7630-5255 years BP (Frison 1998). An overlying Oxbow occupation/component (5500-4200 years BP) (Vickers 1986) in both the Tuscany and Hawkwood sites makes the material culture comparison even more striking between these two sites.

The Gap (DlPo-20)

The Gap site is located in a large opening through which the Oldman River flows eastward between two sections of the Livingstone Range in the southwestern Rocky Mountain Front Ranges in Alberta. The site was located on a postglacial river terrace with stratigraphy that included a buried Mazama tephra horizon with two buried paleosols underlying it (Reeves and Dormaar 1972). Three dates were derived from charcoal and bone remains recovered from stratigraphic contexts at the site.

The oldest 'Ah' soil from 190-200 cm below surface at the site was characterized as a buried degraded alpine eutric brunisol, which is a soil formed under forested conditions (Reeves and Dormaar 1972). Two dates, 9520 ± 40 years BP derived from a charred bison cervical vertebrae and 8000 ± 150 years BP derived from charcoal found in the sediments, help identify a living floor associated with the Mountain-Foothills cultural complex which spans the 10,000-7500 years BP time period (Reeves and Dormaar 1972).

Reeves and Dormaar (1972) associate this soil with a glacial event known as Pinedale III in the area, a time of ice advance and a climate that was cooler and wetter than it was later. The treeline is proposed to have been 600 meters lower at that time (Reeves and Dormaar 1972). An Agate Basin lanceolate projectile point was found at the site, however, it was found on the surface and not in stratigraphic context. Nevertheless, the oldest component of the buried paleosols is proposed to be associated with this cultural evidence, which does fit into the timeframe yielded by the dated material.

Rough Creek, Willow Creek and Bellvue

Three other sites in southern Alberta locations are described by Waters (1972) and Dormaar (1978) though not in great detail. Evaluation of the sediment profiles provided by Waters (1979) for the sites shows some similarities exist between them and the Tuscany site in terms of sequences of deposition. They all contain buried paleosols that are found underlying Mazama tephra deposits.

The Bellvue site dated to 9860 ± 320 years BP had a buried gleyed bisequa gray luvisol with 6 soil horizons present in it (Dormaar 1978). This soil type is associated with forest conditions existing during its formation (Agriculture Canada Expert Committee on Soil Survey 1987). This is a different paleosol compared to the Tuscany site, however.

Willow Creek (9290 ± 260 years BP) and Rough Creek (8030 ± 200 years BP) were both formed in floodplain environments and also contain soils that are dissimilar to the Tuscany profile. Only the dates from the sites may yield comparison in terms of the time period included by them. The date for Willow Creek is in the range of time proposed for the deposition and formation of the reddish paleosols labeled Layer J at the Tuscany site. The date from Willow Creek was derived from freshwater gastropod shells. It is not clear if these might be subject to problems with the incorporation of old carbon, which would make the date suspect. There is the indication that conditions may have been drier and possibly more stable during the timeframe yielded from both dates at these sites. This would have to be the case in order for soils to have time to form in a floodplain environment.

SUMMARY AND INTEGRATION WITH OBSERVATIONS AT THE TUSCANY SITE

Previous chapters have examined the Tuscany habitation site paleosols using a descriptive, comparative approach. As noted, there were no soil chemistry analyses carried out on the paleosols from the site. Nevertheless, there is information that can be compared to other sites in southern Alberta where paleosols have been examined in archaeological and other contexts.

Reddish colored paleosol horizons occur in a number of sites around the Calgary region. There is no doubt that more sites with similar paleosol sequences will be found in the future. The primary interest in the studies presented here was the clustering of dates for these soils in the 10,000-7800 years BP range. However, the dates do range widely.

The examination of buried paleosols in southern Alberta produced some sequences comparable to the Tuscany habitation site. The best comparative sequences are in the Calgary City area and include the Fish Creek paleosols, the Crestmont Estates site (EgPn-428), and the Bowfort Road/85th Street Pit site. While there were no charred botanical remains recovered from the Fish Creek or Crestmont Estates sites, there were some small 5 mm charred twigs recovered from the Bowfort Road site. These small charred twigs (by their description) appear to be identical to the twig-like materials recovered through flotation of the Tuscany bulk soil samples. Two independent dates on charred twigs of 8400 ± 150 and 8710 ± 230 years BP were derived from the Bowfort Road site. Dates from Crestmont Estates and Fish Creek sites were both problematic. Crestmont Estates was evaluated to date to 8000 years BP as discussed above and Fish Creek was compared to the earlier Bowfort Road date of 8400 years BP (available in the literature at that time), which may also now include the other date at 8710 years BP (Wilson's date derived in the year following Kingston's research). The standard deviation may push the oldest date very near 9000 years BP.

One buried paleosol section from Fish Creek (Kingston 1982) is described as a brunisol and appears to resemble the Tuscany buried paleosols more than the other sections. The quantities of charred material in the Tuscany flotation samples indicate the possibility of vegetation cover that included shrubs and possibly trees. The seeds recovered in this study are mainly from shrubby plants including kinnikinnick, juniper, buckbrush, shrubby cinquefoil and prickly rose. These would certainly fit into the same vegetation scenario as noted by Wilson (1983) in his assessment of the Bowfort Road/85th Street pit paleosols. Comparison of the pore structure of two small pieces of charred wood from a Tuscany flotation sample suggests that they are closest to *Salix* sp. (willow) or *Populus* (poplar) species. Recovery of charred cone scales from white spruce cones from the Tuscany flotation samples indicates there may have been such trees present near the site during the formation of the paleosols. The buried paleosols at the Tuscany site and Bowfort Road may be brunisols, which formed under forested vegetation. There is a brunisol section in the Fish Creek paleosol profiles. The soils from these sites appear to be most comparable to the Tuscany paleosols from their descriptions. It would be reasonable to propose that the middle two Tuscany site paleosols are reddish brunisols based on comparison to these other sites analyzed. This proposal is compatible with the information from the faunal assemblages, which require shrubby and treed habitats for browse and cover in the vicinity of the Tuscany habitation site on the north side of the Bow River and the Crestmont Estates site on the south side of the Bow River. The Bowfort Road site fits in between them in the Bow River valley. It also has to be remembered that faunal remains indicate general vegetation cover near an archaeological site or over an extensive area surrounding the site vicinity. Animals may be killed and butchered at some distance from a site and then carried back for consumption with the entire social group. Thus, inferences of vegetation cover from the faunal remains must be evaluated at a very general level.

The difference between the observations for the Tuscany flotation project and other projects that have examined macrobotanical remains, is one of sampling scale. The Tuscany project sampled the paleosols on a scale that has not been undertaken in an archaeological site in this region. Flotation analysis has not been extensively pursued as a methodological means of deriving information from archaeological sites in western Canada. The information that has come out of this study may lead to future studies that can provide comparable assemblages.

There is evidence of a fairly severe burning episode or series of burning episodes within the Tuscany landscape. The evidence for this consists of charred botanical remains that have been recovered via flotation analysis. This burn event is related to the redder soil horizon but it certainly does not mean that the soils themselves were ever burned. The only material that is proposed to have burned is the vegetation that grew within the site depression during the past. The reddish color that characterizes the 3rd pedogenic horizon at the Tuscany site is found in sites in the region with buried paleosols underlying a Mazama tephra layer. At one point, the possibility that the soil hue might be the product of oxidation effects on Fe (iron) by fire was considered. However, a red "B" horizon is characteristic of some Chernozemic and Brunisolic soils due to the accumulation of amorphous Fe compounds and natural oxidation of Fe (Agriculture Canada Expert Committee on Soil Survey 1987).

Mention has already been made of the fact that studies on moist forest soils have shown that the water content plays a very important role during fires. It takes a lot of energy (2.5 MJ/L) to convert the water from a liquid state to vapor, which limits achieved soil temperature to about 100°C until all the water evaporates (Fisher and Binkley 2000:244). Water is very heat conductive as well, and this may limit the temperatures moist soils will attain and how deep heat will penetrate during a burn event (Fisher and Binkley 2000:245). If we assume a mesic to submesic moisture regime that supported the Kinnikinnick and Juniper plant populations along with the other plants, then fire should not have affected the soils with a dramatic color change from oxidation of iron. The color should simply be due to normal soil forming processes observed in many forested landscapes.

There is one other important comment to make on the effects of fire regimes in the Tuscany area during the early Holocene. The fact that the charred plant remains recovered appear to be smaller pieces from twigs and shrubs does not preclude other larger trees being present in the landscape. There are many different kinds of fire regimes that burn vegetation landscapes. These regimes can depend on how many previous burn events have taken place over a space of time. Different forests burn at different rates; the wood provides different fuel types depending upon the species of trees present. There are rapidly advancing fires, which do not use much vegetation fuel and do not raise the soil temperature very much (Fisher and Binkley 2000:244). Other fires burn at slower rates, consuming more fuel and raising soil surface temperatures to 700°C, which drops to 200°C a few centimeters down in a mineral soil and reaches normal temperatures from 15 to 30 cm below the surface (Fisher and Binkley 2000:244). Thus, fires can burn in an area with a low fuel load at ground level and leave the main trees relatively unscathed. The implications of these kinds of fires is that they would burn off the low shrubby plants, forbs and grasses in a vegetated landscape, leaving behind only small shrub twigs as evidence of the fire that took place.

The spatial analysis has revealed that there is a definite drop in the frequencies of charred plant remains and charred seed remains in the second pedogenic zone, just beneath the Mazama tephra. The question to be answered here is what could cause the drop in frequency of the charred macrobotanical remains? The important point here is that the frequencies do not drop to zero—they drop to distinctly lower levels. One scenario that could explain the continued occurrence of the charred remains in pedogenic zone two could be effects of translocation from cryoturbation cycles, which could move the charred remains upward from the third pedogenic zone where they are the most abundant. In other words, translocation could cause a blurring effect, up and down, for the charred plant remains. However, this represents a doubtful explanation, because charred plant remains are also found in the upper chernozem (pedogenic zone 1) above the Mazama tephra boundary, but in very low frequencies.

An alternative and more plausible hypothesis for the continued appearance of charred plant remains can be suggested. In this scenario the landscape which was devastated by the previous fire does not recover with the same abundance of vegetation. It recovers, but is covered with grasses and only occasional shrubs. This landscape continues to burn off at intervals when fires sweep through the area. The results of fires in this more sparsely vegetated landscape would be the production of far fewer charred remains to be incorporated into the soil.

The reddish soil identified as Layer J in the Tuscany wall profiles, contains the burned vegetation fragments, which provide the evidence for a burned landscape and indicate a point where instability may have begun to be introduced to that landscape. Instability occurs when vegetation no longer exists to hold soils in place and wind and rain cause erosion to take place in landscapes. It is suggested here that the vegetation that was burned off did not recover subsequently, or at least not in the same proportions as existed previously. It is suggested that this is indicated by the lower frequencies of both charred plant fragments and charred seed representations from the flotation samples from the second pedogenic zone (immediately under the Mazama tephra). The second pedogenic zone above the burn layer was different in color and texture from the third pedogenic zone below. The second pedogenic zone (immediately under the Mazama tephra) appears to have been a developing chernozem soil, though it was fairly shallow. While it did have time to develop distinct "A" and "B" horizons, it appears that it was arrested in its development by the Mazama tephra deposition event.

An alternative hypothesis can be offered to account for the lack of charred materials in pedogenic zones 1 and 2. In this scenario the vegetation does not burn off the landscape. It simply passes through successful life cycles, dies off and becomes incorporated into the developing soil horizons as it degrades. In this case, there should be no charred plant remains to incorporate into the forming soils. This hypothesis is problematic because it is clear that the charred plant remains continue to occur in pedogenic zones 1 and 2—in very reduced quantities. One means of explaining the continued occurrence of these remains is through wind transport. Wind transport can certainly account for smaller (>2 mm) pieces of charred plant remains, however, the larger remains would have to be transported along the ground surface by saltation (eg. Clark 1988a). This scenario implies a very close association of an area within a few hundred meters of the depression, where a substantial vegetation cover burned and high winds followed that event. It also implies that there could not have been any vegetation ground cover like grass, to prevent the movement of larger charred plant remains along the ground between the burned area and the Tuscany depression. This hypothetical scenario is far more complex than the first proposition. The conclusion here is that the first hypothesis in which reduced amounts of natural vegetation burned in place in the Tuscany depression leaving behind reduced quantities of charred remains, accounts most easily and simply for the observed patterns of distribution within the profiles.

The examination of other sites available in the published literature reveals that many of the dated paleosols encountered in archaeological and geographical related studies are not reddish brunisols as the Tuscany paleosols appear to be. Discussions with a number of local archaeologists have given the impression that these reddish paleosols are encountered fairly regularly. If so, they are not being reported in the literature. This is unfortunate, as patterns of deposition are not being recognized. The distribution of these sites should be mapped and the soils examined and compared. This may represent a research topic that could be pursued by a student interested in a geoarchaeological project in the near future.

This chapter has examined the sediments of the Tuscany site and the processes that led to the formation of a series of paleosols within the locality. The comparison of other site profiles has provided useful information in terms of whether or not similar soil forming processes can be ascertained to have taken place through time in other localities within the greater region of southern Alberta. The fundamental problem with other studies related to the soils/sediments of different localities is the types of information that have been focussed upon. The perspective of this study for macrobotanical remains is relatively new and there is very little related information available on this from other sediment-related research. Thus, a number of important insights have been derived from this discussion, however, it is apparent that research projects with similar goals for paleoenvironmental reconstruction are required for more informative comparisons.

Chapter 9
SUMMARY AND CONCLUSIONS

The research objective of this dissertation has been the examination of macrobotanical remains including charred seeds and charred plant fragments to reconstruct the paleovegetation present at the Tuscany habitation site. The timeframe of interest includes the late Pleistocene between 9950 ± 120 years BP and the Mazama tephra boundary at 6730 ± 40 ^{14}C years BP. The charred botanical remains have provided evidence for a record of paleoenvironmental change at the Tuscany site. This record of change appears to fit in with the perspective of regional paleoenvironmental change that has been derived from examination of pollen and macrobotanical studies from the Rocky Mountains, Foothills and southern Alberta landscapes.

This project represents the first attempt at paleoenvironmental reconstruction from macrobotanical remains from an archaeological site located in the northern Plains in Alberta. As such, it provides the first working model for this type of research for this area. It has been established that paleoenvironmental reconstruction is an inductive process in which a number of interlinked variables are examined to understand and infer what the vegetation was like and what changes may have taken place through time. This is the process that has taken place through this dissertation.

Chapter two summarized the methodological approaches used in the study to extract the macrobotanical remains from bulk sediment samples from the Tuscany site. Two methodologies were utilized to complete the processing of the samples, tub and bucket flotation. A modified bucket flotation method was developed specifically for the project to speed up the processing procedure. The bucket method actually proved to be more rigorous than the tub method and provided data that were of slightly superior quality. The data were not compromised at all by the change in methodologies.

Various other methodologies were explored and employed to facilitate the identification process of the charred plant materials recovered. These included the use of a digital camera along with work on a Scanning Electron Microscope (SEM) and an Environmental Scanning Electron Microscope (ESEM), which all provided useful views of the macrobotanical remains.

The approach to the research included an extensive background summary of the main elements utilized to derive a perspective of a changing environment through time. Chapter three summarized the geology, geomorphology, climate and vegetation of the Calgary area and the Tuscany site today. The primary purpose of the summary was to provide a known vegetation background to compare against information from the macrobotanical analysis. Thus, the background information covered both regional and local site perspectives of the modern vegetation regimes and their relatedness to each other. Interestingly, while the Tuscany site fits into a Parkland environmental designation it appears to be in the transitional zone between the Fescue Grasslands to the east of Calgary and the more forested areas west including the Foothills and Rocky Mountains. This position in the landscape has provided an extra dynamic to the area and the vegetation found therein through time as climatic trends fluctuated. Comparison of the vegetation of the inventories from the present and past shows a trend from more wooded, Parkland type cover to a more open grassland-forb cover that is found today.

A description and discussion of the charred seeds identified from the charred plant remains from the Tuscany site examines the ethnobotanical implications of the assemblage. This information is necessarily descriptive. The charred seeds were identified with the aid of a comparative seed collection from the Provincial Museum of Alberta, Edmonton. A series of figures compares the identified charred seed remains to modern examples from the comparative seed collection.

Kinnikinnick and juniper were the most abundant seeds identified from the Tuscany flotation samples. Their combination in the landscape was found to occur extensively along the Foothills and flanks of the Rocky

Mountains in southwestern Alberta. They are both found today in the greater landscape around the Tuscany site, but do not occur within or close to the site itself. Other plants identified by their charred seed remains include a *Chenopodium* species, a rose species, shrubby cinquefoil, white spruce, northern bedstraw, wild bergamot and a *Castilleja* sp. All the plants have preferences for mesic to submesic moisture regimes. Kinnikinnick and juniper prefer drier and more well drained soils as well as more open cover without much shade. All the plants are found today in the greater Tuscany landscape, especially on the slopes of Twelve Mile Coulee, several hundred meters to the east and north of the site. Thus, Twelve Mile Coulee appears to contain a kind of remnant refugium of the plants found in the Tuscany landscape during the earlier Holocene period.

The spatial distribution of the charred plant remains that appear to be small pieces of burned twigs or roots along with the identified charred seeds and lithic debitage was the topic examined in Chapter five. The charred materials were summed by type or species and the totals were plotted as X, Y, and Z coordinates for each unit in the excavation grid. All of the plots of charred remains reveal an area of low density of 3 x 4 meters around a hearth feature found along the east wall of Unit 37 (N521, E526) Level T. A Lusk projectile point was found in association with the hearth. This low-density area appears to have the dimensions of a trample zone of activity around the hearth feature and may even represent the dimensions of a small structure erected for shelter. The lithic debitage scatter is in association with the charred botanical remains in an area 4 meters to the east of the hearth feature. Bone remains were not plotted in the spatial analysis but are also associated with the charred botanical remains and lithic scatter in a separate activity area to the northeast of the hearth between 5 and 10 meters away. Thus, three activity areas appear to be identified within the space of the excavation grid.

One of the problems resulting from excavation methodology was poor correspondence between the boundaries of the excavated levels from one unit to the next. This problem was resolved to some measure with the development of a series of scaled profile drawings that present the charred macrobotanical data contained within each level of each unit. The profiles do reveal deposition patterning that appears to match up with the sequence of stratigraphy that has been proposed for the site. These profiles reveal four sets of paleosols that developed partially and were subsequently buried. The combination of pedogenesis and macrobotanical remains indicate the probability that there was an open parkland vegetation regime at the Tuscany locale for a lengthy period of time during the early Holocene.

Studies of the forest cover to the west of Calgary provide detailed descriptions on the different types of vegetation patch combinations that occur throughout the landscape. This variability is dependent upon a number of factors like elevation, soil moisture regime, soil type, slope, and aspect. These ecological landscapes, discussed in Chapter 6, have developed through time in southern Alberta and also appear to have long histories, if the Tuscany assemblage is accepted. While the vegetation cover that exists on the landscape has changed with the changing climatic regimes, the suite of plants that thrive in various kinds of communities have survived. They have simply migrated to different areas where soils, elevation and moisture regimes can continue to support them.

These vegetation landscapes have supported cultural landscapes. People camped in the Tuscany site landscape for a short period of time, perhaps up to three weeks during the timeframe around 7800 years BP. They hunted a variety of animal species and perhaps gathered and processed plants from the area. While there is good evidence for hunting endeavors there is, unfortunately, no specific evidence for the use of plants at the site, not even from the hearth feature. Thus, the use of plants by people remains purely conjectural for this region until another site yields more conclusive data. The foraging model proposed for the people that occupied the site is the Plains/Mountain complex as outlined by Reeves (1975) and represents a broad-based foraging pattern where a variety of animals were hunted and killed. The number of different species present in the landscape was a feature of the nature of the vegetation present during those times.

An extensive summary of pollen and macrobotanical studies in Chapter seven provided the background for an account of changing vegetation in the southern Alberta region and the fit of the Tuscany site within the larger trend of changing vegetation through time. The immediate post-glacial period had only just begun when a climatic perturbation is proposed to have occurred with a cool and moist period known as the Younger Dryas from 11,000-10,000 years BP. Rapid climatic change to a period of warming followed and is reflected in dramatic changes in vegetation through the entire southern Alberta region. A spruce forest followed by poplar domination covered the southeastern parts of the Plains region. Pine, spruce and poplar forests appear to have dominated the western side of Alberta along the flanks of the Rocky Mountains and foothills. The Tuscany site has evidence of spruce growing within close range of the depression along with kinnikinnick and juniper as common shrub cover as time progressed. Fire appears to have been a dominant factor affecting vegetation during this time period. All the macrobotanical remains from the Tuscany site are charred and this has proven to be an important factor to their survival within the sediments of the site through time. Evidence of fire has been noted in the pollen and macrobotanical studies across the southern end of the province from east to west and along the eastern flanks of the Rocky Mountains during the early Holocene.

The examination of the buried paleosols underlying the Mazama tephra boundary was the topic of discussion in Chapter eight. While there were no comprehensive soil analyses completed on the Tuscany sediments, a number of important insights were revealed by their description and comparison to other sites in the area around Calgary and greater southern Alberta region. The most comparable soils were those from the Calgary area.

All the lines of evidence summarized throughout the entire dissertation can be used to define a general model for the degree and range of vegetation migration that took place in the Calgary region of southern Alberta during the late Pleistocene and early Holocene. Initially, the Calgary region represented a denuded landscape during the immediate post-glacial period. The plants that colonized the landscape appear to have migrated slowly from refugia to the south in areas that were unaffected by the massive continental glaciations of the late Wisconsin. Arboreal vegetation migrated from the southeastern plains to the north, grassland-forb vegetation migrated from the western plains to the north, and arboreal vegetation migrated from the southern flanks of the eastern Rocky Mountains to the more northerly flanks. The implications for human populations living within these rapidly changing environments are important. The vegetation in the landscape provides shelter and food for animals and human populations. In North America the grassland ecosystems provide a rich environment for grazing animals while the Aspen Parkands, Montane and Boreal ecosystems provide rich environments for browsing animals preferring shrubs and trees. The foraging pattern of a human population occupying a transitional zone between the Plains and Parkland can be broad-scale because of the diversity of animals and plants of both ecosystems that will occupy such zones. This appears to be the case at the Tuscany habitation site where the faunal remains represent animals from both the Parklands and Plains.

The Tuscany site contains evidence of a Lusk cultural component occupation, which is part of the Plains/Mountain complex. This archaeological culture appears well-adapted to the transitional nature of the vegetation landscape and the fauna that it existed within.

One of the important questions that has not been addressed completely in this dissertation is the cause or 'kicker' that has been proposed for the climatic changes that appear in the late Pleistocene and early to mid Holocene. As noted by Campbell and Campbell (1997), orbital perturbations of the earth have induced variations in insolation with a maximum peak of seasonality at 10,000 years BP. At that time winters were cold and dry and summers hot and dry compared to present conditions of cool moist winters and warm moist summers (Campbell and Campbell 1997). These conditions are reflected in the pollen studies that have been reviewed here which reveal changing arboreal vegetation throughout southern Alberta, the foothills and Rocky Mountains during the late Pleistocene-early Holocene timeframe. After 10,000 years BP the trees present on the northern Plains eventually vanished while trees in the foothills and mountains expanded their niches. One of the problems with the interpretation of these changes is the fact that the pollen records indicate the timing of this climatic warming has been variable—it appears to have been occurring at different times in different locations.

This period of warmth is demonstrated to occur around the Pleistocene-Holocene boundary in the Queen Charlotte Islands of British Columbia, northeastern Washington, the Fraser Lowland and Canyon, the interior Douglas Fir zone, southern interior British Columbia, and southwestern British Columbia (Pellat *et al.* 1998:125). This period of warmth is referred to as the "Early Holocene Thermal Maximum" by Pellat *et al.* (1998:125). This period of warmth is noted to occur from approximately 10,000 years to 7000 years BP in all areas excepting northeastern Washington where it ends at 8000 years BP (Pellat *et al.* 1998). Vance (1991:38-39) notes that along the periphery of the northern mixed-grass prairie, the sites of Guardipee Lake and Lost Lake appear to have experienced prolonged low-water levels during the early Holocene until 6000 years BP. MacDonald (1989) assigns the arid period at Toboggan Lake to the timeframe between 8000 and 5000 years BP

Haskell *et al.* (1996:190) note that the "vegetation records from Minnesota and Wisconsin suggest a gradual warming of climate, indicated by the migration of the prairie/forest border to its maximum eastward extension at 6 k.y." and that this same phenomenon eventually spreads into Iowa, but not until 5500 to 3500 years BP. It is noted that three major air masses, the dry westerlies, Arctic flows and moist tropical Gulf air, have a convergence zone that shifts and changes in intensity, producing changing climatic patterns (Haskell *et al.* 1996). It is suggested that this shift of the convergence zone may be a time-transgressive phenomenon (Haskell *et al.* 1996). Thus, it could be proposed that the shifts in the appearance of aridity may be a time-transgressive phenomenon as well.

A period of maximum aridity in the paleoenvironmental record of North America was an idea developed and established by Ernst Antevs between the late 1930's to 1950's (Schweger 1984). Antevs (1955), identified the "Altithermal," a geologic-climatic arid period which he proposed took place between 7500 to 4000 years BP in the western United States. In archaeology, this period has come to represent a chronological marker that has been used to explain why and where change is observed in material culture. According to Schweger (1984:373) this has led to the application of circular arguments when the Altithermal is invoked to explain why material culture has changed. Another term, "Hypsithermal," has also been used in the literature and is the equivalent of "Altithermal" (Schweger 1984:373). Schweger (1984) recommends that Antev's "Altithermal" chronology in archaeology should be dismissed completely because it is no longer applicable as a concept. This implies the concurrent dismissal of the term "Hypsithermal," if we follow the argument to its logical end. I would suggest here that the phrase "Early Holocene Thermal Maximum" is probably not as useful either because of the time-transgressive nature of the period of aridity that is manifested in paleoenvironmental studies. The most useful descriptive phrase for this phenomenon is simply, "the Holocene Thermal Maximum", defined as a period of aridity observed in Holocene paleoenvironmental records and recognized as occurring at different times depending upon location.

The purpose of this discussion is to finalize some of the conclusions derived from the examination of the Tuscany charred plant assemblage. It is proposed that a shrub-herb grass dominated community occupied the landscape during the formation of the lowest paleosol or pedogenic zone 5 (Layer M). This cover gradually changed, filling in with some trees, including spruce and possibly poplar, representative of open parkland vegetation cover. This scenario is derived from the higher representation of charred plant remains recovered from pedogenic zones 3 and 4 (Layers H/I/J and K/L). It has been noted that the paleosol, pedogenic zone 2 (Layers F/G), immediately underlying the Mazama tephra boundary was different in color and texture than the lower three paleosols and also contained far fewer, charred plant remains. A number of arguments were presented to explore this distribution. One of them, which appears to be the most straightforward argument, is that the arboreal and shrubby vegetation cover on the site grew far less abundantly and therefore left behind fewer remains once burned off. The soil that formed and was cut off from further development by the fall of Mazama tephra at 6730 ± 40 years BP is proposed to have been a chernozem, related to grassland vegetation cover.

The question here is, do we see this pattern of changing vegetation cover anywhere else? In fact a similar pattern of landscape vegetation cover is visible in the Toboggan Lake pollen profile, a site which is higher in elevation compared to the Tuscany site and is located only 40 km to the southwest. MacDonald (1989:155) recognizes a shrub and herb dominated landscape at 10,000 years BP. This landscape gradually filled with a pine and spruce forest between 9400 and 8000 years BP (MacDonald 1989:155). Between 8000 and 5000 years BP MacDonald (1989:155) proposes that the tree cover was replaced by grasslands.

The model proposed will include a number of inferences drawn from comparison to the Toboggan Lake pollen analysis. The test of the model can only take place with subsequent studies of sites with similar stratigraphy and similar preservation of macrobotanical remains. Such sites do exist in the greater Calgary area and will present future research opportunities to test the data generated from this research project. This model fits in with the observations from the Tuscany data and the dates derived thus far for the site. It is understood that these dates only provide us with an approximate timeframe for the argument and model. The case could be made more robust with additional comparative dates from the site. At 9950 ± 120, the first date derived from the lowest point in Layer M, the vegetation appears to have been a mixture of shrubs and grasses. By the time of the second set of dates (7840 ± 100, 7610 ± 70) derived from bone and charred plant materials associated with the Lusk cultural component, much more vegetation had gradually occupied the Tuscany locale. After this period, the vegetation cover appears to have changed to predominantly grassland along with a few sporadic shrubs. This sequence appears to tie in with the sequence from Toboggan Lake where aridity is proposed to have played a role in the changes of vegetation cover observed at that locale. It is proposed that aridity was a factor in the changing Tuscany vegetation landscape as well. Fire played an important role at this site as evidenced by the recovery of a completely charred macrobotanical assemblage through time. However, the relationship that may have existed between the people that occupied the site and the vegetation that grew and burned off the landscape is not decipherable from the patterns of distribution of the charred plant assemblage.

The charred plant remains recovered from the paleosols of the Tuscany site include kinnikinnick, juniper, a *Chenopodium* species, a rose species, shrubby cinquefoil, white spruce, northern bedstraw, wild bergamot, a *Castilleja* sp., and small charred woody fragments that appear to represent shrubby species like willows and possibly poplar trees. Each of these species has been discussed from an ethnobotanical perspective and their potential usefulness to a human population occupying the landscape during the past. It is not possible to determine the certain use of any of these plant species by the people who occupied the site over a short period of time during the past. Each of the plants can be argued to represent plants that could have been potentially used by the small group of people who left behind the Lusk point near their small hearth many millenia ago. The potential usefulness of the plants represented in the macrobotanical assemblage can be proposed to represent one of the reasons why this site was chosen for habitation over a short duration of time, probably no more than three weeks in total. This small group appears to have practiced a broadscale foraging strategy within a Parkland vegetation landscape, hunting the animals that were also occupying the more diverse environment available. This model does appear to be different from others because seasonality does not appear to be a determining factor, it is the environment itself that yields an available range of fauna which appears to have been hunted and consumed. What is represented with the macrobotanical assemblage from this research, is a small sub-sample of the entire inventory of plants that existed in the landscape at the time of occupation at approximately 7800 years BP. Certainly, the vegetation scheme was far more complex than represented by the charred plant remains recovered during this research. As well, this knowledge yields a new perspective for archaeologists in the search for understanding the relationships that existed between human populations and the landscapes occupied during the past. The combination of material culture remains with ecological milieu as an approach can provide more insightful models of past behaviors and lifestyles.

The changing paleotopography of the site setting has been explored through a spatial analysis of sediment deposition within the location through time combined with the occurrence of charred plant remains. There does appear to have been a relationship between the topography of the site and the places where charred remains were located. Certainly the small, 3 x 4 meter area around the hearth that appears to lack substantial charred plant remains is one of the most intriguing distribution patterns revealed through the spatial analysis. The suggestion here is that the lack of charred plant remains around the hearth area is due to human activity within a small structure for shelter. Over the remainder of the site, the charred plant remains appear to have been affected by the microtopography within the locale. There does appear to have been downslope movement of charred plant remains toward the northeast, which did represent the area of lowest elevation during the early formation of the landscape following glaciation during the terminal Pleistocene. The area of low elevation may be related to the formation of a channel scar resulting from erosion and drainage of the locale during the formative stages of the landscape evolution of the site after Glacial Lake Calgary drained. What has been demonstrated in the analysis of the microtopography of the site is a definite shift of the area of low elevation in the northeast to a location in the southeast through time. During the time of occupation by the population who left behind the Lusk projectile point, the site appears to have been much more level, but still maintaining a slope oriented toward the northeast. The entire orientation of the depression eventually changed to the southeast along with the type of parent material deposited in the area. Eventually a deep chernozem developed in these sediments over the Mazama tephra deposit and the series of paleosols. The subsequent vegetation that thrived in the depression area appears to have also changed through time, especially in the timeframe following the Mazama tephra deposition, when it became less shrubby and more grassy-herbaceous. Today, the site is predominantly covered with herb-grass vegetation along with some small clumps of buckbrush and shrubby cinquefoil.

REFERENCES CITED

AGENBROAD, L. D. 1978 – *The Hudson-Meng site: an Alberta bison kill in the Nebraska High Plains.* University Press of America, Washington, DC.

Agriculture Canada Expert Committee on Soil Survey 1987 – *The Canadian system of soil classification.* 2nd edition. Agriculture Canada Publication 1646, Ottawa.

ALLEY, N. E. 1972 – The Quaternary History of Part of the Rocky Mountains, Foothills, Plains and Western Porcupine Hills, Southwestern Alberta. Unpublished PhD Dissertation (Geography). University of Calgary, Calgary.

AMMANN, B., H. J. B. BIRKS, S. J. BROOKS, U. EICHER, U. VON GRAFSTEIN, W. HOFMANN, G. LEMDAHL. J. SCHWANDER, K. TOBOLSKI, L. WICK 2000 – Quantification of biotic responses to rapid climatic changes around the Younger Dryas—a synthesis. *Palaeogeography, Palaeoclimatology, Palaeoecology* 159:313-347.

ANTEVS, E. 1955 – Geologic-climatic dating in the west. *American Antiquity* 20(4):317-335.

ARCHER, S., C. A. HASTORF, J. COIL, E. DEAN, R. GODDARD, J. NEAR, M. ROBINSON, W. WHITEHEAD, and E. WOHLGEMUTH 2000 – Paleoethnobotany and Archaeology 2000--The State of Paleoethnbotany in the Discipline. *Society for American Archaeology Bulletin* 18(3):33-38.

ARCHIBALD, J. H., G. D. KLAPPSTEIN and I. G. W. CORNS 1986 – *Field guide to ecosites of southwestern Alberta.* Special Report 8, National Resources Canada, Canadian Forest Service, Northwest Region, Northern Forestry Centre, Edmonton.

ARMSTRONG, S. W. 1993 – Alder complex kitchens: experimental replication of paleoindian cooking facilities. *Archaeology in Montana* 34(2):1-63.

ARNOLD, T. G. 1985 – A comparison of plano complexes. Unpublished MA Thesis (Archaeology), University of Calgary, Calgary.

ASCH, D. L. 1979 – On Sample size problems and the uses of nonprobabilistic sampling. In *Sampling in Archaeology*. Second printing, edited by J. W. Mueller. The University of Arizona Press, Tucson, Arizona.

ASCH, D. L. and N. B. ASCH 1977 – Chenopod as cultigen: a reevaluation of some prehistoric collections from eastern North America. *Mid-Continental Journal of Archaeology* 2(1):3-45.

ASCH, D. L. and N. B. ASCH 1985 – Prehistoric plant cultivation in west-central Illinois. In *Prehistoric Food Production in North America,* edited by R. I. Ford, pp. 149-204. Anthropological Papers, Museum of Anthropology, University of Michigan, No. 75, Ann Arbor.

ASCH, D. L. and N. B. ASCH SIDELL 1988 – Archaeological plant remains: applications to stratigraphic analysis. In *Current Paleoethnobotany: analytical methods and cultural interpretations of archaeological plant remains*, edited by C. A. Hastorf and V. S. Popper, pp. 86-96. University of Chicago Press, Chicago.

BACON, C. R. 1983 – Eruptive history of Mount Mazama and Crater Lakes caldera, Cascade Range, U.S.A. *Journal of Volcanology and Geothermal Research* 18:57-115.

BANFIELD, A. W. F. 1974 – *The Mammals of Canada.* University of Toronto Press, Toronto.

BARNOSKY, C. W. 1989 – Postglacial vegetation and climate in the northwestern Great Plains of Montana. *Quaternary Research* 31:57-73.

BARNOSKY, C. W., E. C. GRIMM and H. E. WRIGHT JR. 1987 – Towards a postglacial history of the northern Great Plains: a review of the Paleoecologic problems. *Annals of the Carnagie Museum* 56:259-273.

BEAUDOIN, A. B. 1992 – Early Holocene paleoenvironmental data preserved in 'non-traditional' sites; Palliser Triangle Global Change Meeting, Regina, Saskatchewan, Program and Abstracts, p. 1-2.

BEAUDOIN, A. B. 1992 – A compendium and evaluation of postglacial pollen records in Alberta. *Canadian Journal of Archaeology* 17:92-113.

BEAUDOIN, A. B. 1996 – An early Holocene macrofossil record from the Fletcher site (DjOw-1) in southern Alberta; Canadian Association of Geographers Meeting, Saskatoon, Saskatchewan, Program and Abstracts, p. 61-62.

BEAUDOIN, A. B. and L. HALSEY 1989 – Some preliminary observations on a 9,000 year palaeoenvironmental record from the Grande Prairie area, Alberta. In *CANQUA 1989 Program and Abstracts*, pp. 26. Edmonton.

BEAUDOIN, A. B. and R. H. KING 1988 – Late Quaternary vegetation history of Wilcox Pass, Jasper National Park, Alberta. *Palaeogeography, Palaeoclimatology, Palaeoecology* 80:129-144.

BEAUDOIN, A. B. and M. WRIGHT 1993 – Links between culture and environment: an example from archaeological and environmental research in the Grande Prairie region, northwestern Alberta, 1987-1991. In *Culture and Environment: A Fragile Co-existence*, edited by R. W. Jamieson, S. Abonyi and N. Mirau, pp. 231-235. Proceedings of the 24th Annual Chacmool Conference, Department of Archaeology, Calgary.

BECKINGHAM, J. D. and J. H. ARCHIBALD 1996 – *Field guide to ecosites of northern Alberta*. Special Report 5, National Resources Canada, Canadian Forest Service, Northwest Region, Northern Forestry Centre, Edmonton.

BINFORD, L. R. 1978 – *Nunamiut ethnoarchaeology*. Academic Press, New York.

BIRKS, H. J. B. and H. H. BIRKS 1980 – *Quaternary Palaeoecology*. University Park Press, Baltimore.

BJÖRCK, S., B. KROMER, S. JOHNSEN, O. BENNIKE, D. HAMMARLUND, G. LEMDAHL, G. POSSNERT, T. L. RASMUSSEN, B. WOHLFARTH, C. U. HAMMER, M. SPURK 1996 – Synchronized terrestrial-atmosphere deglacial records around the north Atlantic. *Science* 274:1155-1160.

BLANKINSHIP, J. W. 1905 – Native economic plants of Montana. *Bulletin Montana Agricultural Experiment Station* 56:1-38.

BOSTOCK, H. S. 1970 – Physiographic subdivisions of Canada. In *Geology and economic minerals of Canada*, edited by R. J. W. Douglas, pp. 9-31. Geological Survey of Canada, Economic Geology Report No. 1.

BROWN, J. A. 1979 – Deep-site excavation strategy as a sampling problem. In *Sampling in Archaeology*, Second printing, edited by J. W. Mueller. The University of Arizona Press, Tucson, Arizona.

CAMPBELL, C. and I. A. CAMPBELL 1997 – Calibration, review, and geomorphic implications of postglacial radiocarbon ages in southeastern Alberta, Canada. *Quaternary Research* 47:37-44.

CAMPBELL, C., I. A. CAMPBELL, C. B. BLYTH and J. H. MCANDREWS 1994 – Bison extirpation may have caused aspen expansion in western Canada. *Ecography* 17(4):360-362.

CANNON, K. P. 1996 – A model for prehistoric economies of the Yellowstone Plateau during the Altithermal. In *The Ecological Implications of Fire in Greater Yellowstone*, edited by J. Greenlee, pp. 1-5. Proceedings of the Second Biennial Conference on the Greater Yellowstone Ecosystem. The International Association of Wildland Fire, Fairfield, Washington.

CARAN, S. C. 1998 – Quaternary paleoenvironmental and paleoclimatic reconstruction: a discussion and critique, with examples from the southern High Plains. *Plains Anthropologist* 43(164):111-124.

CHRISTIANSEN, O. A. and L. V. HILLS 1985 – Part 1: Palynologic and paleoclimatic interpretation of Holocene sediments, Waterton Lakes National Park, Alberta. In *Climatic Change in Canada 5 Critical Periods in the Quaternary Climatic History of Northern North America*, edited by C. R. Harington, pp. 345-354. National Museums of Canada, National Museum of Natural Sciences Project on Climatic Change in Canada During the Past 20,000 Years. Syllogeus No. 55. National Museum of Natural Sciences, National Museums of Canada, Ottawa.

CHURCHER, C. S. 1968 – Pleistocene ungulates from the Bow River gravels at Cochrane, Alberta. *Canadian Journal of Earth Sciences* 5(6):1467-1488.

CHURCHER, C. S. 1975 – Additional evidence of Pleistocene ungulates from the Bow River gravels at Cochrane, Alberta. *Canadian Journal of Earth Sciences* 12(1):68-76.

CLARK, J. S. 1988a – Particle motion and the theory of stratigraphic charcoal analysis: source area, transport, deposition, and sampling. *Quaternary Research* 30:67-80.

CLARK, J. S. 1988b – Stratigraphic charcoal analysis on petrographic thin sections: application to fire history in northwestern Minnesota. *Quaternary Research* 30:81-91.

CONRAD, L. A. and R. C. KOEPPEN 1972 – An analysis of charcoal from the Brewster site (13CK15), Iowa. *Plains Anthropologist* 17(55):52-54.

CORE, H. A., W. A. CÔTÉ and A. C. DAY 1979 – *Wood structure and identification*. 2nd Edition. Syracuse University Press, New York.

CORNS, I. G. W. and R. M. ANNAS 1986 – *Field guide to forest ecosystems of west-central Alberta*. National Resources Canada, Canadian Forest Service, Northern Forestry Centre, Edmonton.

DAIMANT, S. 1979 – A short history of archaeological sieving at Franchthi Cave, Greece. *Journal of Field Archaeology* 6:203-217.

DAVIS, M. B. 1963 – On the theory of Pollen analysis. *American Journal of Science* 261:897-912.

DINCAUZE, D. F. 2000 – *Environmental archaeology: principles and practice*. Cambridge University Press, New York.

DIXON, E. J. 1999 – *Bones, Boats & Bison*. University of New Mexico Press, Albuquerque.

DORMAAR, J. F. 1978 – Paleosol studies in western Canada. In *Quaternary Soils*, edited by W. C. Mahaney, pp. 209-221. Geo Abstracts Ltd., Norwich.

DRASS, R. R. 1993 – Macrobotanical remains from two early Plains village sites in central Oklahoma. *Plains Anthropologist* 38(142):51-64.

DRIVER, J. C., L. V. HILLS and B. O. K. REEVES 1985 – Part 3: palynology of Crowsnest Lake, Alberta, with Comments on Holocene Paleoenvironments of the Southern Alberta Rockies and Surrounding Areas. In *Climatic Change in Canada 5 Critical Periods in the Quaternary Climatic History of Northern North America*, edited by C. R. Harington, pp. 345-354. National Museums of Canada, National Museum of Natural Sciences Project on Climatic Change in Canada During the Past 20,000 Years. Syllogeus No. 55. National Museum of Natural Sciences, National Museums of Canada, Ottawa.

DYKE, A. S. and V. K. PREST 1987 – Late Wisconsinan and Holocene retreat of the Laurentide Ice Sheet. Geological Survey of Canada, Map 1702A, scale 1:5,000,000.

EBELL, S. B. 1982 – The northward movement: trees, rocks and water, or else! *Manitoba Archaeological Quarterly* 6(4):96-103.

FAEGRI, K. and J. IVERSEN 1975 – *Textbook of Pollen Analysis*. 3rd ed. Blackwell, Oxford.

FARRAR, J. L. 1995 – *Trees in Canada*. Fitzhenry & Whiteside Limited and the Canadian Forest Service, Markham.

FEDIRCHUK, MCCULLOUGH & ASSOCIATES Ltd. 1990 – *Historical Resources Impact Assessment, Carma Developers Ltd. West Scenic Acres Residential Subdivision.* Delcan Western Ltd., Calgary.

FEDJE, D. W., J. M. WHITE, M. C. WILSON, D. E. NELSON, J. S. VOGEN and J. R. SOUTHON 1992 – Vermilion Lakes site: adaptations and environments in the Canadian Rockies during the latest Pleistocene and early Holocene. *American Antiquity* 60(1):81-108.

FENWICK, I. 1985 – Paleosols: problems of recognition and interpretation. In *Soils and Quaternary Landscape Evolution*, edited by J. Boardman, pp. 3-21. John Wiley & Sons, Chichester.

FISHER, R. F. and D. BINKLEY 1998 – *Ecology and management of forest soils*. John Wiley & Sons, Inc., New York.

FISHER, T. G. 1999 – Sedimentology, stratigraphy, and geomorphology of glacial Lake Calgary at Cochrane, Alberta, Canada. *Canadian Journal of Earth Sciences* 36:791-803.

FLADMARK, K. R., J. C. DRIVER and D. ALEXANDER 1988 – The paleoindian component at Charlie Lake cave (HbRf 39), British Columbia. *American Antiquity* 53:371-384.

FORD, R. I. 1979 – Paleoethnobotany in American archaeology. In *Advances in Archaeological Method and Theory*, Vol. 2, M. B. Schiffer (ed.), pp. 285-336.

FRISON, G. C. 1983 – Stone circles, stone-filled firepits, grinding stones and High Plains archaeology. Memoir 19, *Plains Anthropologist*, 28(102-2):81-91.

FRISON, G. C. 1991 – *Prehistoric hunters of the high plains*. Second Edition, Academic Press, San Diego.

FRISON, G. C. 1998 – The Northwestern and Northern Plains Archaic. In *Archaeology on the Great Plains*, edited by W. R. Wood, pp. 140-172. University Press of Kansas, Lawrence.

FRITZ, G. J. 1984 – Identification of cultigen amaranth and chenopod from rockshelter sites in northwest Arkansas. *American Antiquity* 49(3):558-572.

FRITZ, G. J. and B. D. SMITH 1988 – Old collections and new technology: documenting the domestication of *Chenopodium* in eastern North America. *Midcontinental Journal of Archaeology* 13(1):3-27.

GEIST, V. 1971 – The relation of social evolution and dispersal in ungulates during the Pleistocene, with emphasis on the Old World deer and the genus *Bison*. *Quaternary Research* 1:283-315.

GIBBON, G. 1989 – *Explanation in archaeology*. Basil Blackwell Inc., New York.

GIFFORD, D. P. 1981 – Taphonomy and Paleoecology: a critical review of archaeology's sister disciplines. In *Advances in Archaeological Method and Theory*, Vol. 4, M. Schiffer (ed.), pp. 365-438.

GLENDINNING, G. R. 1974 – The glacial geomorphology of an area near Calgary, Alberta, between the Bow River and Fish Creek valleys. Unpublished MSc Thesis (Geography), University of Calgary, Calgary.

GREMILLION, K. J. 1993 – Crop and weed in prehistoric eastern North America: the *Chenopodium* example. *American Antiquity* 58(3):496-508.

GREMILLION, K. J. 1994 – Evidence of plant domestication from Kentucky caves and rockshelters. In *Agricultural Origins and Development in the Midcontinent*, edited by W. Green, pp. 105-126. Report 19. Office of the State Archaeologist, The University of Iowa, Iowa.

GRYBA, E. M. 1983 – Sibbald Creek: 11,000 years of human use of the Alberta Foothills. Archaeological Survey of Alberta, Occasional Paper 22, Alberta Culture, Edmonton.

GUTHRIE, R. D. 1980 – Bison and man in North America. *Canadian Journal of Anthropology* 1:55-73.

HALLET, D. J., L. V. HILLS, and J. J. CLAGUE 1997 – New accelerator mass spectrometry radiocarbon ages for the Mazama tephra layer from Kootenay Park, British Columbia, Canada. *Canadian Journal of Earth Sciences* 34:1202-1209.

HALLWORTH, B. (editor) 1988 – *Nose Hill*. Calgary Field Naturalists' Society, Calgary.

HALLY, D. J. 1981 – Plant Preservation and the content of paleobotanical samples: a case study. *American Antiquity* 46(4):723-742.

HARRISON, J. F. 1973 – Dated paleosol from below Mazama (?) tephra. *Geological Survey of Canada*, Paper 73-1, Part B: 128-128.

HASKELL, B. J., D. R. ENGSTROM and S. C. FRITZ 1992 – Late Quaternary paleohydrology in the North American Great Plains inferred from the geochemistry of endogenic carbonate and fossil ostracodes from Devils Lake, North Dakota, USA. *Palaeogeography, Palaeoclimatology, Palaeoecology* 124:179-193.

HASTORF, C. A. 1999 – Recent research in paleoethnobotany. *Journal of Archaeological Research* 7(1):55-103.

HILLS, L. V. 1969 – Problems in Palynology. In *Proceedings of the 1st annual Paleoenvironmental Workshop of the University of Calgary Archaeological Association*, edited by R. G. Forbis, L. B. Davis, O. A. Christensen and G. Fedirchuk, pp. 67-108. The Students' Press, University of Calgary, Calgary.

HILLS, L. V., B. KOOYMAN, P. MCNEIL, and S. TOLMAN 1999 – The St. Mary Reservoir paleontological site. Paper presented at the 4th Annual Workshop and Poster Session, Alberta Paleontological Society, Calgary.

HOFMAN, J. L. and R. W. GRAHAM 1998 – The paleo-Indian cultures of the Great Plains. In *Archaeology on the Great Plains*, edited by W. R. Wood, pp. 87-139. University Press of Kansas, Lawrence.

HOUSTON, D. B. 1973 – Wildfires in northern Yellowstone National Park. *Ecology* 54(5):1111-1117.

IRISH, E. J. W. 1972 – Physiographical and geologic setting of the city of Calgary. *Bulletin of Canadian Petroleum Geology* 20(2):362-369.

IRWIN-WILLIAMS, C., H. IRWIN, G. AGOGINO and C. V. HAYNES, JR. 1973 – Hell Gap: Paleo-Indian occupation on the High Plains. *Plains Anthropologist* 18(59):40-53.

JACKSON, L. E., JR. 1977 – Quaternary Stratigraphy and terrain inventory of the Alberta portion of the Kananaskis Lakes 1:250,000 Sheet (82-J). Unpublished PhD Dissertation (Geology), University of Calgary, Calgary.

JACKSON, L. E., JR., F. M. PHILLIPS and E. C. LITTLE 1997 – Cosmogenic ^{36}Cl dating of the Foothills erratics train, Alberta, Canada. *Geology* 25(3):195-198.

JACKSON, L. E., JR., F. M. PHILLIPS and E. C. LITTLE 1999 – Cosmogenic ^{36}Cl dating of the maximum limit of the Laurentide Ice Sheet in southwestern Alberta. *Canadian Journal of Earth Sciences* 36:1347-1356.

JOHNSON, D., L. KERSHAW, A. MACKINNON and J. POJAR 1995 – *Plants of the western boreal forest and aspen parkland.* Lone Pine Publishing and the Canadian Forest Service, Edmonton.

JOHNSON, E. A., D. A. GREGORY, C. LARSEN, and M. A. PRUDEN 1987 – Climatically-induced change in fire frequency from 1586 to 1984 in the southern Canadian Rockies. In *Fire Frequency Studies in the Kananaskis River Watershed*, edited by E. A. Johnson, pp. 3-45. Final Report to Alberta Environment Trust, Calgary.

JOHNSON, E. A., and C. P. S. LARSEN 1991 – Climatically induced change in fire frequency in the southern Canadian Rockies. *Ecology* 72(1):194-201.

JOHNSON, E. A. and K. MIYANISHI 2001 – Strengthening fire ecology's roots. In *Forest Fires: Behavior and Ecological Effects*, edited by E. A. Johnson and K. Miyanishi, pp. 1-9. Academic Press, San Diego.

JOHNSTON, A. 1987 – *Plants and the Blackfoot.* Lethbridge Historical Society, Historical Society of Alberta, Occasional Paper No. 15, Lethbridge.

KEARNEY, M. S. and B. H. LUCKMAN 1983 – Postglacial vegetational history of Tonquin Pass, British Columbia. *Canadian Journal of Earth Sciences* 20:776-786.

KEARNEY, M. S. and B. H. LUCKMAN 1987 – A mid-Holocene vegetational and climatic record from the subalpine zone of the Maligne Valley, Jasper National Park, Alberta (Canada). *Palaeogeography, Palaeoclimatology, Palaeoecology 59*:227-242.

KELLEY, J. H. and M. P. HANEN 1988 – *Archaeology and the methodology of science.* University of New Mexico Press, Albuquerque.

KINGSTON, M. S. 1982 – An evaluation of the recommended methods for paleosol investigations: a case study of a Holocene paleosol, Calgary, Alberta. Unpublished MSc Thesis (Geography), University of Western Ontario, London.

KLIVOKIOTIS, P. and R. B. THOMSON 1984 – *The Climate of Calgary.* Climatological Studies Number 38, The Climate of Canadian Cities Number 3. The Canadian Climate Program, Ottawa.

KOOYMAN, B. P. 2000 – *Understanding stone tools and archaeological sites.* University of Calgary Press, Calgary.

KOOYMAN, B. P., M. E. NEWMAN, C. CLUNEY, M. LOBB, S. TOLMAN, P. MCNEIL, and L. V. HILLS 2001 – Identification of horse exploitation by Clovis hunters based on protein analysis. *American Antiquity* 66(4):686-691.

KOOYMAN, B. P. and D. SANDGATHE 2001

KOOYMAN, B.P., S. TOLMAN, L. V. HILLS, and P. MCNEIL 1999 – Extinct mammals and early human adaption in the St. Mary's Basin. Paper presented at the 24th Annual Conference of the Archaeological Society of Alberta, Edmonton.

KUHNLEIN, H. V. and N. J. TURNER 1991 – *Traditional Plant Foods of Canadian Indigenous Peoples.* Gordon and Breach Science Publishers, Philadelphia.

LANDALS, A. J. 1998 – Lake Minnewanka site 1997 mitigation program. Consultant's Report on file, Parks Canada, Calgary.

LARSEN, C. P. S. 1992 – Fire and climate dynamics in the boreal forest of northern Alberta, Canada from AD 1850 to 1989. *The Holocene* 6:449-456.

LEIGHTON, A. L. 1985 – *Wild plant use by the Woods Creek (Nihithawak) of east-central Saskatchewan.* National Museum of Man Mercury Series. Canadian Ethnology Service Paper No. 101, Ottawa.

LEMMEN, D. S. and R. E. VANCE 1999 – Introduction. In *Holocene Climate and Environmental Change in the Palliser Triangle: A Geoscientific Context for Evaluating the Impacts of Climate Change on the Southern Canadian Prairies*, edited by D. S. Lemmen and R. E. Vance, pp. 1-6. Geological Survey of Canada Bulletin 534, Ottawa.

LENNSTROM, H. A. and C. A. HASTORF 1992 – Testing old wives' tales in paleoethnobotany: a comparison of bulk and scatter sampling schemes from Pancán, Peru. *Journal of Archaeological Science* 19:205-229.

LENNSTROM, H. A. and C. A. HASTORF 1995 – Interpretation in context: sampling and analysis in paleoethnobotany. *American Antiquity* 60(4):701-721.

LEWIS, H. T. 1981 – *A Time for Burning.* Occasional Publication No. 17, Boreal Institute for Northern Studies, The University of Alberta, Edmonton.

LICHTI-FEDEROVICH, S. 1970 – The pollen stratigraphy of a dated section of late Pleistocene lake sediment core from central Alberta. *Canadian Journal of Earth Sciences* 7:938-945.

LIMP, F. W. 1974 – Water separation and flotation processes. *Journal of Field Archaeology* 1:337-342.

LONGLEY, R. W. 1967 – Climate and weather patterns. In *Alberta, a Natural History.* Edited by W. G. Hardy, Pp. 53-67. Hurtig, Edmonton.

LOOMAN, J. and K. F. BEST 1994 – *Budd's flora of the Canadian prairie provinces.* Revised 1987. Reprinted 1994. Canada Communication Group, Ottawa.

LUCKMAN, B. H. 2000 – The Little Ice Age in the Canadian Rockies. *Geomorphology* 32(2000):357-384.

LUCKMAN, B. H. and M. S. KEARNEY 1984 – Reconstruction of Holocene changes in alpine vegetation and climate in the Maligne Range, Jasper National Park, Alberta. *Quaternary Research* 26:244-261.

LYMAN, R. L. 1994 – *Vertebrate Taphonomy.* Cambridge University Press, New York.

MACDONALD, G. M. 1982 – Late Quaternary paleoenvironments of the Morley Flats and Kananaskis Valley of southwestern Alberta. *Canadian Journal of Earth Sciences* 19:23-35.

MACDONALD, G. M. 1984 – Postglacial plant migration and vegetation development in the western Canadian boreal forest. Unpublished PhD Dissertation (Botany). University of Toronto, Toronto.

MACDONALD, G. M. 1987a – Postglacial development of the subalpine-boreal transition forest of western Canada. *Journal of Ecology* 75:303-320.

MACDONALD, G. M. 1987b – Postglacial vegetation history of the Mackenzie River Basin. *Quaternary Research* 28:245-262.

MACDONALD, G. M. 1989 – Postglacial palaeoecology of the subalpine forest—grassland ecotone of southwestern Alberta: new insights on vegetation and climate change in the Canadian Rocky Mountains and adjacent foothills. *Palaeogeography, Palaeoclimatology, Palaeoecology* 73:155-173.

MACDONALD, G. M. 1993 – Methodological falsification and the interpretation of paleoecological records: the cause of the early Holocene birch decline in western Canada. *Review of Palaeobotany and Palynology.* 79(1993):83-97.

MACDONALD, G. M., R. P. BEUKENS, W. E. KIESER and D. H. VITT 1987 – Comparative radiocarbon dating of terrestrial plant macrofossils and aquatic moss from the "Ice-Free Corridor" of western Canada. *Geology* 15:837-840.

MACDONALD, G. M. and L. C. CWYNAR 1985 – A fossil pollen based reconstruction of the late Quaternary history of lodgepole pine (*Pinus contorta* ssp. *latifolia*) in the western interior of Canada. *Canadian Journal of Forest Research* 15:1039-1044.

MACDONALD, G. M., C. P. S. LARSEN, J. M. SZEICZ, and K. A. MOSER 1991 – The reconstruction of boreal forest fire history from lake sediments: a comparison of charcoal, pollen, sedimentological, and geochemical indices. *Quaternary Science Reviews* 10:53-71.

MACDONALD, G. M. and J. C. RITCHIE 1986 – Modern pollen spectra from the western interior of Canada and the interpretation of late Quaternary vegetation development. *New Phytologist* 103:245-268.

MACMILLIAN, R. A. 1986 – *Soil survey of the Calgary urban perimeter.* Alberta Soil Survey Report No. 45, Alberta Research Council, Terrain Services Department, Edmonton.

MARKGRAF, V. and T. LENNON 1986 – Paleoenvironmental history of the last 13,000 years of the eastern Powder River basin, Wyoming, and its implications for prehistoric cultural patterns. *Plains Anthropologist* 31(111):1-12.

MARLES, R. J., C. CLAVELLE, L. MONTELEONE, N. TAYS, and D. BURNS 2000 – *Aboriginal plant use in Canada's northwest boreal forest.* UBC Press, Vancouver.

MARSH, W. F. 1984 – *Earthscape: a physical geography.* John Wiley & Sons, Inc., New York.

MENTIS, M. T. 1984 – Hypothetico-Deductive and Inductive approaches in ecology. *Functional Ecology* 2(1):5-14.

MEYBOOM, P. 1961 – Groundwater resources of the City of Calgary and vicinity. Alberta Research Council Bulletin 8.

MIKSICEK, C. H. 1987 – Formation processes of the archaeobotanical record. In *Advances in Archeological Method and Theory*, Vol. 10, edited by M. B. Schiffer, pp. 211-247.

MINNIS, P. E. 1981 – Seeds in archaeological sites: sources and some interpretive problems. *American Antiquity* 46(1):143-152.

MIYANISHI, K. 2000 – Duff consumption. In *Forest Fires: Behavior and Ecological Effects*, edited by E. A. Johnson and K. Miyanishi, pp. 437-475. Academic Press, New York.

MONTGOMERY, F. H. 1977 – *Seed and fruits of plants of eastern Canada and northeastern United States.* University of Toronto Press, Toronto.

MOORE, P. D., J. A. WEBB and M. E. COLLINSON 1991 – *Pollen Analysis*. Blackwell Scientific Publications, Oxford.

MORAN, S. R. 1986 – *Surficial Geology of the Calgary Urban Area.* Alberta Research Council, Terrain Services Department, Edmonton.

MORGAN, A. V. 1966 – Lithological and Glacial Geomorphological Studies near the Erratics train, Calgary area, Alberta. MSc Thesis (Geography), University of Calgary, Calgary.

MORRISON, K. D. 1994 – Monitoring regional fire history through size-specific analysis of microscopic charcoal: the last 600 years in south India. *Journal of Archaeological Science* 21:675-685.

MORLAN, R. 1999 – Canadian Archaeological radiocarbon database: establishing conventional ages. *Canadian Journal of Archaeology* 23(1 and 2):3-10.

MOSS, E. H. 1983 – *Flora of Alberta.* Second edition, revised by J. G. Packer. University of Toronto Press, Toronto.

MOTT, R. J. and L. E. JACKSON, Jr. 1982 – An 18,000 year palynological record from the southern Alberta segment of the classical Wisconsinan "Ice-free Corridor". *Canadian Journal of Earth Sciences* 19:504-513.

NELSON, J. G. 1969 – Some observations on animals, landscape and man in the Bow Valley area; c. 1760-1885. In *Vegetation, Soils and Wildlife*, edited by J. G. Nelson and M. J. Chambers, pp. 219-237. Methuen, Toronto.

OETELAAR, G. A. 1982 – An analysis of microremains in an Area A column sample. In *The Carrier Mills Archaeological Project: Human Adaptation in the Saline Valley, Illinois*, Volume 2, edited by R. W. Jefferies and B. M. Butler, pp. 939-1007. Center for Archaeological Investigations, Research Paper No. 33. Southern Illinois University, Carbondale.

OETELAAR, G. A. 1990 – Final report of the 1989 University of Calgary Archaeological Field School at the Strathcona site (FjPi-29) (Permit No. 89-10). Report on file, Archaeological Survey of Alberta, Edmonton.

OETELAAR, G. A. 1997 – Interim report on the 1995 and 1996 archaeological excavations at Tuscany (EgPn-377) (Permits 95-40 and 96-29). Consultant's Report (Department of Archaeology, University of Calgary) for Carma Developers Ltd., Calgary. Copy on file, Archaeological Survey of Alberta, Edmonton.

OETELAAR, G. A. and S. ZAYCHUK 1997 – Landscape evolution and human occupation during the Archaic on the northern Plains: a view from Tuscany. Paper presented to the 55th Annual Plains Anthropological Conference, Boulder.

OSBORN, G. and B. H. LUCKMAN 1988 – Holocene glacier fluctuations in the Canadian Cordillera (Alberta and British Columbia). *Quaternary Science Review* 7:115-128.

PATTERSON, W. A. III, K. J. EDWARDS and D. J. MAGUIRE 1987 – Microscopic charcoal as a fossil indicator of fire. *Quaternary Science Reviews* 6:3-23.

PATTIE, D. L. and I. FISHER 1999 – *Mammals of Alberta.* Lone Pine Publishing, Edmonton.

PEACOCK, S. 1992 – Piikáni ethnobotany: traditional plant knowledge of the Piikáni Peoples of the northwestern Plains. Unpublished MA Thesis (Archaeology). The University of Calgary, Calgary.

PEARSALL, D. M. 1988 – Interpreting the meaning of macroremain abundance: the impact of source and context. In *Current paleoethnobotany: analytical methods and cultural interpretations of archaeological plant remains*, C. A. Hastorf and V. S. Popper (eds.), pp. 97-118. University of Chicago Press, Chicago.

PEARSALL, D. M. 1999 – *Paleoethnobotany: a handbook of procedures.* Second edition, Academic Press, Inc., San Diego.

PELLATT, M. G., M. J. SMITH, R. W. MATHEWES and I. R. WALKER 1997 – Palaeoecology of postglacial treeline shifts in the northern Cascade Mountains, Canada. *Palaeogeography, Palaeoclimatology, Palaeoecology* 141:123-138.

PICHA, P. R., S. A. AHLER, R. D. SAYLER and R. W. SEABLOOM 1990 – Effects of prairie fire on selected artifact classes. *Archaeology in Montana* 32(2):15-28.

POPPER, V. S. and C. A. HASTORF 1988 – Introduction. In Current Paleoethnobotany, edited by C. A. Hastorf and V. S. Popper, pp. 1-16. University of Chicago Press, Chicago.

PORSILD, A. E. and W. J. CODY 1980 – *Vascular plants of continental Northwest Territories, Canada.* National Museums of Canada, Ottawa.

POWERS, J. S. 1994 – Late Quaternary paleoenvironmental reconstruction of the Beauvais Lake area, Alberta. Unpublished MSc Thesis (Geology and Geophysics), University of Calgary, Calgary.

PYNE, S. J. 1981 – *Fire in America. A Cultural History of Wildland and Rural Fire.* Princeton University Press, Princeton.

RAMSEY, B. 2000 – OxCal v3.5. http://www.rlaha.ox.ac.uk/orau/06_ind.htm

REASONER, M. A. and M. HICKMAN 1989 – Late Quaternary environmental change in the Lake O'Hara region, Yoho National Park, British Columbia. *Palaeogeography, Palaeoclimatology, Palaeoecology* 72:291-316.

REASONER, M. A. and U. M. HUBER 1999 – Postglacial palaeoenvironments of the upper Bow Valley, Banff

National Park, Alberta, Canada. *Quaternary Science Reviews* 18:475-492.

REEVES, B. O. K. 1975 – Early Holocene (ca. 8000 to 5500 B.C.) prehistoric land resource utilization patterns in Waterton Lakes National Park, Alberta. *Arctic and Alpine Research* 7(3):237-248.

REEVES, B. O. K. and J. F. DORMAAR 1972 – A partial Holocene pedological and archaeological record from the southern Alberta Rocky Mountains. *Arctic and Alpine Research* 4(4):325-336.

REINHARD, K. J., J. R. AMBLER and M. MCGUFFIE 1983 – Diet and parasitism Dust Devil Cave. *American Antiquity* 50:819-824.

RENECKER, L. A. and R. J. HUDSON 1993 – Morphology, bioenergetics and resource use: patterns and processes. In *Hoofed mammals of Alberta*, edited by J. B. Stelfox, pp. 141-163. Lone Pine Publishing, Edmonton.

RITCHIE, J. C. 1976 – The late-Quaternary vegetational history of the Western Interior of Canada. *Canadian Journal of Botany* 54:1793-1818.

RITCHIE, J. C. and G. A. YARRANTON 1978 – The Late-Quaternary history of the Boreal Forest of central Canada, based on standard pollen stratigraphy and principle components analysis. *Journal of Ecology* 66:199-212.

RONAGHAN, B. M. and A. B. BEAUDOIN 1988 – An archaeological survey in the upper North Saskatchewan River valley. In *Archaeology in Alberta 1987*, edited by M. Magne, pp. 25-45. Archaeological Survey of Alberta, Occasional Paper No. 32. Alberta Culture, Edmonton.

ROWE, J. S. 1959 – *Forest regions of Canada*. Bulletin 123, Forestry Branch, Canadian Department of Northern Affairs and Natural Resources, Ottawa.

ROYER, F. and R. DICKINSON 1999 – *Weeds of Canada and the northern United States*. University of Alberta Press and Lone Pine Publishing, Edmonton.

RUHE, R. V. 1956 – Geomorphic surfaces and the nature of soils. *Soil Science* 82:441-455.

RUHE, R. V. 1967 – Quaternary paleopedology. In *The Quaternary of the United States*, edited by H. Wright and D. Frey, pp. 755-764. Princeton University Press.

RUTTER, N. W. 1980 – *Special AMQUA Issue, the ice-free corridor and peopling the New World, Proceedings of the 5th Biennial Meeting of the American Quaternary Association 2-4, 1978, Edmonton, Alberta, Canada*. Canadian Journal of Anthropology 1(1), University of Alberta Press, Edmonton.

SAUCHYN, D. J. 1990 – A reconstruction of Holocene geomorphology and climate, western Cypress Hills, Alberta and Saskatchewan. *Canadian Journal of Earth Sciences* 27:1504-1510.

SAUCHYN, M. A. and D. J. SAUCHYN 1991 – A continuous record of Holocene pollen from Harris Lake, southwestern Saskatchewan, Canada. *Palaeogeography, Palaeoclimatology, Palaeoecology* 88:13-23.

SCHIFFER, M. B. 1987 – *Formation processes of the archaeological record*. University of New Mexico Press, Albuquerque.

SCHOCK, J. M. 1971 – Indoor water flotation – a technique for the recovery of archaeological materials. *Plains Anthropologist* 16:228-231.

SCHWEGER, C. E. 1984 – A 24,000 year pollen record from Goldeye Lake, Alberta – implications to the paleoecology of the ice-free corridor. In *CANQUA Symposium on the Paleoenvironmental Reconstruction of the Late Wisconsin Deglaciation and the Holocene, Lethbridge, Program-with-Abstracts and Field Guide*, p. 55.

SCHWEGER, C. E. and M. HICKMAN 1989 – Holocene paleohydrology, central Alberta: Testing the General-Circulation-Model climate simulations. *Canadian Journal of Earth Sciences* 26:1826-1833.

SCOTT-CUMMINGS, L. 1997 – Bucket flotation methods. AOL.com

SIEGFRIED, E. V. 1994 – Ethnobotany of the Northern Cree of Wabasca/Desmarais. Unpublished MA Thesis (Archaeology). University of Calgary, Calgary.

SIEGFRIED, E. V. and G. A. OETELAAR 2000 – A place to call home: the Organization and use of space at a paleoindian encampment. Paper presented at the 65th Annual Meeting of the Society for American Archaeology, Philadelphia.

SMITH, B. D. 1984 – *Chenopodium* as a prehistoric domesticate in Eastern North America: evidence from Russell Cave, Alabama. *Science* 226:165-167.

SMITH, B. D. 1985a – The role of *Chenopodium* as a domesticate in pre-maize garden systems of the eastern United States. *Southeastern Archaeology* 4(1):51-72.

SMITH, B. D. 1985b – *Chenopodium berlandieri* ssp. *jonesianum*: evidence for a Hopewellian domesticate from Ash Cave, Ohio. *Southeastern Archaeology* 4(2):107-133.

SMITH, B. D. 1987 – Origins of agriculture in eastern North America. *Science* 246:366-371.

SMITH, B. D. 1995 – *The Emergence of Agriculture*. Scientific American Library, New York.

SMITH, B. D. and C. W. COWAN 1987 – Domestication *Chenopodium* in prehistoric eastern North America: new accelerator dates from eastern Kentucky. *American Antiquity* 52(2):355-357.

SMITH, C. S. 1987 – Seeds, weeds and prehistoric hunters and gatherers: the plant macrofossil evidence from southwest Wyoming. *Plains Anthropologist* 33(120):141-158.

STEINHAUSER, R. A. 1991 – Recovery of fine fractions: the Fletcher site (DjOw-1). In *Archaeology in Alberta, 1988 and 1989*, edited by M. Magne, pp. 135-144.

Archaeological Survey, Provincial Museum of Alberta, Occasional Paper No. 33. Alberta Culture and Multiculturalism, Edmonton.

STRAUSS, A., and J. CORBIN 1990 – *Basics of Qualitative Research: Grounded Theory Procedures and Techniques*, Sage Publication, Newbury.

STRAUSS, A., and J. CORBIN 1994 – Grounded theory. In *Handbook of Qualitative Research*, N. K. Denzin and Y. S. Lincoln (eds.), Sage Publication, Thousand Oaks.

STRONG, W. L. 1977 – Pre- and post-settlement palynology of southern Alberta. *Review of Palaeobotany and Palynology* 23:1373-1387.

STRUEVER, S. 1968 – Flotation techniques for the recovery of small-scale archaeological remains. American Antiquity 33:353-362.

SWAIN, A. M. 1973 – A history of fire and vegetation in northeastern Minnesota as recorded in Lake sediments. *Quaternary Research* 3:383-396.

SWAIN, A. M. 1978 – Ennvironmental changes during the past 2000 years in north-central Wisconsin: analysis of pollen, charcoal and seeds from varved lake sediments. *Quaternary Research* 10:55-68.

SWANSON, E. H., Jr. 1961 – Early cultures in northwestern North America. *American Antiquity* 28:151-158.

THARIN, J. C. 1960 – Glacial Geology of the Calgary, Alberta Area. Unpublished Ph.D. Thesis (Geology), University of Illinois, Urbana.

THOMAS, D. H. 1998 – *Archaeology*. 3rd Edition. Harcourt Brace & Company, Troy.

TURNER, N. J. 1990 – "Burning mountain sides for better crops:" Aboriginal landscape burning in British Columbia. *Archaeology in Montana* 32(2):57-73.

VANCE, F. R., J. R. JOWSEY, J. S. MCLEAN and F. A. SWITZER 1999 – *Wildflowers across the prairies*. Greystone Books, Vancouver.

VANCE, R. E. 1986 – Late Quaternary vegetation and climate of the eastern slopes – the record from pollen studies. In *Eastern slopes prehistory: selected papers*, B. Ronaghan (ed.), pp. 247-268, Archaeological Survey of Alberta Occasional Paper No. 30. Alberta Culture Historical Resources Division, Edmonton.

VANCE, R. E. 1991 – A paleobotanical study of Holocene drought frequency in southern Alberta. Unpublished PhD Dissertation (Biological Sciences), Simon Fraser University, Burnaby.

VANCE, R. E., A. B. BEAUDOIN, B. H. LUCKMAN 1993 – The paleoecological record of 6 ka BP climate in the Canadian prairie provinces. *Géographie physique et Quaternaire* 49(1):81-98.

VANCE, R. E., R. W. MATHEWES and J. J. CLAGUE 1995 – 7000 year record of lake-level change on the northern Great Plains: A high-resolution proxy of past climate. *Geology* 20:879-882.

VAN DYKE, S. 1982 – Final Report, Archaeological conservation studies at EfPm-104, The Bow Bottom Site. (ASA Permit 78-5). Consultants Report (Lifeways of Canada Limited) for The City of Calgary. Copy on file, Archaeological Survey of Alberta, Edmonton.

VAN DYKE, S. and S. STEWART 1985 – *Hawkwood Site (EgPm-179): a multicomponent prehistoric campsite on Nose Hill.* Occasional Paper No. 7, Archaeological Survey of Alberta, Alberta Culture, Edmonton.

VICKERS, J. R. 1983 – *Alberta Plains prehistory: a review*. Archaeological Survey of Alberta. Occasional Paper No. 27, Alberta Culture, Historical Resources Division, Edmonton.

VICKERS, J. R. 1989 – *The Ross Site (DlPd-3) 1980 Research.* Archaeological Survey of Alberta Manuscript Series 14. Edmonton.

VICKERS, J. R. and A. B. BEAUDOIN 1989 – A limiting AMS date for the Cody Complex occupation at the Fletcher site, Alberta, Canada. *Plains Anthropologist* 34(123):261-264.

VIVIAN, B. C., K. BOSCH, and B. REEVES 1998 – Final Report, Historical resource conservation excavations EgPn-230. (ASA Permit 97-83). Consultant's Report (Lifeways of Canada Limited) for The City of Calgary. Copy on file, Archaeological Survey of Alberta, Edmonton.

VIVIAN, B. C. and B. REEVES 1999 – Final Report, Historical resource conservation excavations EgPn-428 Crestmont Estates. (ASA Permit 97-30, 98-131). Consultant's Report (Lifeways of Canada Limited) for Crestmont Developments Inc. Copy on file, Archaeological Survey of Alberta, Edmonton.

WATERS, P. L. 1977 – Postglacial paleoenvironment of southern Alberta. Unpublished MSc Thesis, University of Alberta, Edmonton.

WATSON, P. J. 1975 – In pursuit of prehistoric subsistence: A comparative account of some contemporary flotation systems. Mid-Continental Journal of Archaeology 1:77-100.

WHEAT, J. B. 1979 – The Jurgens site. *Plains Anthropologist*, Memoir 15. Lincoln.

WHITE, J. M. 1987 – *Late Pleistocene and Recent Environments of the Bow Valley, Banff National Park.* Report to Parks Canada, Western Regional Office, Calgary, Alberta.

WHITE, J. M. and R. W. MATHEWES 1992 – Postglacial vegetation and climatic change in the upper Peace River District, Alberta. *Canadian Journal of Botany* 64:2305-2318.

WHITLEY, D. S. 1998 – New approaches to old problems. In *Reader in Archaeological Theory: Post-processual and Cognitive Approaches*, edited by D. S. Whitley, pp. 1-28. Routledge, New York.

WHITNEY-SMITH, E. 1997 – Late Pleistocene extinctions by second order overkill or why the bison made it. In *International Symposium on Bison Ecology and Management in North America*, edited by L. Irby and J. Knight, pp. 337-345. Montana State University, Boseman.

WILKINSON, K. 1990 – *Trees and Shrubs of Alberta.* Lonepine Publishing, Edmonton.

WILKINSON, K. 1999 – *Wildflowers of Alberta.* The University of Alberta Press, Edmonton.

WILLOCK, T. 1990 – *A prairie coulee.* Lone Pine Publishing, Edmonton.

WILSON, H. D. 1981 – Domesticated *Chenopodium* of the Ozark Bluff Dwellers. *Economic Botany* 35:233-239.

WILSON, H. D. 1989 – Quinua and relatives (Chenopodium sect. Chenopodium subsect. Cellulata). In New Perspectives on the Origin and Evolution of New World Domesticated Plants, edited by P. Bretting. *Economic Botany* 44(3):92-110.

WILSON, M. 1982 – *Once Upon a River: Archaeology and Geology of the Bow River Valley at Calgary, Alberta, Canada.* Archaeological Survey of Canada, Paper No. 114. National Museums of Canada, Ottawa.

WILSON, M. and C. S. CHURCHER 1977 – Late Pleistocene *Camelops* from the Galleli Pit, Calgary, Alberta: morphology and geologic setting. *Canadian Journal of Earth Sciences* 15(5):729-740.

WILSON, M. C. and J. A. BURNS 1999 – Searching for the earliest Canadians: wide corridors, narrow doorways, small windows. In *Ice Age People of North America: Environments, Origins and Adaptations*, edited by R. Bonnichsen and K. L. Turnmire, pp. 213-248. Oregon State University Press, Corvallis.

W. L. Strong Ecological Land Surveys Ltd. 1992 – *Ecoregions and Ecodistricts of Alberta, Volume 1.* Alberta Forestry, Lands and Wildlife, Land Information Services Division, Edmonton.

WRIGHT, J. V. 1976 – *The Grant Lake site, Keewatin District, NWT.* National Museum of Man, Archaeological Survey of Canada, Mercury Series 47.

WRIGHT, J. V. 1995 – *A history of the Native people of Canada.* Canadian Museum of Civilization, Hull.

YANSA, C. H. 1991 – An early postglacial record of vegetation change in southern Saskatchewan, Canada. Unpublished MSc Thesis. University of Saskatoon, Saskatoon.

YANSA, C. H. 1998 – Holocene paleovegetation and paleohydrology of a prairie pothole in southern Saskatchewan, Canada. *Journal of Paleolimnology* 19:429-441.

YANSA, C. H. and J. F. BASINGER 1998 – A postglacial plant macrofossil record of vegetation and climate change in southern Saskatchewan. In *Holocene Climate and Environmental Change in the Palliser Triangle: A Geoscientific Context for Evaluating the Impacts of Climate Change on the Southern Canadian Prairies*, edited by D. S. Lemmen and R. E. Vance, pp. 139-172. Geological Survey of Canada Bulletin 534, Ottawa.

YUDCOVITCH, N. 1967 – Factors influencing the temperature variation in Calgary. *The Albertan Geographer* 3:11-19.

www.ingramcontent.com/pod-product-compliance
Lightning Source LLC
LaVergne TN
LVHW071213210926
848482LV00015B/801